NEUROSCIENCE RESEARCH PROGRESS

METABOTROPIC GLUTAMATE RECEPTORS

MOLECULAR MECHANISMS, ROLE IN NEUROLOGICAL DISORDERS AND PHARMACOLOGICAL EFFECTS

Neuroscience Research Progress

Additional books in this series can be found on Nova's website under the Series tab.

Additional E-books in this series can be found on Nova's website under the E-book tab.

NEUROSCIENCE RESEARCH PROGRESS

METABOTROPIC GLUTAMATE RECEPTORS

MOLECULAR MECHANISMS, ROLE IN NEUROLOGICAL DISORDERS AND PHARMACOLOGICAL EFFECTS

M. FOSTER OLIVE
EDITOR

New York

For permission to use material from this book please contact us:
Telephone 631-231-7269; Fax 631-231-8175
Web Site: http://www.novapublishers.com

Library of Congress Cataloging-in-Publication Data

Library of Congress Control Number: 2014933508

ISBN: 978-1-63117-451-3

Published by Nova Science Publishers, Inc. † New York

Contents

Preface

This book provides a collection of up-to-date reviews and novel findings, written by authors representing 4 different continents, on metabotropic glutamate receptors (mGluRs) and their role in basic brain functioning as well as various neurological and neuropsychiatric diseases. The first of the eight known mGluRs was identified over 25 years ago, and since that time the field of mGluR research and drug development has grown immensely. This growth has witnessed groundbreaking discoveries towards identification of novel treatments for numerous afflictions of the nervous system, and many of these notable advances in the field are highlighted in this book. Such advances include identification of a role for mGluR receptor signaling in regulating amyloidogenesis and aberrant dendritic protein translation in Fragile X syndrome. In addition, the field has witnessed advancement of mGluR ligands into clinical trials for the treatment of this common form of mental retardation as well as other afflictions including anxiety disorders, gastroesophageal reflux, chronic pain, and L-dopa-induced dyskinesias in Parkinson's disease.

In Chapter 1, Díaz-Casares and colleagues from the Universidad de Málaga review the current literature on the role of glutamate signaling via mGluRs in the modulation of cardiovascular and respiratory responses within the brainstem. The authors detail the effects of local pharmacological manipulation of mGluR function in regions such as the nucleus of the solitary tract, medial parabrachial region, intertrigeminal region, and A5 region on autonomic regulation of blood pressure, heart and respiratory rate, and defensive responses. The authors also summarize the potential of targeting of brainstem mGluRs for the treatment of different diseases affecting vegetative output such as essential hypertension or chronic pain.

In Chapter 2, Jun-Ichi Goto and colleagues from Yamagata University and the RIKEN Brain Institute review the intracellular signaling mechanisms of Group I mGluRs (mGluR1 and mGluR5), namely the phospholipase C (PLC) and inositol 1,4,5-triphosphate receptor (IP3R) pathways. The authors detail these signaling pathways as they contribute to forms of synaptic plasticity and metaplasticity such as long-term potentiation and synaptic depotentiation, respectively. Goto and colleagues conclude by presenting evidence supporting their hypothesis that mGluR-PLC-IP3R signaling acts as a synaptic and neural network stabilizer that exceeds the duration of N-methyl-D-aspartate (NMDA) receptor-mediated neuronal regulation.

Supporting the notion that Group I mGluRs mediate synaptic plasticity and metaplasticity, as summarized by Goto and colleagues in Chapter 2, in Chapter 3 Lucas

Watterson and M. Foster Olive from Arizona State University summarize behavioral data from rodent studies that support the notion that Group I mGluRs mediate various forms of learning and memory including spatial learning, behavioral flexibility, and the extinction of fear-related and drug-seeking behaviors. The authors also review evidence that positive allosteric modulation of Group I mGluRs, particular mGluR5, facilitates extinction learning may represent a novel therapeutic avenue for the treatment of abberant learning in anxiety disorders and drug addiction.

In Chapter 4, Cara Westmark from the University of Wisconsin at Madison summarizes recent findings identifying amyloid-beta protein precursor mRNA as a translationally regulated target of mGluR5 signaling. Thus, dysfunctional mGluR5 signaling may underlie the pathological accumulation of beta amyloid protein in amyloidogenic disorders including Alzheimer's disease, Down syndrome, Fragile X syndrome, and autism. Westmark hypothesizes that mGluR5 inhibition is a novel potential therapeutic strategy to modulate amyloid-beta protein accumulation and catabolite levels in these neurological disorders.

Chapter 5, authored by Kelly Newell and colleagues from the University of Wollongong, provide the first of several chapters devoted to reviewing the literature on mGluR involvement in schizophrenia and major depressive disorder. Newell and colleagues emphasize not only the importance of considering the status of mGluRs in the pathophysiology of these disorders, which may in turn influence the therapeutic potential of mGluR ligands, but also whether current therapeutics such as antipsychotics and antidepressants indirectly influence mGluR function as novel mechanisms of action of these pharmaceuticals.

In Chapter 6, Gentaroh Suzuki and colleagues from Japan Tobacco Inc. and Banyu Pharmaceutical Co. focus on the potential therapeutic value of targeting mGluR1, an often understudied mGluR subtype, in the treatment of schizophrenia. The authors describe a series of studies in which they have identified, characterized, and optimized novel mGluR1-selective negative allosteric modulators, which show potential antipsychotic activity in established rodent models with minimal potential for extrapyramidal side effects. Suzuki and colleagues also describe unique patterns of immediate early gene expression induced by these novel mGluR1 ligands as well as the development of a functional positron emission tomography tracer for this receptor. The development of this radiotracer represents a highly valuable advancement in the ability to assess mGluR1 receptor protein and occupancy levels in human studies.

In Chapter 7, Elizabeth Holly of Tufts University, Amber LaCrosse of Arizona State University, and Todd Hillhouse of Virginia Commonwealth University build on emphasizing of the importance of mGluR status in neuropsychiatric disease to major depressive disorder, as previously discussed in the context of schizophrenia in Chapter 5. Here, Holly and colleagues underscore the importance of potential glutamate dysfunction in major depressive disorder, as well as potential influences of existing antidepressants on mGluR function, both of which can potentially influence the therapeutic potential of Group I and II mGluR ligands as novel antidepressants. Holly and colleagues also summarize the existing preclinical literature on efficacy of various mGluR ligands in animal models of depression.

Chapter 8 turns to the subject of drug abuse and addiction, where Alessandra Matzeu and colleagues from the Scripps Research Institute in La Jolla, CA, summarize the substantial and still growing preclinical literature on the therapeutic potential of mGluR5 antagonists and mGluR2/3 agonists for the treatment of drug addiction. As most existing pharmacological and non-pharmacological approaches to treating substance use disorders are inadequate, Matzeu and colleagues emphasize in detail how targeting these specific receptors may potentially reduce on-going drug and alcohol intake, withdrawal symptoms, and prevent relapse.

Finally, in Chapter 9, Peter Kufahl and Sara Taylor from Arizona State University pursue a similar vein in which they summarize the potential for Group I, II, and III mGluR ligands as potential novel treatments for substance use disorders. The authors frame their review in the context of stress reactivity, as substance abuse is frequently comorbid with other forms of maladaptive stress reactivity such as depression and anxiety. Kufahl and Taylor propose that the use of mGluR agents in the treatment of addiction is potential tempered by, and therefore must take under consideration, the extent and nature of maladaptive changes in the brain brought about by stress dysregulation, mood disorders, and the addiction cycle itself.

It has been a great pleasure to serve as editor of this book, as mGluRs have been the central focus of my research for over 10 years. I would like to dedicate this book to the memory of my father, Ronald Lee Olive, who passed away on March 18, 2013. He was a life-long inspiration to me for strength, kindness, selflessness, humor, and patience. His passing has created a void in my life that can never be filled, and can only be lessened by both time and carrying on his legacy of hard work, creativity, and dedication to family.

M. Foster Olive, Ph.D.
Arizona State University
December 2013

In: Metabotropic Glutamate Receptors
Editor: M. Foster Olive

ISBN: 978-1-63117-451-3

Chapter 1

Brainstem Metabotropic Glutamate Receptors and Regulation of Autonomic Responses

A. Díaz-Casares[1], M. V. López-González[1] and M. S. Dawid-Milner[1]

[1]Universidad de Málaga, Málaga, Spain

Abstract

A considerable body of evidence has established that glutamate, the major excitatory neurotransmitter of the central nervous system, plays a key role in the modulation of cardiovascular and respiratory responses at the level of the brainstem [1-7]. The involvement of glutamate in central autonomic regulation, has been defined via microinjection of selective agonist and antagonists of ionotropic subtypes of glutamate (NMDA and non-NMDA) receptors within regions of the brainstem [8-10]. Recent evidence has established a role for a second major family of glutamate receptors, the metabotropic receptors (mGluR), which can regulate synaptic signalling both pre and postsynaptically by influencing intracellular processes via G-proteins [11, 12]. The existence of this family of glutamate receptors results in additional physiological mechanisms in central autonomic regulation. The mGluR agonist, tACPD, when microinjected into the nucleus of the solitary tract (NTS) produces dose-dependent depressor and bradycardic responses similar to those elicited by glutamate itself [13, 14]. Moreover, these responses are insensitive to the ionotropic receptor antagonist kynurenate, indicating the involvement of mGluRs. This family of glutamate receptors is additionally likely to have cardiorespiratory [15] and chemoreceptor [16, 17] roles within the NTS suggesting a wider involvement in central autonomic mechanisms. Moreover, electrophysiological studies, demonstrate that mGluRs located in pontine areas have an important role on the modulation of cardiorespiratory parameters at rest [5]. Microinjections of MCPG, an antagonist of mGluRs, within the medial parabrachial (mPB) or the A5 region produce an increase in blood pressure and heart rate in resting conditions. The respiratory rate is also decreased with the microinjection of MCPG into the A5 region (unpublished observations). There is evidence suggesting that mGluRs, located in the parabrachial complex (PBc) and the A5 region, are also involved in the modulation of autonomic changes associated with the defense response [5]. Studies involving specific metabotropic, specific ionotropic and nonspecific glutamate receptor antagonists, suggest that an interaction between ionotropic and metabotropic receptors from PBc and A5 region is necessary for the modulation of the cardiorespiratory response evoked by hypothalamic defense area (HDA) stimulation. The stimulation of the HDA in the rat produces a pattern of autonomic and somatic changes characteristic of the defense reaction. The cardiorespiratory response includes tachypnoea, tachycardia and a marked pressor response. This pattern of response is clearly altered after the microinjections of glutamate receptor antagonists into the PBc [5] and A5 region (unpublished observations). As brainstem glutamate is involved in central autonomic responses, mGluRs are potential drug targets for the therapy of different diseases affecting vegetative output, as for example essential hypertension or chronic pain [11, 18-22]. Further research is needed to further understand the role of mGluRs in the vegetative manifestation of these diseases. This chapter provides a review of the most important evidences that confirm the implication of mGluRs in the control of autonomic responses.

Keywords: Autonomic regulation, brainstem, hypertension, respiration, defense reaction, pressor response

Introduction

In the mid to late 1980s, studies providing the first evidence for the existence of glutamate receptors that are not ligand-gated cation channels but are coupled to intracellular effector systems through GTP-binding proteins were published. Since those initial reports, tremendous progress has been made in characterizing these metabotropic glutamate receptors (mGluRs). These include cloning and characterization of cDNA that encodes a family of eight mGluRs subtypes, several of which have multiple splice variants. Also, progress has been made in developing new highly selective mGluR agonists and antagonists, towards determining the physiological roles of the different mGluR subtypes in mammalian brain. Three groups of mGluRs have been defined according to their respective transduction mechanisms, sequence homologies and pharmacological characteristics: type I receptors (mGluR1,5) linked to stimulation of the phosphoinositide cascade, and type II (mGluR2,3) and type III receptors (mGluR4,6,7,8), both of which inhibit adenylyl cyclase when activated although they display different agonist and antagonist selectivities. The existence of this family of mGluRs offers additional physiological mechanisms, presumably also functional in central autonomic regulation, whereby the brain's major excitatory transmitter can subtly modulate cellular communication [11, 12].

Glutamate is the transmitter of the vast majority of the fast excitatory synapses in the mammalian central nervous system (CNS) and plays an important role in a wide variety of central autonomic functions. Until recently, the actions of glutamate in mammalian brain were thought to be mediated exclusively by activation of glutamate-gated cation channels termed ionotropic glutamate receptors (iGluRs). Modulation of neurotransmission through glutamatergic circuits was thought to require neuromodulators from extrinsic afferents (e.g., dopamine, acetylcholine, serotonin and norepinephrine) that modulate synaptic transmission and cell excitability by activating GTP-binding protein (G-protein)-linked receptors. In 1980s, however, evidence for the existence of glutamate receptors directly coupled to second messenger systems via G-proteins began to be discovered when they were found to be coupled to activation of phosphoinositide hydrolysis [23, 24]. Since that time, it has become clear that glutamate activates a large family of these receptors, termed metabotropic glutamate receptors (mGluRs), that are coupled to effector systems through GTP-binding proteins. The existence of this family of mGluRs offers additional physiological mechanisms in central autonomic regulation.

The discovery of mGluRs changed the traditional view of glutamatergic neurotransmission since activation of mGluRs can modulate the activity of glutamatergic circuits in a manner that was previously associated with neuromodulators arising from non-glutamatergic afferents. Unlike receptors for monoamines and other neuromodulators, mGluRs provide a mechanism by which glutamate can modulate the activity of the same synapses at which it elicits fast excitatory postsynaptic responses. Due to the extensive distribution of glutamatergic synapses in the CNS, mGluRs have the potential to participate in a wide variety of autonomic functions within the brainstem. In addition, the wide diversity and heterogeneous distribution of mGluRs subtypes provide an opportunity for developing pharmacologic agents that selectively interact with mGluRs involved in only one or a limited number of brainstem functions. Thus, gaining a detailed understanding of the specific roles of mGluRs in brainstem functions could allow to develop novel treatment strategies for a variety of autonomic manifestations of neurological disorders that could have a significant impact on neuropharmacology.

The involvement of glutamate receptors in central autonomic regulation has been investigated using pharmacological agents, especially with selective agonists and antagonists that are active at glutamate receptor sites. Here we review important evidence that confirm the implication of mGluRs in the control of a variety of cardiovascular and respiratory responses thus indicating that these receptors are important for mediating autonomic functions.

Metabotropic Glutamate Receptors and Brainstem Cardiovascular Functions

Whilst the involvement of iGluRs in central cardiovascular regulation has been well-studied, the role of mGluRs is less well understood. A considerable body of evidence has established that glutamate plays a key role in the coordination of brain's responses to changes in blood pressure (BP) and heart rate (HR) [8-10]. Thus, glutamate is a transmitter of vagal baroreceptors, chemoreceptors and cardiopulmonary afferent fibers which project within the different regions of the NTS.

The potential role of NTS mGluRs in cardiovascular regulation was examined following the microinjection of a number of specific selective mGluR compounds in anaesthetized rats within the NTS. The mGluR selective agonist 1S,3R-1amino-cyclopentane dicarboxylate (1S,3R–ACPD) elicits depressor and bradycardic actions following microinjections. This hypotension and decrease in HR is similar to that produced by L-glumamate [13, 15]. Similarly, decreases in BP and HR are observed upon administration of the type I and II selective mGluR agonists, (R,S)-3,5-dihydroxyphenylglycine (DHPG) and 2R, 4R-4-

aminopyrrolidine-2,4-dicarboxylate (APDC), respectively [13] These actions of DHPG are selectively attenuated by ±-1-aminoindane-1,5-dicarboxylate (AIDA), a type I mGluR antagonist. Cardiovascular responses to APDC were unaffected by this antagonist [13]. Interestingly, the type II mGluR antagonist, (2S,4S)-2-animo-4-(4, 4-diphenylbut-1yl)-pentane-1,5-doic acid (LY307452), reduces the cardiovascular responses to intra-NTS administration of both APDC and DHPG [13]. These studies provide evidence for a role of type I and II mGluRs in modulating cardiovascular responses within the NTS.

Glutamate is also present as a transmitter in other neuronal pathways involved in the regulation of sympathetic nerve activity and cardiovascular functions. These pathways include projections from the NTS to the caudal ventrolateral medulla (CVLM), and to reticulospinal neurons descending from the rostral ventrolateral medulla (RVLM) to the intermediolateral cell column (IML) of the spinal cord. These three glutamatergic pathways (plus the GABAergic projection from CVLM to RVLM) are the main components of the baroreceptor reflex circuitry, which plays a major role in resetting cardiovascular reflexes and hence determining normal BP and HR values [8-10].

Given that the IML is the location of preganglionic sympathetic neurons, it represents an important site for regulation of sympathetic nerve activity, the cardiovascular responses to intrathecal administration of mGluR agonists and antagonists conscious rats via telemetry has been examined [25]. Stimulation of type I or II mGluRs causes an increase in BP and HR, providing evidence that both mGluR subtypes in the IML are involved in the modulation of sympathetic flow and thus of vasomotor tone [25]. Within the IML, L-glutamate elicits marked, transient pressor and tachycardic responses [25]. In addition, (1S,3R)-ACPD evokes similar effects. Three selective mGluR agonists DHPG, APDC and CCG-1 ((2S,3S,4S)-2-(carboxycyclopropyl)glycine), a type I/II mGluR agonist, increase BP and HR, although the timing of their effects is different, the rank order being DHPG (type I) > CCG-1 (type I/II) > APDC (type II) (Table 1) [25]. These data support the view that both type I and II mGluR subtypes are involved in cardiovascular responses at the spinal level, but that the activation of type I mGluR exerts a greater influence on sympathetic cardiovascular outflow.

The effects of activation of mGluR subtypes on sympathetic preganglionic neurons of the IML differ by mGluR class. Type I mGluRs mediate excitation while type II mGluRs mediate inhibition.

Table 1. Maximum changes on BP and HR and duration of effect following intrathecal injection of mGluR compounds

	GLU	*(1S,3R)-ACPD*	*DHPG*	*CGC*	*APDC*
BP	Increase **	Increase **	Increase **	Increase **	ns
Duration BP (min)	5	20	<90	5	-
HR	Increase **	Increase **	Increase **	Increase **	Increase *
Duration HR (min)	20	20	>90	5	3

BP, blood pressure; HR, heart rate; ns, no significant changes. Asterisks show differences vs. baseline. * $p<0.05$, ** $p<0.01$.

Modified from [25].

Type I mGluR excitatory effects are predominant [26]. In vivo, type I excitation may represent normal synaptic transmission at the spinal level, but an additional type II-mediated inhibitory action of APDC and, to a lesser extent, CCG-1 modulates the initial excitatory effect.

The involvement of type I and II mGluRs in the IML of the spinal cord in response to DHPG and APDC has been evaluated using microinjections of their selective antagonists, AIDA, a selective antagonist of type I mGluR, and LY307452, respectively. Unexpectedly, both AIDA and LY307452 alone also caused pressor and tachycardic responses as their agonists. The reason for this is still not clear [27]. One explanation might be that these selective subtype mGluR antagonists exhibit partial agonistic activity, although this finding has not previously been reported [27]. As expected, AIDA markedly attenuates the overall cardiovascular effects of DHPG, although the maximum increase in BP is unaffected. Similarly, the type II mGluR antagonist, LY307452, seems to blunt the cardiovascular effects of APDC.

Interestingly, the effects of intrathecal mGluR antagonists described here are in contrast to the effects of the administration of the iGluR antagonist, kynurenate (kynurenic acid). Kynurenate decreases BP and HR in conscious rats [28], an affect due to the inhibition of sympathetic tone. These opposing actions of iGluR and mGluR antagonists are consistent with a growing body of evidence for contrasting actions of iGluR and mGluR agonists and antagonists [11, 29, 30]. Moreover, these cardiovascular effects are site-specific. AIDA and LY307452 exert no effects when injected into the NTS and decrease the cardiovascular effects of DHPG and APDC [13]. An interesting observation is that intrathecal administration of the type I mGluR agonist, DHPG and the type I/II mGluR agonist, CCG-1, induce associated forepaw licking, grooming, face washing, rearing, head jerk, and vocalization. These effects are very similar to those observed in the so called spontaneous nociceptive behaviour, which has recently been reported [31]. Type II mGluR agonists do not cause spontaneous nociceptive behaviour, which infers a type I mGluR-mediated response [31]. It is interesting to point that the cardiovascular actions persist longer than the behavioural events thus not contributing to the cardiovascular changes.

In summary, intrathecal administration of the mGluR agonists DHPG (type I), CCG-1 (type I/II) and APDC (type II) markedly increase BP and HR. Prior intrathecal administration of the type I and type II mGluR antagonists, AIDA and LY307452, attenuate the cardiovascular responses to DHPG and APDC [31]. These results therefore suggest that functional type I and II mGluR exist in the spinal cord, and that their activation may play an important role in the regulation of sympathetic outflow and thus in autonomic cardiovascular homeostasis.

There is evidence suggesting that the activity of IML is modulated directly or indirectly by other projections arising from different sites of the brainstem. One of these regions is the PBc, located within the dorsolateral pons, which plays an important role in central cardiovascular control [32, 33]. PBc participates in a variety of visceral regulatory functions that range from BP and respiratory control to taste and feeding [7, 33-37]. The functional heterogeneity of the PBc is reflected in its anatomy; at least 11 parabrachial subnuclei have been identified on the basis of their unique cytoarchitecture, chemical phenotype, and circuitry [1, 38, 39]. lutamate microinjection within the PBc evokes and increase in BP with a small increase in HR, a similar effect is observed when glutamate is microinjected within this region (Table 2) [33].

Table 2. Cardiovascular and respiratory changes before (control) and 4 minutes after the glutamate, kynurenic acid, MK-801, CNQX or MCPG microinjections within the lateral parabrachial (lPB), medial parabrachial (mPB) and A5 region

	Glutamate	Kynurenic Acid	MK-801	CNQX	MCPG
lPB					
BP	Increase **	Increase **	Increase ***	Increase **	ns
HR	Increase *	ns	Increase **	Increase *	ns
RR	Increase *	Decrease *	ns	ns	ns
mPB					
BP	Increase ***	Increase *	ns	Increase ***	Increase *
HR	Increase *	ns	ns	ns	Increase *
RR	Decrease **	Decrease *	ns	ns	ns
A5					
BP	Increase *	Increase *	ns	Increase **	Increase **
HR	Increase *	ns	ns	ns	ns
RR	Decrease **	Decrease **	ns	Decrease *	Decrease *

(BP, blood pressure; HR, heart rate; RR, respiratory; ns, no significant changes). Asterisks show differences between Control vs. unspecific and specific antagonist glutamate receptors. Data for A5 region glutamate antagonist receptors are unpublished observations. * $p<0.05$, ** $p<0.01$, *** $p<0.001$.

Modified from [5, 7].

Recent studies demonstrate the implications of different glutamate receptors subtypes within the different PBc subnuclei in this cardiovascular modulation [5]. Kynurenic acid microinjected within the lateral parabrachial region (lPB) or medial parabrachial region (mPB), produces, 4 min after its administration, a significant increase in resting BP with no changes in resting HR. An interesting observation is that no cardiovascular effects are observed when (RS)-α-Methyl-4-carboxyphenylglycine, MCPG, a nonspecific antagonist of mGluRs, is microinjected into the lPB; however, an increase in BP and HR are observed when MCPG is microinjected into the mPB [5] (Table 2). The cardiovascular effects of mGluR inhibition depend on the exact location of mGluR within the different PBc subnuclei, indicating the complexity of the PBc in the modulation of cardiovascular parameters at rest.

Similar effects are observed with MCPG microinjections within the A5 region (Table 2) (unpublished observations). The A5 region is located within the ventrolateral pons, and is known to provide the major component of the noradrenergic input to the sympathetic preganglionic neurons of the IML [40-42]. Whereby the A5 region has been implicated in cardiovascular control [4, 7, 43-47], it also contains non-catecholaminergic neurons which are mainly located at the level of the most caudal part of the A5 region [48, 49].

Kynurenic acid and MCPG microinjected within the A5 region produces, 4 min after its administration, a significant increase in resting BP with no changes in resting HR (unpublished observations). The similarity of the cardiovascular responses observed between mPB and A5 region to kynurenic acid and MCPG microinjections suggests a possible interaction between these two regions for the modulation of the cardiovascular activity in resting conditions. This modulation is influenced by mGluRs.

Metabotropic Glutamate Receptors and Brainstem Respiratory Functions

Some studies show evidence that confirm that mGluRs are also involved in the generation of the respiratory rhythm within the primary brainstem respiratory network and thus mGluRs are involved in the regulation of respiration. Glutamate microinjection within the retrotrapezoid nucleus, which is an integrative region of the brainstem network regulating cardiorespiratory reflexes, increases phrenic amplitude in a dose-dependent manner [50, 51]. The response duration depends on the timing of glutamate microinjection: responses to 3-second microinjections last a few minutes, and responses to 60-second microinjections last for > 30 minutes. This long-lasting effect is reproduced with microinjection of the mGluR agonist 1S,3R–ACPD at 0.01-0.02 times the glutamate dose. These data suggests that the retrotrapezoid nucleus has an endogenous source of glutamate to maintain eucapnic phrenic output and that its stimulation increases phrenic output via mGluR processes with the latter resulting in long-lasting phrenic stimulation.

Sites within the medulla oblongata where mGluRs are involved in regulating respiration have been identified [15]. A good example is the NTS. Unilateral microinjections of glutamate into the NTS of anaesthetized rats elicit apnea. Similarly, 1S,3R–ACPD also induces apnea, suggesting that mGluRs are involved in NTS regulation of respiratory functions.

Some pontine areas have also been studied to analyze their potential role in the control of respiration. A series of studies have established that respiratory disturbances, including apnea, can be produced by microinjections of glutamate into the intertrigeminal region (ITR) and the neighbouring PBc and Kölliker-Fuse nucleus of the pons in rats [2, 52-54]. In 1998, Chamberlin and Saper [52] extended this observation with neuroanatomical evidence postulating that ITR neurons may participate in apneic airway protective reflexes and may integrate afferent inputs from different portions of the airway, relaying this information to the central respiratory pattern generator. Further evidence for a function of this region in respiratory control derives from the demonstration that small ibotenic acid lesions of the ITR increase sleep-related apneas in rats [55]. These observations suggest the importance of defining the mechanism underlying glutamatergic neurotransmission or neuromodulation within the ITR, in terms of their impact on respiration. Further research is needed to solve this issue.

Microinjections of glutamate within the ITR induces apnea which can be completely abolished by pre-injecting the same site with kynurenic acid [56]. Blocking endogenous ITR glutamatergic neurotransmission with kynurenic acid also increases the duration of vagally mediated reflex apnea induced by intravenous injection of serotonin (5-HT) [53]. However, neither the specific iGluR antagonists, AP-5 nor NBQX, when injected into the ITR, affect the timing of 5-HT induced reflex apnea [54, 57]. These differential effects between kynurenic acid, AP5 and NBQX on 5-HT induced apnea suggest that iGluRs in the ITR are not involved in modulation of vagally mediated reflex apnea. Therefore, the modulation of this reflex seems to be modulated by mGluRs within the ITR.

It has been described that glutamate microinjections within ITR induce a reflex evoked apnea [58] which is also induced by intravenous serotonin administration. Although it has no effects on ITR-glutamate-induced apnea, AIDA, a selective antagonist at type I mGluRs, markedly increases total apnea duration and respiratory instability following intravenous 5-HT [58]. Respiratory disturbances caused by serotonergic activation of the peripheral vagus nerves are increased by ITR injection of the mGluR antagonist AIDA in a dose-dependent manner. This finding is in sharp contrast to the previously reported effects of iGluR antagonists, which block or attenuate ITR-glutamate-induced apnea, but have little or no impact on vagal reflex apnea [54, 57]. This finding is of interest because it suggests a role of type I mGluR within the ITR in dampening respiratory disturbances. Type I mGluRs within the ITR appear to exert a potent effect on respiratory reflexes, which is a generally accepted view of mGluR function in neuromodulation.

It can be speculated that endogenous glutamate release within the ITR modulates the ongoing respiratory pattern acting at type I mGluR localized in one or more of three anatomical domains: somato-dendritic sites, resulting in direct modulation of ITR neurons; presynaptic nerve terminals, resulting in modulation of inputs to ITR neurons, or axonal sites, resulting in modulation of ITR outputs or outputs from other areas with projection axons passing through the ITR. It is known that type I mGluRs are expressed in each of these anatomical neuronal domains [59]. In 2004, Radulovacki et al. [55] showed that unilateral lesions of the ITR increases sleep-related apnea in unanesthetized rats, a finding correlated with a significant cell loss in the ITR with no destruction of axons of passage. In view of this, it can be considered that AIDA-induced modulation of axons of passage within the ITR is the least likely explanation for these effects; with either direct modulation of ITR neurons or modulation of inputs to ITR neurons as more likely mediators of the effects observed in this study. Also, AIDA microinjections within the ITR results in spontaneous intermittent apneas occurring over at least a 10 min interval following the initial serotonin-evoked apnea [58]. Presynaptic type I mGluR inhibitory autoreceptors are well-known and widely distributed in the brainstem [60]. Furthermore, the primary vagal afferent relay neurons of the NTS are largely glutamatergic and send projections to the ITR [52]. Thus it is possible that AIDA may act on presynaptic autoreceptors to inhibit glutamate release in the ITR, leaving vagal modulation of respiration less dampened.

In summary, there is evidence that AIDA, an mGluR antagonist, when injected into the ITR significantly alters vagal reflex apnea, an important physiological phenomenon which may be a part of the protective mechanism of reflex apneas. These findings are consistent with the possible existence of functional mGluR1 receptors in respiratory modulating areas of the ITR and suggest that the observed effects may have been mediated by type I mGluRs.

As mentioned before, there are other areas within the brainstem modulating respiratory activity at rest. Two of them are well studied: the PBc and the A5 region [5, 49]. In rat, the PBc modulates respiration in two different ways. Glutamate microstimulation of neurons located within the mPB evokes a decrease of respiratory rate due to a prolongation of expiration, while stimulation of neurons located within the lPB increases respiratory rate due to a decrease in the duration of expiration together with a facilitation of inspiratory activity (Table 2) [2, 33, 52]. These results support the suggestions of Chamberlin and Saper in 1994 [2]. Using very small chemical stimuli, these investigators described three types of respiratory responses within the mPB, Kölliker-Fuse nucleus and its ventral–lateral boundaries, suggesting a complex organization of the PBc modulating inspiratory, postinspiratory and expiratory mechanisms. Parabrachial neurons are also involved in a topographic organized control of bulbar laryngeal motoneurons [7].

Recent studies, microinjecting nonspecific and specific glutamate receptor antagonists within the PBc, demonstrate the role of the different groups of glutamate receptors in the regulation of ongoing respiratory activity at rest [5]. The microinjection of kynurenic acid, within the mPB or lPB, 4 min after its administration, decreased respiratory rate due to an increase in expiratory time.

However, the microinjection of MK-801 ((+)-5-methyl-10,11-dihydro-5H-dibenzo[a,d]cyclohepten-5, 10-imine maleate, a specific NMDA receptor antagonist), CNQX (6-cyano-7-nitroquinoxaline-2,3-dione, specific non NMDA receptor antagonist) and MCPG within the mPB or lPB did not produce changes in respiratory rate, although, an increase in inspiratory time was observed. The increase in inspiratory time observed was compensated with a decrease in expiratory time. Since inspiratory time is shorter and its duration is more constant than expiratory time at rest, the probability of finding significant differences in this parameter increases, but its influence on respiratory rate is less effective than the changes in expiratory time. Thus, it is common to find changes in inspiratory time with no differences in respiratory rate. The integrity of the PBc seems to be needed to produce the expiratory facilitatory response, since PBc glutamatergic inhibition with kynurenic acid produced the same effect as glutamate activation, a decrease in respiratory rate due to an increase in expiratory time.

In previous reports it has been demonstrated that the activation of lPB somata with glutamate evokes an inspiratory facilitatory response [7]. The response consisted of an increase of respiratory rate due to a decrease in expiratory time. In a study by Díaz-Casares et al. [5] the inhibition of glutamate receptors located within the PBc with kynurenic acid was able to produce the opposite effect. Kynurenic acid decreased respiratory rate due to an increase in expiratory time (Table 2). The fact that microinjections of kynurenic acid within all the regions of the PBc decreases respiratory rate but that this effect is not observed with specific glutamate receptor antagonists suggests that glutamate receptors are involved in respiratory rate at rest. An integration of all types of receptors, ionotropic and metabotropic, is necessary for this effect, thus suggesting the complexity of the PBc in respiratory regulation. Further research on specific role of mGluR subtypes is needed to understand its role on respiratory modulation.

Another pontine area, the A5 region, also seems to be involved in the respiratory control at rest. The A5 region has connections with the NTS in the medulla oblongata, the Kölliker-fuse nucleus and lPB in the pons.

These connections suggest a role in the control of sympathetic activity and imply also a role in respiratory control. Glutamate microinjections within this region, which specifically activates perikarya, result mainly in a decrease in respiratory rate due to an increase in expiratory time [4]. Microinjections of kynurenic acid, CNQX and MCPG within the A5 region produce, 4 min after its administration, a decrease in respiratory rate due to an increase in expiratory time (table 2) (unpublished observations), thus suggesting that both groups of glutamate receptors are mediating this respiratory response and thus indicating the complexity of the A5 region in respiratory regulation at rest and implies a role for mGluRs in respiratory control at rest.

Role of Metabotropic Glutamate Receptors in the Defensive Behaviour

Recent studies demonstrate that pontine mGluRs have an important role in the modulation of the cardiorespiratory response evoked during defensive behaviour [5, 61]. Stimulation of the HDA in the rat produces a pattern of visceral and somatic changes characteristic of the defense reaction. Blood pressure and heart rate rise to meet the increased metabolic demands of fight or flight. The response includes tachypnoea, vasodilation of the femoral vascular bed and vasoconstriction of the renal vasculature, exophthalmos, retraction of the palpebrae, piloerection, twitching of the vibrissae and retroflexion of the tail [62].

The cardiorespiratory changes observed during the defense response are partially mediated by a facilitation of the chemoreceptor reflex and an attenuation of baroreceptor [63] and laryngeal reflexes [7, 64] involving a GABAergic mechanism in the NTS [65]. The cardiovascular response is also mediated by direct descending connections from the hypothalamic paraventricular nucleus to sympathetic preganglionic neurons of the intermediate lateral cell column in the thoracic spinal cord [66], the RVLM [67] and the A5 catecholaminergic region of the pons [68].

As mentioned before, HDA evoked response includes tachypnoea, tachycardia and a marked pressor response [62]. It is known that the PBc is a major secondary relay within the pons for transmission of cardiovascular and respiratory information from the NTS [69]. In fact, the PBc and mainly the lPB have been shown, using neuroanatomical techniques, to be reciprocally connected with forebrain structures involved in cardiorespiratory regulation [38, 70, 71]. Specific stimulation with glutamate of cell bodies located within lPB evokes a pressor response associated with tachycardia and tachypnoea, while stimulation of cell bodies of the mPB evokes the same cardiovascular response, hypertension and tachycardia, associated with bradypnea [33]. The similarity of the responses to HDA and PBc stimulation suggests a possible interaction between these cardiorespiratory regions. It has been demonstrated that microinjection of muscimol, a GABA-A agonist that inhibits cell somata by sustained hyperpolarization, within the PBc modifies the cardiorespiratory response evoked by HDA stimulation [61]. Muscimol microinjection within the lPB abolished the respiratory response to HDA stimulation and decreased the pressor response. Muscimol microinjected within the mPB decreased the magnitude of the pressor and tachycardic responses to HDA stimulation without modifying the respiratory response.

Previous studies have demonstrated the presence of both mGluR and iGluR in different nuclei of the PBc and Kölliker-Fuse nucleus [72, 73]. Activation of vagal afferent fibers releases glutamate within the PBc [74]. An ascending excitatory pathway involving glutamate from the NTS to the PBc has been described [75]. In vitro studies also show that glutamate agonists depolarize neurons in the PBc [76]. Therefore, glutamate receptors within the different PBc subnuclei are involved with the cardiorespiratory response to HDA stimulation. This hypothesis has been confirmed in a recent study [5]. The cardiovascular response evoked during the stimulation of the HDA was modified after the microinjection of kynurenic acid into the lPB and mPB. In both regions, the increase in BP is reduced and the tachycardia observed during HDA stimulation is abolished (Figure 1A,B, 2B,C and 3B,C) [5]. These results confirm that HDA-evoked cardiovascular responses are partially mediated through a glutamatergic acti-vation of lPB and mPB regions.

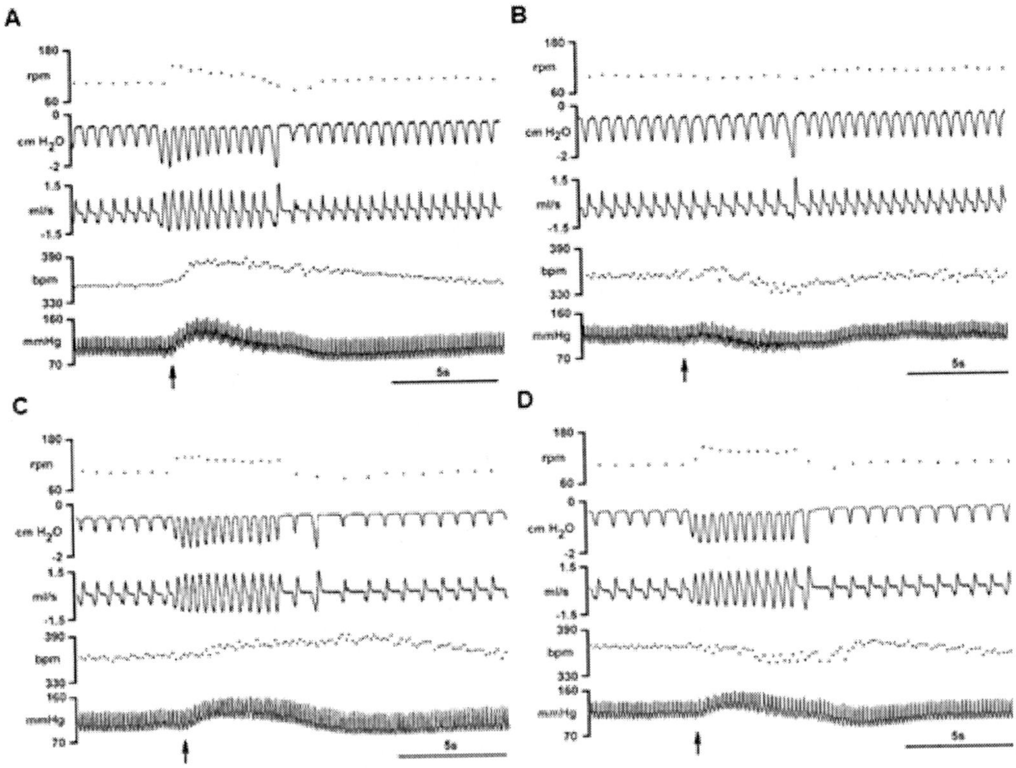

From ref. [5].

Figure 1. Instantaneous respiratory rate (upper trace, respirations per minute -rpm-), respiratory flow (ml/s), pleural pressure (cm H_2O), instantaneous heart rate (beats per minute -bpm-) and blood pressure (mmHg) showing the cardiorespiratory response evoked on HDA stimulation before (A, C) and after (B, D) the microinjection of kynurenic acid (50 nl) in the lPB (A, B) and in the mPB region (50 nl) (C, D). The arrows show the onset of the HDA electrical stimulation.

MK-801 and CNQX microinjections into the lPB produce a decrease in the pressor response and abolish the tachycardia evoked by HDA stimulation. Similarly, the microinjection of MK-801 and CNQX within the mPB also reduces pressor and HR responses. No changes are observed in the cardiovascular response after the microinjection of MCPG within the lPB or mPB (Figure 2B,C and 3B,C) [5].

These results suggest that the pressor response evoked during the stimulation of the HDA may involve a iGluR activation and recruitment of neurons in both the lPB and mPB subdivisions producing an indirect activation of the sympathetic preganglionic neurons of the IML [5]. The inhibition of the baroreceptor response seems to be more dependent on lPB iGluR activation than mPB since the tachycardia associated with the pressor response is only abolished with microinjections within the lPB. However, mGluRs do not seem to be involved in this modulation.

The respiratory response evoked by HDA stimulation is abolished after the microinjection of kynurenic acid into the lPB (Figure 1A and 2A). No changes are observed in the respiratory response after kynurenic acid microinjection within the mPB (Figure 1B, 3A) [5]. These data suggest that glutamate receptors located into the lPB are involved in the modulation of the respiratory response evoked to HDA stimulation.

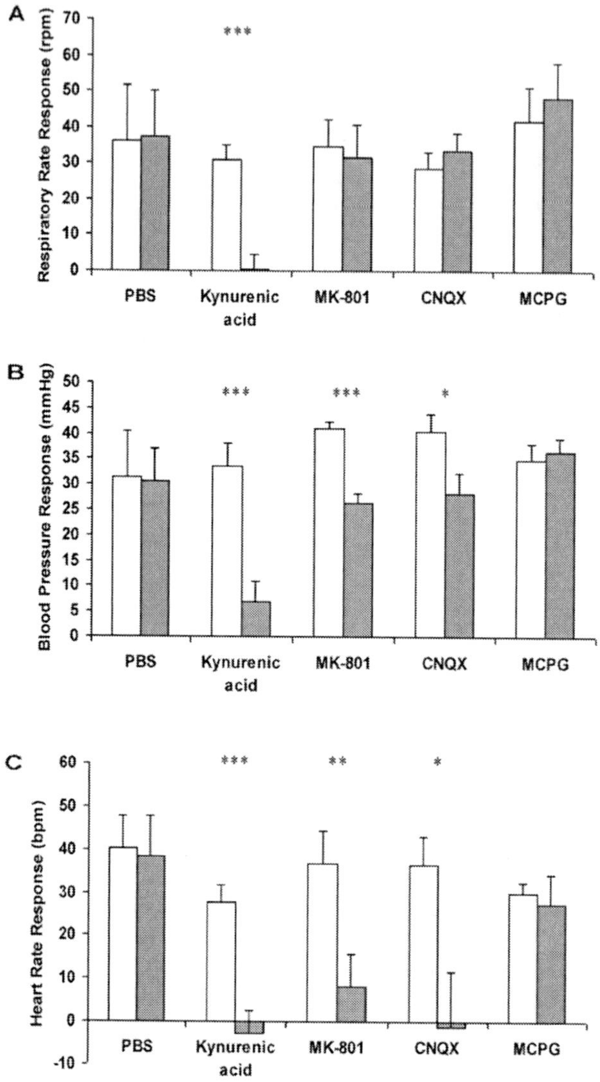

From ref. [5].

Figure 2. Changes in the intensity of the cardiorespiratory response evoked on HDA stimulation (respiratory rate response in respirations per minute -rpm-, blood pressure response -mmHg-and heart rate response in beats per minute -bpm-) before (empty bar) and after (filled bar) microinjection of PBS, kynurenic acid, MK-801, CNQX and MCPG within the lPB. *p<0.05, **p<0.01, *** p<0.001.

There are indications that HDA stimulation may facilitate the chemoreceptor reflex at specific cells located within the NTS [77]. These neurons are activated by HDA-NTS direct excitatory connections and are also the main targets of excitatory inputs from the lPB [78]. The observations of Díaz-Casares et al. [5] suggest that glutamate activates these excitatory inputs. The inhibition of the activation of these lPB projections with kynurenic acid leads to the abolishment of the tachypnoea evoked by HDA stimulation.

However, no changes in the respiratory response to HDA stimulation are observed after the blockade of iGluRs or mGluRs in the lPB (Figure 2A) [5]. One may speculate that there has to be an interaction between cells with different subtypes of iGluRs or mGluRs to produce this effect. Similar interactions have been previously reported [79]. Therefore, we can conclude that, although the mGluRs do not seem to be involved in the modulation of the cardiovascular response to HDA stimulation, iGluRs and mGluRs together may be involved in the respiratory response evoked by HDA stimulation.

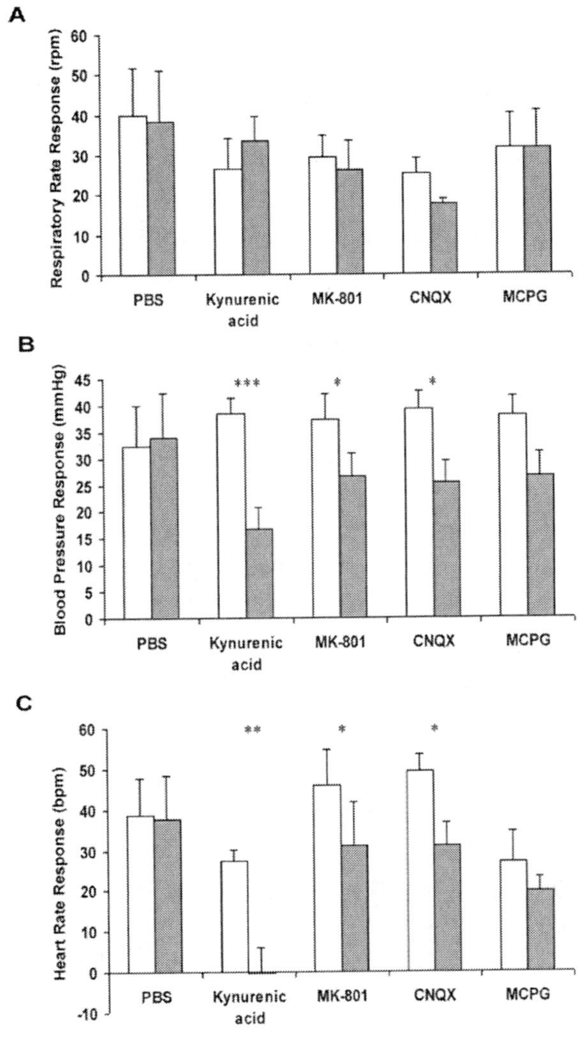

From ref. [5].

Figure 3. Changes in the intensity of the cardiorespiratory response evoked on HDA stimulation (respiratory rate response in respirations per minute -rpm-, blood pressure response –mmHg-and heart rate response in beats per minute -bpm-) before (empty bar) and after (filled bar) microinjection of PBS, kynurenic acid, MK-801, CNQX and MCPG within the mPB-KF. *$p<0.05$, ** $p<0.01$, *** $p<0.001$.

On the contrary, in another pontine area, the A5 region, microinjection of kynurenic acid decreases the cardiovascular response evoked to HDA stimulations. The significant increase in BP was still present but the amplitude of the increase in BP is significantly smaller after the kynurenic acid microinjection. Tachycardia was almost abolished. MK-801 and CNQX microinjection into the A5 region produce a decrease in the pressor response and reduced the tachycardia evoked by HDA stimulation. MCPG produces only a reduction in the tachycardia but does not affect the pressor response. The increase in BP and HR were still present, although only the amplitude of HR was smaller than before MCPG microinjection into the A5 region (unpublished observations). These results confirm that HDA-evoked cardiovascular responses are partially mediated through glutamatergic activation of the A5 region. Moreover, MCPG microinjections show a modulation and thus a role for mGluRs in the tachycardia evoked by HDA stimulation.

In conclusion, the role of mGluRs within the brainstem is dependent on their exact location, and thus is very important to define the functional role of these receptors in the modulation of the autonomic functions. Further research is needed to resolve this issue.

Metabotropic Glutamate Receptors as a Potential Drug Target for the Therapy of Diseases

Because of the wide diversity and heterogeneous distribution of mGluRs subtypes within the brain, the opportunity exists for developing highly selective drugs that affect autonomic functions. mGluRs therefore provide a novel target for development of therapeutic agents for the treatment of a wide variety of CNS disorders. It is important to note that drugs acting at mGluRs would have more subtle effects on transmission at glutamatergic synapses than would iGluR agonists and antagonists, and would probably have fewer side effects.

We know that agonists of group II and group III mGluRs have neuroprotective actions [80-83] while group I mGluR agonists can potentiate excitotoxicity [80, 84]. With these results it can be concluded that mGluR agonists and antagonists can reduce excitotoxic neuronal damage in neurodegenerative disorders. mGluRs are involved in motor circuits of the basal ganglia and cerebellum [11]. So, mGluR compounds have been suggested to be useful for the treatment of cerebellar ataxia, chorea and Parkinson disease [11]. The role of mGluRs in synaptic plasticity has suggested a possible role of mGluR ligands in cognitive functions [85-87]. Their modulation could be useful for memory and learning and thus have the potential to reduce the manifestations of Alzheimer disease [11].

Moving to a more close location and focusing on the structures of the brainstem involved in vegetative responses, mGluRs have a potential role modulating these autonomic manifestations associated with more general diseases such as hypertension or pain.

Treatment of Hypertension

The importance of mGluRs in cardiovascular control has been previously evidenced by studies demonstrating that the resting sympathetic vasomotor flow is maintained by tonic activation of group I mGluRs in the paraventricular nucleus of the hypothalamus, the so-called defense region, an important location for the control of brainstem sympathetic output and thus HR and BP [18]. This information is important for understanding the mechanisms involved in the control of the increased sympathetic outflow observed in some types of central essential hypertension. Group I mGluRs may represent a new target for treatment of certain types of central hypertension.

Moreover, the mGluR agonist 1S,3R-ACPD directly depolarizes neurons in the NTS and virtually abolishes evoked inhibitory postsynaptic potentials [11, 88]. This would be expected to elicit cardiovascular responses associated with the activation of the baroreceptor reflex. Consistent with this, microinjection of *trans*-ACPD into the NTS induces an increase in HR and BP [14].

In contrast, activation of mGluRs in the raphe obscurus induces a decrease in BP [29], and within the mPB [5] and A5 region induces an increase of BP. Thus, antagonists and agonists of mGluRs involved in regulating cardiovascular function in the NTS, raphe obscurus, PBc and A5 region respectively, could provide novel centrally acting agents for treatment of arterial hypertension.

Treatment of Pain

We know that the PBc, NTS, ventrolateral medulla and raphe nuclei receive converging nociceptive and visceral inputs from the spinal and trigeminal dorsal horns and initiate arousal, affective, autonomic, motor and pain modulatory responses to painful stimuli [89].

The analysis of the effects of mGluR agonists and antagonists on the responses of thalamic [19, 21], periaqueductal grey matter (PAG) [90] and spinal cord neurons [20, 22] to noxious and non-noxious stimuli suggest that mGluRs are involved in transmitting noxious sensory information to the thalamus and thus responsible for the vegetative response to noxious stimuli evoked by activation of the different cardiorespiratory effector regions of the pons and brainstem. It is known that postsynaptic group I mGluRs have the potential to directly modulate the analgesic, behavioural and autonomic functions of the PAG [90]. mGluRs in the spinal cord are involved in the induction of long-term enhancement of responses to noxious stimuli. The administration of selective antagonists for mGluR subtypes involved in transmitting noxious sensory information could be useful in the treatment of acute and chronic pain.

We know also that patients with cardiac diseases often experience cardiac-related secondary hyperalgesia, including deep muscle-like pain, persistent postoperative musculoskeletal pain and chronic myofascial pain [91]. The NTS plays a descending facilitatory role in cardiac-evoked muscle hyperalgesia, and NMDA receptors and mGluRs mediate this facilitatory effect [92]. Different mGluR subtypes display different roles in NTS nociception [93].

Group III mGluRs in the NTS plays an inhibitory role in cardiac nociception, while mGluR7 and mGluR8 in the NTS are oppositely involved in modulating cardiac nociception. Activation of mGluR7 in the NTS has anti-nociceptive effects, whereas activation of mGluR8 in the NTS has pro-nociceptive effects on cardiac-evoked muscle hyperalgesia. A presynaptic mechanism may be involved in the facilitatory effect of mGluR8. Similar results indicating a potential role of mGluR for the treatment of NTS pain modulation have been published [94-96].

Conclusion

A great deal of progress has been made in isolating and characterizing the family of mGluRs and in determining the physiologic roles of these receptors in the mammalian brainstem. This chapter reviews the most important evidences that confirm the implication of brainstem mGluRs in the control of a variety of autonomic, cardiovascular and respiratory responses indicating that these receptors are important for mediating autonomic functions. Evidence described above demonstrate that mGluRs located in different regions within the brainstem have an important role in the modulation of cardiovascular and respiratory parameters at rest. Moreover, this evidence shows the importance of these receptors in the modulation of the cardiorespiratory response evoked during hypothalamic defense area stimulation.

In conclusion, these observations provide valuable information for the understanding of the role of brainstem mGluRs, giving clear indications for the potential use of selective mGluR agonists or antagonists as neuroprotective agents. Based on the modulatory function of mGluRs located within the brainstem, agonists or antagonists of mGluRs are potentially relevant for the development of novel treatment strategies for a large number of central disorders presenting vegetative brainstem modulated symptoms.

References

[1] Chamberlin N. L., Saper C. B. (1992) Topographic organization of cardiovascular responses to electrical and glutamate microstimulation of the parabrachial nucleus in the rat. *J. Comp. Neurol.*, 326:245-262.

[2] Chamberlin N. L., Saper C. B. (1994) Topographic organization of respiratory responses to glutamate microstimulation of the parabrachial nucleus in the rat. *J. Neurosci.,* 14:6500-6510.

[3] Dawid-Milner M. S., Lara J. P., Gonzalez-Baron S., Spyer K. M. (2001) Respiratory effects of stimulation of cell bodies of the A5 region in the anaesthetised rat. *Pflugers Arch.,* 441:434-443.

[4] Dawid Milner M. S., Lara J. P., Lopez de Miguel M. P., Lopez-Gonzalez M. V., Spyer K. M., Gonzalez-Baron S. (2003) A5 region modulation of the cardiorespiratory responses evoked from parabrachial cell bodies in the anaesthetised rat. *Brain Res.,* 982:108-118.

[5] Diaz-Casares A., Lopez-Gonzalez M. V., Peinado-Aragones C. A., Gonzalez-Baron S., Dawid-Milner M. S. (2012) Parabrachial complex glutamate receptors modulate the cardiorespiratory response evoked from hypothalamic defense area. *Auton Neurosci.,* 169:124-134.

[6] Foley C. M., Moffitt J. A., Hay M., Hasser E. M. (1998) Glutamate in the nucleus of the solitary tract activates both ionotropic and metabotropic glutamate receptors. *Am. J. Physiol.*, 275:R1858-1866.

[7] Lara J. P., Dawid-Milner M. S., Lopez M. V., Montes C., Spyer K. M., Gonzalez-Baron S. (2002) Laryngeal effects of stimulation of rostral and ventral pons in the anaesthetized rat. *Brain Res.,* 934:97-106.

[8] Dampney R. A. (1994) Functional organization of central pathways regulating the cardiovascular system. *Physiol. Rev.,* 74:323-364.

[9] Lawrence A. J., Jarrott B (1996) Neurochemical modulation of cardiovascular control in the nucleus tractus solitarius. *Prog. Neurobiol.,* 48:21-53.

[10] Sun M. K. (1995) Central neural organization and control of sympathetic nervous system in mammals. *Prog. Neurobiol.*, 47:157-233.

[11] Conn P. J., Pin J. P. (1997) Pharmacology and functions of metabotropic glutamate receptors. *Annu. Rev. Pharmacol. Toxicol.,* 37:205-237.

[12] Nicoletti F., Bruno V., Copani A., Casabona G., Knopfel T. (1996) Metabotropic glutamate receptors: a new target for the therapy of neurodegenerative disorders? *Trends Neurosci.,* 19:267-271.

[13] Jones N. M., Beart P. M., Monn J. A., Widdop R. E. (1999) Type I and II metabotropic glutamate receptors mediate depressor and bradycardic actions in the nucleus of the solitary tract of anaesthetized rats. *Eur. J. Pharmacol.*, 380:129-135.

[14] Pawloski-Dahm C., Gordon F. J. (1992) Evidence for a kynurenate-insensitive glutamate receptor in nucleus tractus solitarii. *Am. J. Physiol.*, 262:H1611-1615.

[15] Vitagliano S., Berrino L., Pizzirusso A., D'Amico M., Calderaro V., Maione S., Rossi F. (1994) Metabotropic glutamate receptors are involved in the control of breathing at the medulla oblongata level of anaesthetized rats. *Neuropharmacology,* 33:859-864.

[16] Vardhan A., Kachroo A., Sapru H. N. (1993a) Excitatory amino acid receptors in commissural nucleus of the NTS mediate carotid chemoreceptor responses. *Am. J. Physiol.,* 264:R41-50.

[17] Vardhan A., Kachroo A., Sapru H. N. (1993b) Excitatory amino acid receptors in the nucleus tractus solitarius mediate the responses to the stimulation of cardio-pulmonary vagal afferent C fiber endings. *Brain Res.,* 618:23-31.

[18] Broekamp C. L., Phillips A. G., Cools A. R. (1979) Facilitation of self-stimulation behavior following intracerebral microinjections of opioids into the ventral tegmental area. *Pharmacol. Biochem. Behav.,* 11:289-295.

[19] Eaton S. A., Jane D. E., Jones P. L., Porter R. H., Pook P. C., Sunter D. C. et al. (1993) Competitive antagonism at metabotropic glutamate receptors by (S)-4-carboxyphenylglycine and (RS)-alpha-methyl-4-carboxyphenylglycine. *Eur. J. Pharmacol.,* 244:195-197.

[20] Neugebauer V., Lucke T., Schaible HG (1994) Requirement of metabotropic glutamate receptors for the generation of inflammation-evoked hyperexcitability in rat spinal cord neurons. *Eur. J. Neurosci.*, 6:1179-1186.

[21] Salt T. E., Eaton S. A. (1994) The function of metabotropic excitatory amino acid receptors in synaptic transmission in the thalamus: studies with novel phenylglycine antagonists. *Neurochem. Int.,* 24:451-458.

[22] Young M. R., Fleetwood-Walker S. M., Mitchell R., Dickinson T. (1995) The involvement of metabotropic glutamate receptors and their intracellular signalling pathways in sustained nociceptive transmission in rat dorsal horn neurons. *Neuropharmacology,* 34:1033-1041.

[23] Sladeczek F., Pin J. P., Recasens M., Bockaert J., Weiss S. (1985) Glutamate stimulates inositol phosphate formation in striatal neurones. *Nature,* 317:717-719.

[24] Sugiyama H., Ito I., Hirono C. (1987) A new type of glutamate receptor linked to inositol phospholipid metabolism. *Nature,* 325:531-533.

[25] Li X. C., Beart P. M., Monn J. A., Jones N. M., Widdop R. E. (1999) Type I and II metabotropic glutamate receptor agonists and antagonists evoke cardiovascular effects after intrathecal administration in conscious rats. *Br. J. Pharmacol.*, 128:823-829.

[26] Nolan M. F., Logan S. D. (1998). Metabotropic glutamate receptor-mediated excitation and inhibition of sympathetic preganglionic neurones. *Neuropharmacology,* 37:13-24.

[27] Wermuth C. G., Mann A., Schoenfelder A., Wright R. A., Johnson B. G., Burnett J. P. et al. (1996) (2S,4S)-2-amino-4-(4,4-diphenylbut-1-yl)-pentane-1,5-dioic acid: a potent and selective antagonist for metabotropic glutamate receptors negatively linked to adenylate cyclase. *J. Med. Chem.*, 39:814-816.

[28] Verberne A. J., Widdop R. E., Maccarrone C., Jarrott B., Beart P. M., Louis W. J. (1990) Intrathecal kynurenate reduces arterial pressure, heart rate and baroreceptor-heart rate reflex in conscious rats. *Neurosci. Lett.,* 114:309-315.

[29] D'Amico M., Berrino L., Pizzirusso A., de Novellis V., Rossi F. (1996) Opposing effects on blood pressure following the activation of metabotropic and ionotropic glutamate receptors in raphe obscurus in the anaesthetized rat. *Naunyn. Schmied. Arch. Pharmacol.*, 353:302-305.

[30] Leyva J., Maione S., Pallotta M., Berrino L., Rossi F. (1995) Metabotropic and ionotropic glutamate receptors mediate opposite effects on periaqueductal gray matter. *Eur. J. Pharmacol.,* 285:123-126.

[31] Fisher K., Coderre T. J. (1996) Comparison of nociceptive effects produced by intrathecal administration of mGluR agonists. *Neuroreport,* 7:2743-2747.

[32] Coote J. H., Hilton S. M., Zbrozyna A. W. (1973) The ponto-medullary area integrating the defence reaction in the cat and its influence on muscle blood flow. *J. Physiol.,* 229:257-274.

[33] Lara J. P., Parkes M. J., Silva-Carvhalo L., Izzo P., Dawid-Milner M. S., Spyer K. M. (1994) Cardiovascular and respiratory effects of stimulation of cell bodies of the parabrachial nuclei in the anaesthetized rat. *J. Physiol.*, 477:321-329.

[34] Biondolillo J. W., Williams L. A., King M. S. (2009) Blocking glutamate receptors in the waist area of the parabrachial nucleus decreases taste reactivity behaviors in conscious rats. *Chem. Senses.*, 34:221-230.

[35] De Gobbi J. I., Beltz T. G., Johnson R. F., Menani J. V., Thunhorst R. L., Johnson A. K. (2009) Non-NMDA receptors in the lateral parabrachial nucleus modulate sodium appetite. *Brain Res.,* 1301:44-51.

[36] Hayward L. F. (2007) Midbrain modulation of the cardiac baroreflex involves excitation of lateral parabrachial neurons in the rat. *Brain Res.,* 1145:117-127.

[37] Norgren R., Pfaffmann C. (1975) The pontine taste area in the rat. *Brain Res.,* 91:99-117.

[38] Fulwiler C. E., Saper C. B. (1984) Subnuclear organization of the efferent connections of the parabrachial nucleus in the rat. *Brain Res.,* 319:229-259.

[39] Herbert H., Moga M. M., Saper C. B. (1990) Connections of the parabrachial nucleus with the nucleus of the solitary tract and the medullary reticular formation in the rat. *J. Comp. Neurol.*, 293:540-580.

[40] Bruinstroop E., Cano G., Vanderhorst V. G., Cavalcante J. C., Wirth J., Sena-Esteves M., Saper C. B. (2012) Spinal projections of the A5, A6 (locus coeruleus), and A7 noradrenergic cell groups in rats. *J. Comp. Neurol.*, 520:1985-2001.

[41] Byrum C. E., Stornetta R., Guyenet P. G. (1984) Electrophysiological properties of spinally-projecting A5 noradrenergic neurons. *Brain Res.,* 303:15-29.

[42] Loewy A. D., Gregorie E. M., McKellar S., Baker R. P. (1979) Electrophysiological evidence that the A5 catecholamine cell group is a vasomotor center. *Brain Res.,* 178:196-200.

[43] Dampney R. A., Polson J. W., Potts P. D., Hirooka Y., Horiuchi J. (2003) Functional organization of brain pathways subserving the baroreceptor reflex: studies in conscious animals using immediate early gene expression. *Cell Mol. Neurobiol.*, 23:597-616.

[44] Guyenet P. G. (2006) The sympathetic control of blood pressure. *Nat. Rev. Neurosci.*, 7:335-346.

[45] Huangfu D. H., Koshiya N., Guyenet P. G. (1991) A5 noradrenergic unit activity and sympathetic nerve discharge in rats. Am. *J. Physiol.*, 261:R393-402.

[46] Loewy A. D., McKellar S., Saper C. B. (1979) Direct projections from the A5 catecholamine cell group to the intermediolateral cell column. *Brain Res.,* 174:309-314.

[47] Pilowsky P. M., Goodchild A. K. (2002) Baroreceptor reflex pathways and neurotransmitters: 10 years on. *J. Hypertens.*, 20:1675-1688.

[48] Goodchild A. K., Phillips J. K., Lipski J., Pilowsky P. M. (2001) Differential expression of catecholamine synthetic enzymes in the caudal ventral pons. *J. Comp. Neurol.,* 438: 457-467.

[49] Lopez-Gonzalez M. V., Diaz-Casares A., Peinado-Aragones C. A., Lara J. P., Barbancho M. A., Dawid-Milner M. S. (2013) Neurons of the A5 region are required for the tachycardia evoked by electrical stimulation of the hypothalamic defence area in anaesthetized rats. *Exp. Physiol.*, 98:1279-1294.

[50] Li A, Nattie EE (1995) Prolonged stimulation of respiration by brain stem metabotropic glutamate receptors. J Appl Physiol 79:1650-1656.

[51] Nattie E. E., Li A. (1995) Rat retrotrapezoid nucleus iono- and metabotropic glutamate receptors and the control of breathing. *J. Appl. Physiol.*, 78:153-163.

[52] Chamberlin N. L., Saper C. B. (1998) A brainstem network mediating apneic reflexes in the rat. *J. Neurosci.*, 18:6048-6056.

[53] Radulovacki M., Pavlovic S., Saponjic J., Carley D. W. (2003) Intertrigeminal region attenuates reflex apnea and stabilizes respiratory pattern in rats. *Brain Res.,* 975:66-72.

[54] Radulovacki M., Stoiljkovic M., Saponjic J., Carley D. W. (2007) Effects of intertrigeminal region NMDA and non-NMDA receptors on respiratory responses in rats. *Respir. Physiol. Neurobiol.*, 156:40-46.

[55] Radulovacki M., Pavlovic S., Carley D. W. (2004) Pontine intertrigeminal region attenuates sleep apneas in rats. *Sleep*, 27:383-387.

[56] Perkins M. N., Stone T. W. (1982) An iontophoretic investigation of the actions of convulsant kynurenines and their interaction with the endogenous excitant quinolinic acid. *Brain Res.,* 247:184-187.

[57] Isenovic E. R., Radulovacki M., Carley D. W. (2007) Impact of intertrigeminal region AMPA receptor blockade on respiratory responses in rats. *Respir. Physiol. Neurobiol.,* 158:39-44.

[58] Stoiljkovic M., Radulovacki M., Carley D. W. (2009) Local antagonism of intertrigeminal region metabotropic glutamate receptors exacerbates apneic responses to intravenous serotonin. *Respir. Physiol. Neurobiol.*, 165:137-142.

[59] Ohishi H., Neki A., Mizuno N. (1998) Distribution of a metabotropic glutamate receptor, mGluR2, in the central nervous system of the rat and mouse: an immunohistochemical study with a monoclonal antibody. *Neurosci. Res.*, 30:65-82.

[60] Raiteri M. (2008) Presynaptic metabotropic glutamate and $GABA_B$ receptors. *Handb. Exp. Pharmacol.*, 184:373-407.

[61] Diaz-Casares A., Lopez-Gonzalez M. V., Peinado-Aragones C. A., Lara J. P., Gonzalez-Baron S., Dawid-Milner M. S. (2009) Role of the parabrachial complex in the cardiorespiratory response evoked from hypothalamic defense area stimulation in the anesthetized rat. *Brain Res.,* 1279:58-70.

[62] Hilton S. M., Redfern W. S. (1986) A search for brain stem cell groups integrating the defence reaction in the rat. *J. Physiol.*, 378:213-228.

[63] Silva-Carvalho L., Dawid-Milner M. S., Goldsmith G. E., Spyer K. M. (1995) Hypothalamic modulation of the arterial chemoreceptor reflex in the anaesthetized cat: role of the nucleus tractus solitarii. *J. Physiol.*, 487:751-760.

[64] Dawid-Milner M. S., Silva-Carvalho L., Goldsmith G. E., Spyer K. M. (1995) Hypothalamic modulation of laryngeal reflexes in the anaesthetized cat: role of the nucleus tractus solitarii. *J. Physiol.*, 487:739-749.

[65] Jordan D., Mifflin S. W., Spyer K. M. (1988) Hypothalamic inhibition of neurones in the nucleus tractus solitarius of the cat is GABA mediated. *J. Physiol.*, 399:389-404.

[66] Hosoya Y., Sugiura Y., Okado N., Loewy A. D., Kohno K. (1991) Descending input from the hypothalamic paraventricular nucleus to sympathetic preganglionic neurons in the rat. *Exp. Brain Res.,* 85:10-20.

[67] Pyner S., Coote J. H. (1999) Identification of an efferent projection from the paraventricular nucleus of the hypothalamus terminating close to spinally projecting rostral ventrolateral medullary neurons. *Neuroscience,* 88:949-957.

[68] Hosoya Y., Sugiura Y., Ito R., Kohno K. (1990) Descending projections from the hypothalamic paraventricular nucleus to the A5 area, including the superior salivatory nucleus, in the rat. *Exp. Brain Res.*, 82:513-518.

[69] Spyer K. M. (1990) The central nervous organisation of reflex circulatory control. In: Central regulation of autonomic functions (Loewy AD, Spyer KM, eds). Oxford University Press, New York, pp 168–188.

[70] Moga M. M., Herbert H., Hurley K. M., Yasui Y., Gray T. S., Saper C. B. (1990) Organization of cortical, basal forebrain, and hypothalamic afferents to the parabrachial nucleus in the rat. *J. Comp. Neurol.*, 295:624-661.

[71] Moga M. M., Saper C. B., Gray T. S. (1990) Neuropeptide organization of the hypothalamic projection to the parabrachial nucleus in the rat. *J. Comp. Neurol.,* 295: 662-682.

[72] Chamberlin N. L., Saper C. B. (1995) Differential distribution of AMPA-selective glutamate receptor subunits in the parabrachial nucleus of the rat. *Neuroscience,* 68: 435-443.

[73] Guthmann A., Herbert H. (1999) Expression of N-methyl-D-aspartate receptor subunits in the rat parabrachial and Kolliker-Fuse nuclei and in selected pontomedullary brainstem nuclei. *J. Comp. Neurol.,* 415:501-517.

[74] Saleh T. M., Bauce L. G., Pittman Q. J. (1997) Glutamate release in parabrachial nucleus and baroreflex alterations after vagal afferent activation. *Am. J. Physiol.*, 272:R1631-1640.

[75] Jhamandas J. H., Harris K. H. (1992) Excitatory amino acids may mediate nucleus tractus solitarius input to rat parabrachial neurons. *Am. J. Physiol.,* 263:R324-330.

[76] Zidichouski J. A., Jhamandas J. H. (1993) Electrophysiological characterization of excitatory amino acid responses in rat lateral parabrachial neurons in vitro. *Brain Res.,* 611:313-321.

[77] Silva-Carvalho L., Dawid-Milner M. S., Goldsmith G. E., Spyer K. M. (1993) Hypothalamic-evoked effects in cat nucleus tractus solitarius facilitating chemoreceptor reflexes. *Exp. Physiol.,* 78:425-428.

[78] Felder R. B., Mifflin S. W. (1988) Modulation of carotid sinus afferent input to nucleus tractus solitarius by parabrachial nucleus stimulation. *Circ. Res*., 63:35-49.

[79] Barnard E. A. (1997) Ionotropic glutamate receptors: new types and new concepts. *Trends Pharmacol. Sci.,* 18:141-148.

[80] Bruno V., Battaglia G., Copani A., Giffard R. G., Raciti G., Raffaele R. et al. (1995) Activation of class II or III metabotropic glutamate receptors protects cultured cortical neurons against excitotoxic degeneration. *Eur. J. Neurosci.,* 7:1906-1913.

[81] Buisson A., Yu S. P., Choi D. W. (1996) DCG-IV selectively attenuates rapidly triggered NMDA-induced neurotoxicity in cortical neurons. *Eur. J. Neurosci.,* 8:138-143.

[82] Copani A., Bruno V., Battaglia G., Leanza G., Pellitteri R., Russo A. et al. (1995) Activation of metabotropic glutamate receptors protects cultured neurons against apoptosis induced by beta-amyloid peptide. *Mol. Pharmacol*., 47:890-897.

[83] Maiese K., Greenberg R., Boccone L., Swiriduk M. (1995) Activation of the metabotropic glutamate rec:eptor is neuroprotective during nitric oxide toxicity in primary hippocampal neurons of rats. *Neurosci. Lett*., 194:173-176.

[84] Bruno V., Copani A., Knopfel T., Kuhn R., Casabona G., Dell'Albani P. et al. (1995) Activation of metabotropic glutamate receptors coupled to inositol phospholipid hydrolysis amplifies NMDA-induced neuronal degeneration in cultured cortical cells. *Neuropharmacology,* 34:1089-1098.

[85] Ohno M., Watanabe S. (1996) Concurrent blockade of hippocampal metabotropic glutamate and N-methyl-D-aspartate receptors disrupts working memory in the rat. *Neuroscience*, 70:303-311.

[86] Riedel G., Wetzel W., Reymann K. G. (1994) (R,S)-alpha-methyl-4-carboxyphenylglycine (MCPG) blocks spatial learning in rats and long-term potentiation in the dentate gyrus in vivo. *Neurosci. Lett.,* 167:141-144.

[87] Van der Staay F. J., Antonicek H., Helpap B., Freund W. D. (1995) Effects of the selective metabotropic glutamate receptor agonist, L-CCG-I, on acquisition of a Morris task by rats. *Eur. J. Pharmacol*., 294:361-365.

[88] Glaum S. R., Miller R. J. (1994) Inhibition of phosphoprotein phosphatases blocks metabotropic glutamate receptor effects in the rat nucleus tractus solitarii. *Mol. Pharmacol*., 45:1221-1226.

[89] Benarroch E. E. (2006) Pain-autonomic interactions. *Neurol. Sci.,* 27 Suppl 2:S130-133.

[90] Wilson-Poe A. R., Mitchell V. A., Vaughan C. W. (2013) Postsynaptic mGluR mediated excitation of neurons in midbrain periaqueductal grey. *Neuropharmacology,* 66:348-354.

[91] Levine P. R., Mascette A. M. (1989) Musculoskeletal chest pain in patients with "angina": a prospective study. *South Med. J.,* 82:580-585.

[92] Liu X. H., Sun N., Du J. Q., Tang J. S., Han M., Zhu J. X., Huo F. Q. (2012) Chemical lesioning and glutamate administration reveal a major role for the nucleus tractus solitarius in the cardiac-somatic reflex in rats. *Neuroscience,* 207:326-332.

[93] Goudet C., Magnaghi V., Landry M., Nagy F., Gereau R. W. 4th, Pin J. P. (2009) Metabotropic receptors for glutamate and GABA in pain. *Brain Res. Rev.,* 60:43-56.

[94] Marabese I., de Novellis V., Palazzo E., Mariani L., Siniscalco D., Rodella L. et al. (2005) Differential roles of mGlu8 receptors in the regulation of glutamate and gamma-aminobutyric acid release at periaqueductal grey level. *Neuropharmacology*, 49 Suppl 1:157-166.

[95] Marabese I., de Novellis V., Palazzo E., Scafuro M. A., Vita D., Rossi F., Maione S. (2007) Effects of (S)-3,4-DCPG, an mGlu8 receptor agonist, on inflammatory and neuropathic pain in mice. *Neuropharmacology,* 52:253-262.

[96] Marabese I., Rossi F., Palazzo E., de Novellis V., Starowicz K., Cristino L. et al. (2007) Periaqueductal gray metabotropic glutamate receptor subtype 7 and 8 mediate opposite effects on amino acid release, rostral ventromedial medulla cell activities, and thermal nociception. *J. Neurophysiol.*, 98:43-53.

In: Metabotropic Glutamate Receptors
Editor: M. Foster Olive

ISBN: 978-1-63117-451-3

Chapter 2

Activation of Group I Metabotropic Glutamate Receptors Negatively Regulates Synaptic Plasticity and Network Activity in the Central Nervous System via Inositol 1,4,5-Trisphosphate Receptor Signaling

Jun-Ichi Goto[1,2], *Satoshi Fujii*[1,2] *and Katsuhiko Mikoshiba*[2]
[1]Department of Physiology, Yamagata University, School of Medicine, Yamagata, Japan
[2]Laboratory for Developmental Neurobiology, Brain Science Institute, RIKEN, Japan

Abstract

Stimulation of group I metabotropic glutamate receptors (mGluRs) activates two distinct signaling pathways involving protein kinase C (PKC) or inositol 1,4,5-trisphosphate receptors (IP_3Rs). The IP_3R transduces the second messenger, IP_3, into another second messenger, intracellular Ca^{2+} release from internal store sites, and thus exerts effects on various cellular functions, such as synaptic plasticity. Targeted ablation of the type 1 IP_3R results in epilepsy-like seizures and cerebellar ataxia in mice, suggesting that mGluR-phospholipase C-IP_3R signaling plays a pivotal role in the normal functions of the nervous system. Synaptic plasticity, such as long-term potentiation (LTP), is one of the best characterized physiological functions in the central nervous system. It is widely accepted that the N-methyl D-aspartate receptor (NMDAR) plays a central role in LTP induction at hippocampal and cortical synapses and together with mGluRs, contributes to induction of synaptic plasticity. In addition to a direct effect on LTP induction, mGluR stimulation is also involved in metaplastic changes in synaptic efficacy, such as depotentiation or suppression of LTP. Interestingly, both depotentiation and suppression of LTP exert their effects by negatively regulating synaptic transmission, consistent with the stabilizing effect of mGluR-IP_3R signaling on network activity demonstrated in knockout mice experiments. Diacylglycerol-PKC signaling, the other branch of the group I mGluR signaling pathway, is also involved in the induction of synaptic plasticity. Pharmacological studies have shown that the direction of plasticity is determined by the interaction between NMDARs and mGluR-IP$_3$R and mGluR-PKC signaling, probably by the phosphorylation and dephosphorylation of intracellular proteins, such as Ca^{2+}/calmodulin-dependent kinase II and calcineurin. Accordingly, we propose that mGluR-PLC-IP$_3$R signaling primarily acts as a stabilizer for each synapse as well as the whole network and that this effect has a longer time constant than NMDAR-mediated regulation of neuronal functions.

Keywords: Synaptic plasticity, long-term potentiation, long-term depression, depotentiation, suppression of LTP, inositol 1,4,5-trisphosphate receptor

1. Metabotropic Glutamate Receptors in Neurons

Excitatory synapses at cortical and hippocampal neurons utilize glutamate as both a transmitter of spiking activity and a signaling molecule to induce synaptic plasticity. Glutamate activates pre- and postsynaptic glutamate receptors, which can be roughly divided into two groups, ionotropic glutamate receptors (iGluRs), consisting of α-amino-3-hydroxy-5-methyl-4-isoxazolepropionic acid receptors (AMPARs), NMDARs, and kainate receptors (KARs), and metabotropic glutamate receptors (mGluRs) [1, 2]. The iGluRs are ion channels and induce ionic fluxes into and/or out of cells, while mGluRs are coupled to G_q or $G_{i/o}$ G-protein signaling which initiate a range of cellular responses [3]. For the purposes of this chapter, we will focus our discussion to group I mGluRs (see below), and will therefore utilize the term mGluR to infer group I mGluR unless otherwise noted.

The physiological function of iGluRs depends on the movement of ions between the extracellular and intracellular spaces. AMPARs are responsible for the majority of inward ionic current in response to glutamate release from presynaptic sites. Sodium ions (Na^+) carry a significant proportion of the inward current passing through AMPAR channels, while calcium ions (Ca^{2+}) also contribute to the current through certain AMPAR subtypes [4]. NMDARs provide the major pathway for Ca^{2+} influx from the extracellular to the intracellular space. The opening of NMDARs requires both the binding of glutamate to NMDARs and the depolarization of the postsynaptic membrane, which releases NMDARs from magnesium block. This enables NMDARs to detect the conjunctive spiking of pre- and postsynaptic neurons using glutamate binding and postsynaptic depolarization [5, 6]. The KAR is a less well understood family of glutamate receptors [7], but, in terms of ion conductance, is similar to the AMPAR, as it usually permeable to Na^+ (and potassium), but less permeable to Ca^{2+}, although certain subtypes of AMPARs and KARs are more permeable to Ca^{2+} than others.

Group I mGluRs are mainly expressed on postsynaptic elements of neurons, whereas group II/III mGluRs are more abundant in presynaptic sites [3]. Stimulation of group I mGluRs (mGluR1/5), which are coupled to G_q, activates phospholipase C (PLC), which hydrolyses phosphatidylinositol 4,5-bisphosphate into inositol 1,4,5-trisphosphate (IP$_3$) and diacylglycerol (DAG). IP$_3$ diffuses into the intracellular space and activates the IP$_3$ receptor (IP$_3$R), a calcium channel in the membrane of the endoplasmic reticulum (ER), and the opening of IP$_3$R channels results in release of Ca^{2+} from the ER calcium store. DAG remains on the plasma membrane and activates protein kinase C (PKC). The first finding suggesting the involvement of IP_3Rs in synaptic plasticity was obtained by targeted ablation of type 1 IP_3Rs. Type1 IP_3R null mice exhibit epilepsy-like seizures, and diazepam treatment relieves the symptoms [8]. Cerebellar long-term depression, which is thought to be a cellular basis of cerebellum-dependent motor learning [9-15] (but see [16]), is abolished in type 1 IP_3R null mice [17]. Furthermore, IP_3R+/- mice exhibit impairments in motor learning [18]. These results suggest that mGluR-IP_3R signaling might control the excitation/inhibition balance, presumably by modulating synaptic plasticity.

2. Metabotropic Glutamate Receptors, NMDA Receptors, and Synaptic Plasticity

Long-term potentiation (LTP) is the persistent enhancement of synaptic transmission, and is probably the best characterized type of synaptic plasticity in terms of the induction paradigm, underlying molecular mechanisms, and correlation with animal behaviors [19-21]. Early studies revealed that NMDARs play a dominant central role in the induction of LTP. Activity-dependent LTP is induced by a brief high-frequency stimulation (HFS) of the input pathway and is generally considered to be triggered by influx of Ca^{2+} into the post-synaptic neuron through NMDA receptor channels. The increase in the postsynaptic Ca^{2+} concentration leads to activation of Ca^{2+}/calmodulin-dependent kinase II (CaMKII) and other downstream targets responsible for LTP induction and expression (Figure 1A). Another form of synaptic plasticity, homosynaptic long-term depression (LTD), induced by low-frequency stimulation (LFS) at 1 Hz, requires the activation of NMDARs and an increase in the postsynaptic $[Ca^{2+}]_i$ caused by influx of Ca^{2+} [22]. This Ca^{2+} increase is believed to activate the calcium-dependent protein phosphatase calcineurin, which is critical in LTD induction (Figure 1B) [23, 24].

Although NMDARs are indispensable in LTP/LTD induction, mGluRs are indispensable for induction of typical LTP by a standard tetanus of 100 Hz and 100 pulses [21, 25, 26], but not for atypical LTP [27]. In the case of LTD, there are at least two forms, one of which NMDAR-dependent (Figure 1B) and the other mGluR-dependent (Figure 1C) [28, 29]. Thus, for both LTP and LTD, mGluR activation is indispensable for some types of LTP/LTD, but not others. If activation of mGluRs is completely blocked by an antagonist during tetanic stimulation, LTP induction is inhibited [25]. Bortolotto et al. [30] reported that LTP induction requires the activation of both NMDARs and mGluRs and that mGluR activation during a preceding stimulation overcomes the need for mGluR stimulation during subsequent LTP induction. These results show that mGluR activation is indispensable for typical LTP induction. The requirement for mGluR stimulation indicates that downstream target(s) of mGluRs are involved in LTP induction. PKC is probably the main target of mGluRs in the facilitation of LTP induction [31]. A role for IP_3R in LTP induction seems to be more controversial. Several lines of studies suggest that IP_3R activation may facilitate LTP induction [32-34], while, under certain conditions, it inhibits LTP induction [35, 36].

3. Activation of mGluRs Reverses LTP (Depotentiation)

Stimulation of mGluRs not only plays a role in LTP, but is also indispensable for depotentiation (DP). DP (also known as reversal of LTP) is induced by LFS applied 20-60 minutes after LTP induction and reverses a pre-established LTP [37-39].

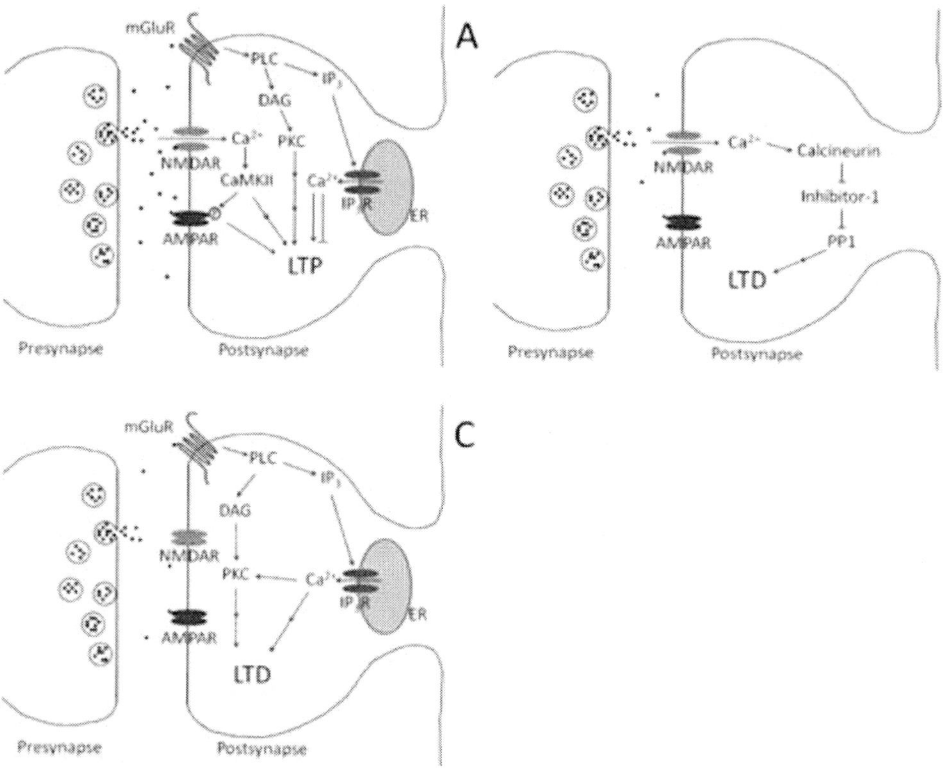

Figure 1. Schematic diagrams of postsynaptic signalings in the induction of synaptic plasticity. (A) LTP induced by standard tetanus involves activation of NMDAR and mGluR. Glutamate released from the presynaptic terminal activates AMPA and NMDA receptor. Activation of NMDA receptor induces Ca^{2+} influx through NMDA receptor channel and results in the increase of postsynaptic Ca^{2+}. Postsynaptic Ca^{2+} increase activates CaMKII, which phosphorylates AMPA receptor (such as Ser831 of GluA1), and induces LTP. Glutamate also activates mGluR, which activates PLC-DAG-PKC signaling pathway. The activation of PKC, together with CaMKII, is also required for the induction of standard LTP. Stimulation of mGluR also activates PLC-IP_3-IP_3R signaling pathway, which may facillitate or inhibit LTP indcution. (B) NMDAR-dependent LTD is mediated by Ca^{2+} influx through NMDAR channel. Postsynaptic Ca^{2+} rise activates calcineurin, which induces LTD via protein phosphatase signaling including Inhibitor-1 and PP1. (C) MGluR-dependent LTD is induced by activation of IP_3R and PKC. Stimulation of mGluR activates PLC, which catalyses hydrolysis of PIP_2 into DAG and IP_3. DAG activates PKC, and IP_3 activates IP_3R which mediates Ca^{2+} release from ER. Both of PKC and Ca^{2+} release through IP_3R facillitate mGluR-LTD induction.

The mossy fiber-CA3 pyramidal neuron synapse is a good system for investigating the role of mGluRs in DP induction. LTP at the mossy fiber-CA3 neuron synapse is induced in the field EPSP by delivery of HFS (a tetanus of two trains of 100 pulses at 100 Hz with a 10 s interval) and is reversed by a train of LFS of 1000 pulses at 2 Hz applied 60 minutes later [40]. This DP is triggered by activation of mGluR-IP_3R signaling during HFS, which may increase the postsynaptic $[Ca^{2+}]_i$, leading to a cellular process responsible for modification of LTP expression. In DP, activation of CaMKII during and/or after the HFS is necessary for both LTP induction and LTP reversal in the field EPSP, which are mediated by postsynaptic AMPARs. Thus, it is suggested that the mGluR-IP_3R-CaMKII signaling pathway, which is triggered by HFS, may be involved in the mechanism of DP induction at mossy fiber-CA3 neuron synapses. Furthermore, activation of IP_3Rs or protein phosphatase during LFS is required for the DP expressed at mossy fiber-CA3 neuron synapses. These results suggest that, at these synapses, activation of IP_3Rs by an LTP-inducing HFS results in IP_3R modulation and/or postsynaptic protein phosphorylation during a subsequent LFS, leading to a decrease in the field EPSP and erasure of LTP. Thus, it is possible that a postsynaptic $[Ca^{2+}]_i$ increase during/after the LFS, resulting from activation of IP_3Rs during LTP induction produced by a signaling cascade involving group I mGluRs, induces DP at mossy fiber-CA3 neuron synapses.

DP at Schaffer collateral (SC)-CA1 neuron synapses has a similar, but slightly different, mechanism to that at mossy fiber-CA3 neuron synapses [27, 37, 40]. In hippocampal CA1 neurons, activation of NMDARs during LTP induction is critical for LTP induction and thus the expression of DP. In contrast, in DP at mossy fiber-CA3 neuron synapses, induction of LTP is NMDAR-independent. To elucidate the role of NMDARs in DP induction, LTP was induced by HFS at the SC-CA1 neuron synapse, then NMDARs were blocked during the application of a subsequent LFS, and this was found to inhibit DP induction [37]. Activation of group I mGluRs during LTP induction is also necessary for DP at SC-CA1 neuron synapses, as application of an mGluR antagonist during LTP induction inhibits DP induction [27].

4. mGluR Activation Inhibits Certain Forms of LTP (Suppression of LTP)

In addition to direct effects on synaptic efficacy, mGluR activation also regulates metaplastic changes at synapses. A typical example is suppression of LTP, in which LFS at 1-2 Hz given prior to HFS delivery suppresses LTP induction [41].

Suppression of LTP was first reported at SC-CA1 neuron synapses [41], but the detailed physiological and pharmacological characteristics were first reported at mossy fiber-CA3 neuron synapses [42]. Induction of LTP in the field EPSP by delivery of HFS (a tetanus of two trains of 100 pulses at 100 Hz with a 10 s interval) to mossy fiber-CA3 neuron synapses is suppressed when the synapses are preconditioned by LFS of 1000 pulses at 2 Hz, and this effect is inhibited when the LFS preconditioning is performed in the presence of an IP_3R antagonist or a protein phosphatase inhibitor. Furthermore, activation of group I mGluRs during HFS cancels the effects of an IP_3R antagonist applied during the preconditioning LFS on subsequent LTP induction at mossy fiber-CA3 neuron synapses. These results suggest that, in mossy fiber-CA3 neuron synapses, activation of IP_3Rs during a preconditioning LFS results in dephosphorylation events that lead to failure of HFS to induce a subsequent LTP.

Delivery of HFS to the inputs of mossy fiber-CA3 neuron synapses increases the $[Ca^{2+}]_i$ in the presynaptic terminals [43, 44] and also evokes Ca^{2+} release in postsynaptic pyramidal neurons via the mGluR-PLC-IP_3R pathway during LTP induction [45-48]. In hippocampal CA3 neurons, the Ca^{2+} released from postsynaptic intracellular Ca^{2+} stores by activation of group I mGluRs and/or IP_3Rs can stimulate the Ca^{2+}/calmodulin and CaMKII signaling pathways [49] to phosphorylate postsynaptic non-NMDARs [24, 50], resulting in long-lasting enhancement of synaptic transmission [21].

Immunoreactivity for the Ca^{2+}/calmodulin-dependent protein phosphatase, calcineurin, is strong in the stratum lucidum and moderate in the pyramidal cells in the CA3 subfield [51]. As LFS generates LTD in CA1 pyramidal neurons through activation of calcineurin [52], it is possible that a postsynaptic $[Ca^{2+}]_i$ increase due to activation of IP_3Rs induces either phosphorylation or dephosphorylation of postsynaptic group I mGluRs [53, 54] in CA3 neurons.

The ability of the preconditioning LFS to turn on the molecular switch for suppression of the HFS-induced LTP in postsynaptic CA3 neurons involves dephosphorylation [42]. Thus, it is suggested that Ca^{2+}, released through IP_3Rs by the preconditioning LFS, stimulates calcineurin and activates protein phosphatase signaling pathways [52], resulting in modification of LTP expression at mossy fiber-CA3 neuron synapses. We also suggest that the activation of group I mGluRs by the preconditioning LFS triggers dephosphorylation events that prevent the HFS-induced activation of mGluRs that is necessary for LTP induction.

The mechanism by which suppression of LTP is induced at SC-CA1 synapses is less understood, but activation of NMDARs during the preconditioning LFS is required [41].

5. Negative Regulatory Effect of mGluR-IP3R Signaling in Synaptic Plasticity

Activation of mGluR-IP_3R signaling not only reverses LTP [27, 37, 40], but also suppresses a subsequent LTP [41, 42], implying a general inhibitory role for the mGluR-IP_3R pathway in synaptic plasticity. This is of great interest, since there might be a link between the negative regulatory effect on LTP [35, 36] and the epilepsy-like seizures observed in IP_3R1 knockout mice [8]. In other words, constitutive negative feedback on synaptic plasticity via mGluR-IP_3R signaling might be an underlying mechanism by which neuronal circuits are protected from hyperactivity and epilepsy-like phenotype. Downstream targets of mGluR-IP_3R signaling are also involved in the modulation of neuronal network activity. Stimulation of IP_3Rs induces release of Ca^{2+} from the ER, resulting in depletion of the ER calcium store. To replenish the depleted store, store-operated calcium entry (SOCE) from the extracellular space has to be activated [55]. The SOCE channel consists of two proteins, stromal interacting molecule (STIM) and ORAI. STIM, an ER transmembrane protein with Ca^{2+}-binding EF hand domains, serves as a sensor for the ER luminal Ca^{2+} concentration [56-58], while ORAI is the pore-forming channel protein [59-61]. When STIM senses that ER Ca^{2+} is low, it forms dimers via its coiled-coil domains and translocates below the plasma membrane, thus activating ORAIs. Recent studies have shown that inhibition of IP_3Rs or SOCE affects the excitability of neurons and thus the stability of neuronal networks. Inhibition of IP_3Rs by an antagonist reduces IP_3R-mediated intracellular Ca^{2+} release and increases neuronal excitability in cortical and hippocampal acute brain slices [62, 63].

6. Blockade of either NMDARs or mGluRs Converts LTD to LTP

We have mentioned that typical LTP induction requires the activation of both NMDARs and mGluRs [21, 25, 26], while mGluR-IP_3R signaling may negatively regulate LTP under some conditions [27, 35, 36, 40, 42]. This is controversial, as, depending on experimental conditions, mGluR activation can facilitate or inhibit LTP induction. To explain this paradox, we need to consider what determines whether a synapse undergoes LTP or LTD.

To explain this, the Bienenstock-Cooper-Munro (BCM) theory [64] is probably one of the most widely accepted regarding the mechanism by which the sign of the plasticity (i.e. LTP vs. LTD) is controlled.

A classical paradigm for inducing LTP is a brief high frequency tetanic stimulation (e.g. 100 pulses at 100 Hz) [19-21], whereas LTD is induced by LFS for a longer time (e.g. 1000 pulses at 1 Hz) [22].

One form of LTD is mGluR-dependent; in this case, a high Ca^{2+} concentration over a short period tends to induce LTP, while a moderate Ca^{2+} concentration for a longer time facilitates LTD. The BCM theory postulates that a lower Ca^{2+} increase induces LTD and a higher Ca^{2+} increase induces LTP [64]. However, it is not always the case that a higher or lower Ca^{2+} concentration leads, respectively, to LTP or LTD.

One of the examples that are not consistent with the BCM theory is NMDAR-dependent LTD in hippocampal CA1 neurons. Pharmacological blockade of NMDARs during LFS of 1000 pulses at 1 Hz not only inhibits LTD induction, but converts LTD to LTP; interestingly, mGluR or IP_3R antagonists also convert LTD to LTP [65]. These results suggest that, if one of the Ca^{2+} sources is blocked during LFS of 1000 pulses at 1 Hz, the direction of synaptic plasticity changes from LTD to LTP. Thus, under certain conditions, a lower Ca^{2+} increase causes LTP, rather than LTD.

Homeostatic synaptic plasticity also exhibits synaptic depression in response to stronger synaptic inputs and synaptic potentiation in response to weaker stimulation [66]. This is quite similar to the situation in which LTD induced by LFS is converted to LTP if the Ca^{2+} increase is partially blocked by drug application [65]. However, these results conflict with BCM theory-based synaptic plasticity, in which stronger stimulation induces LTP and weaker stimulation induces LTD.

It is of great interest that, in situations explained by the BCM theory, the duration of synaptic inputs is relatively short, while, in those in which homeostatic plasticity applies and in LFS-induced LTD, the duration of inputs is long. The amplitude and direction of synaptic plasticity can also be modulated by previous synaptic inputs.

Yamazaki et al. [27] showed that LTP induced at SC-CA1 neuron synapses by delivery of LFS of 80 pulses at 1 Hz is depotentiated by an identical LFS applied 20 minutes later. In this DP, activation of mGluRs and/or IP_3Rs during the first LFS is required for the reversal of LTP by the second LFS, although the activation of these receptors is not involved in LTP induction itself.

In addition, activation of NMDARs during the second LFS plays an important role in determining the direction of synaptic plasticity; when LTP at SC-CA1 neuron synapses is induced in standard solution, but the second LFS is applied in the presence of an NMDAR antagonist, this LTP is not reversed (depotentiated), but increased [27]. The excitatory postsynaptic current through NMDARs recorded from CA1 pyramidal neurons increases significantly 20 minutes after a single LFS and this increase is inhibited when the LFS is delivered in the presence of a group I mGluR inhibitor, an IP_3R inhibitor, or a CaMKII inhibitor. These results suggest that, in hippocampal CA1 neurons, activation of IP_3Rs by a previous LFS results in postsynaptic protein phosphorylation and/or enhancement of NMDAR activation during a subsequent LFS, leading to reversal of LTP in the field EPSP.

7. LTP and LTD May Involve Both Facilitatory and Inhibitory Changes and the Direction and Amplitude of Synaptic Plasticity Is Determined by the Balance of Facilitatory and Inhibitory Mechanisms Downstream of Receptor Activation

As discussed above, the difference between LTP and LTD is relative, since synaptic plasticity may change to the opposite direction if the strength of the input stimuli becomes stronger or weaker, depending on the stimulation paradigm [27, 65]. In other words, synaptic potentiation and depression may coexist.

LTP induced by a short tetanus (e.g. 100 Hz for 10 pulses) involves both potentiation and depression, as revealed by inhibition of mGluR-IP_3R signaling [36]. This LTP is enhanced if mGluR-IP_3R signaling is blocked by antagonists or gene targeting of IP_3R1 [35]. Interestingly, with a longer tetanus (e.g. 100 Hz for 100 pulses), LTP is no longer sensitive to mGluR-IP_3R inhibition [35, 36]. This implies that mGluR-IP_3R-mediated LTP inhibition is effective if the LTP-inducing HFS has a short duration (~100 msec), while HFS of longer duration (~1 sec) has stronger LTP inducing activity and LTP inhibition by mGluR-IP3R signaling is overwhelmed.

In the course of the induction of synaptic plasticity, an increase in the $[Ca^{2+}]_i$ is one of the earliest steps in intracellular signaling. Many explanations for the mechanisms of Ca^{2+}-dependent plasticity, such as the BCM theory [64], regard the strength of the input stimulus or the amplitude of the Ca^{2+} transient as the major determinant for the plasticity outcome. However, other aspects of Ca^{2+} signaling, such as duration or frequency of the Ca^{2+} transient, should be taken into account.

Besides the amplitude of the Ca^{2+} transient, probably the most important properties of Ca^{2+} signaling are its temporal features, i.e. duration and frequency. CaMKII and calcineurin, major downstream targets of Ca^{2+} signaling in plasticity induction, can be differentially activated by a Ca^{2+} transient, depending on its duration and frequency. Li et al. [67] reported that a Ca^{2+} increase activates both CaMKII and calcineurin, whereas a higher input frequency is more effective in activating CaMKII than calcineurin. They also showed that CaMKII activation depends on the total amount of the Ca^{2+} increase, estimated as the product of the amplitude and duration, and that, if the total amount of the Ca^{2+} increase is fixed, a higher frequency with a smaller amplitude is more effective in activating CaMKII than calcineurin. This result is intriguing, since it might be in accordance with the conversion of LTD to LTP by blockade of either NMDARs or mGluRs [65]. Partial block of the Ca^{2+} increase reduces the amplitude of each Ca^{2+} increase, but leaves the frequency intact. Given that CaMKII is more efficiently activated by a Ca^{2+} increase of higher frequency and smaller amplitude [67], there might be a certain range of frequencies in which a reduction in the amplitude of the Ca^{2+} increase decreases calcineurin activation more than that of CaMKII. It is generally considered that, in hippocampal pyramidal neurons, activation of CaMKII leads to LTP [68, 69], while activation of calcineurin is likely to induce LTD or DP [23, 39]. Thus, it is possible that partial inhibition of both CaMKII and calcineurin, with less inhibition of CaMKII and more inhibition of calcineurin, converts LTD to LTP.

Activation of CaMKII or calcineurin and induction of LTP or LTD is not a simple one-to-one relationship. Recent studies have shown that CaMKII, as a kinase, activated by synaptic inputs can be rapidly inactivated within a couple of minutes [70], and that it plays a structural role in a later phase of LTP [71]. Activation of calcineurin is also required for the induction of LTP [72]. Thus, CaMKII and calcineurin may not be mutually counteracting factors for LTP/LTD. Instead, intracellular signaling components, including CaMKII, calcineurin, and other kinases and phosphatases and phosphatase inhibitors, may constitute signaling pathways for the induction/reversal of LTP/LTD and the balance of the activity of these kinases/phosphatases might be important for the sign of the plasticity. In the cerebellum, LTD at parallel fiber-Purkinje cell synapses is CaMKII-dependent [15, 73], suggesting that different brain regions may use different intracellular signaling systems. Accordingly, synaptic plasticity might be induced by a mixture of several phosphorylation/dephosphorylation signals and the direction of plasticity might be determined by the balance of these signaling pathways, rather than merely the amplitude of the Ca^{2+} increase.

8. mGluR-IP$_3$R Signaling Acts as a Negative Regulator of LTP and Stabilizes Network Activity

We have mentioned that activation of both NMDARs and mGluRs is involved in the induction of LTP and LTD. Since activation of mGluRs stimulates two signaling pathways, the mGluR-PKC and mGluR-IP$_3$R pathways, it can be either facilitatory or inhibitory for LTP induction, depending on the stimulation paradigm, but, in most cases, activation of the mGluR-IP$_3$R pathway is inhibitory.

An increase in the $[Ca^{2+}]_i$ can induce either LTP or LTD, but the relationship between the $[Ca^{2+}]_i$ and the direction of plasticity is not straightforward. Synaptic inputs for a relatively short period (1 second to less than 1 minute) cause LTP at a higher $[Ca^{2+}]_i$ and LTD at a lower $[Ca^{2+}]_i$, as suggested by the BCM theory [64]. On the other hand, a chronic change in network activity leads to homeostatic plasticity, which shows the opposite Ca^{2+}-plasticity relationship [62, 63, 66]. Gene targeting of the IP$_3$R1 revealed that IP$_3$Rs regulate network activity [8, 35], a result consistent with homeostatic plasticity. Recent studies have shown that synaptic inputs applied for an intermediate period of time (1 - 10 minutes) induces LTP or LTD at a lower or higher $[Ca^{2+}]_i$, respectively [65]. These results suggest that mGluR-IP$_3$R signaling negatively regulates synaptic plasticity when synaptic inputs are delivered on an intermediate (~1 minute) or a longer (10 min) or chronic (hours to days) time scale. The roles of mGluR-IP_3R signaling in learning and memory are still largely elusive, but present understanding of this signaling pathway provide clues for investigating animal behaviors, such as erasure of memory, which are presumably regulated by negative synaptic plasticity.

References

[1] Nakanishi S. (1992) Molecular diversity of glutamate receptors and implications for brain function. *Science,* 258:597-603.

[2] Dingledine R., Borges K., Bowie D., Traynelis S. F. (1999) The glutamate receptor ion channels. *Pharmacol. Rev.*, 51:7-61.

[3] Niswender C. M., Conn P. J. (2010) Metabotropic glutamate receptors: physiology, pharmacology, and disease. *Annu. Rev. Pharmacol. Toxicol.*, 50:295-322.

[4] Jonas P., Burnashev N. (1995) Molecular mechanisms controlling calcium entry through AMPA-type glutamate receptor channels. *Neuron,* 15:987-990.

[5] Mayer M. L., Westbrook G. L., Guthrie P. B. (1984) Voltage-dependent block by Mg^{2+} of NMDA responses in spinal cord neurones. *Nature,* 309:261-263.

[6] Nowak L., Bregestovski P., Ascher P., Herbet A., Prochiantz A. (1984) Magnesium gates glutamate-activated channels in mouse central neurones. *Nature,* 307:462-465.

[7] Lodge D. (2009) The history of the pharmacology and cloning of ionotropic glutamate receptors and the development of idiosyncratic nomenclature. *Neuropharmacology,* 56:6-21.

[8] Matsumoto M., Nakagawa T., Inoue T., Nagata E., Tanaka K., Takano H., Minowa O., Kuno J., Sakakibara S., Yamada M., Yoneshima H., Miyawaki A., Fukuuchi Y., Furuichi T., Okano H., Mikoshiba K., Noda T. (1996) Ataxia and epileptic seizures in mice lacking type 1 inositol 1,4,5-trisphosphate receptor. *Nature,* 379:168-171.

[9] Aiba A., Kano M., Chen C., Stanton M. E., Fox G. D., Herrup K., Zwingman T. A., Tonegawa S., (1994) Deficient cerebellar long-term depression and impaired motor learning in mGluR1 mutant mice. *Cell,* 79:377-388.

[10] Kim J. J., Thompson R. F. (1997) Cerebellar circuits and synaptic mechanisms involved in classical eyeblink conditioning. *Trends Neurosci.*, 20:177-181.

[11] De Zeeuw C. I., Hansel C., Bian F., Koekkoek S. K., van Alphen A. M., Linden D. J., Oberdick J. (1998) Expression of a protein kinase C inhibitor in Purkinje cells blocks cerebellar LTD and adaptation of the vestibulo-ocular reflex. *Neuron*, 20:495-508.

[12] Feil R., Hartmann J., Luo C., Wolfsgruber W., Schilling K., Feil S., Barski J. J., Meyer M., Konnerth A., De Zeeuw C. I., Hofmann F. (2003) Impairment of LTD and cerebellar learning by Purkinje cell-specific ablation of cGMP-dependent protein kinase I. *J. Cell Biol.*, 163:295-302.

[13] Koekkoek S. K., Hulscher H. C., Dortland B. R., Hensbroek R. A., Elgersma Y., Ruigrok T. J., De Zeeuw C. I. (2003) Cerebellar LTD and learning-dependent timing of conditioned eyelid responses. *Science,* 301:1736-1739.

[14] Boyden E. S., Katoh A., Pyle J. L., Chatila T. A., Tsien R. W., Raymond J. L. (2006) Selective engagement of plasticity mechanisms for motor memory storage. *Neuron,* 51:823-834.

[15] Hansel C., de Jeu M., Belmeguenai A., Houtman S. H., Buitendijk G. H., Andreev D., De Zeeuw C. I., Elgersma Y. (2006) alphaCaMKII Is essential for cerebellar LTD and motor learning. *Neuron,* 51:835-843.

[16] Schonewille M., Gao Z., Boele H. J., Veloz M. F., Amerika W. E., Simek A. A., De Jeu M. T., Steinberg J. P., Takamiya K., Hoebeek F. E., Linden D. J., Huganir R. L., De Zeeuw C. I. (2011) Reevaluating the role of LTD in cerebellar motor learning. *Neuron,* 70:43-50.

[17] Inoue T., Kato K., Kohda K., Mikoshiba K. (1998) Type 1 inositol 1,4,5-trisphosphate receptor is required for induction of long-term depression in cerebellar Purkinje neurons. *J. Neurosci.,* 18:5366-5373.

[18] Ogura H., Matsumoto M., Mikoshiba K. (2001) Motor discoordination in mutant mice heterozygous for the type 1 inositol 1,4,5-trisphosphate receptor. *Behav. Brain Res.,* 122:215-219.

[19] Bliss T. V., Gardner-Medwin A. R. (1973) Long-lasting potentiation of synaptic transmission in the dentate area of the unanaestetized rabbit following stimulation of the perforant path. *J. Physiol.*, 232: 357-374.

[20] Bliss T. V., Lømo T. (1973) Long-lasting potentiation of synaptic transmission in the dentate area of the anaesthetized rabbit following stimulation of the perforant path. *J. Physiol.,* 232:331-356.

[21] Bliss T. V., Collingridge G. L. (1993) A synaptic model of memory: long-term potentiation in the hippocampus. *Nature,* 361:31-39.

[22] Dudek S. M., Bear M. F. (1992) Homosynaptic long-term depression in area CA1 of hippocampus and effects of N-methyl-D-aspartate receptor blockade. *Proc. Natl. Acad. Sci. USA,* 89:4363-4367.

[23] Mulkey R. M., Endo S., Shenolikar S., Malenka R. C. (1994) Involvement of a calcineurin/inhibitor-1 phosphatase cascade in hippocampal long-term depression. *Nature,* 369:486-488.

[24] Lee H. K., Barbarosie M., Kameyama K., Bear M. F., Huganir R. L. (2000) Regulation of distinct AMPA receptor phosphorylation sites during bidirectional synaptic plasticity. *Nature,* 405:955-959.

[25] Bashir Z. I., Bortolotto Z. A., Davies C. H., Berretta N., Irving A. J., Seal A. J., Henley J. M., Jane D. E., Watkins J. C., Collingridge G. L. (1993) Induction of LTP in the hippocampus needs synaptic activation of glutamate metabotropic receptors. *Nature,* 363:347-350.

[26] Musgrave M. A., Ballyk B. A., Goh J. W. (1993) Coactivation of metabotropic and NMDA receptors is required for LTP induction. *Neuroreport,* 4:171-174.

[27] Yamazaki Y., Fujii S., Aihara T., Mikoshiba K. (2012) Activation of inositol 1,4,5-trisphosphate receptors during preconditioning low-frequency stimulation leads to reversal of long-term potentiation in hippocampal CA1 neurons. *Neuroscience*, 207: 1-11.

[28] Oliet S. H., Malenka R. C., Nicoll R. A. (1997) Two distinct forms of long-term depression coexist in CA1 hippocampal pyramidal cells. *Neuron,* 18:969-982.

[29] Domenici M. R., Berretta N., Cherubini E. (1998) Two distinct forms of long-term depression coexist at the mossy fiber-CA3 synapse in the hippocampus during development. *Proc. Natl. Acad. Sci. USA*, 95: 8310-8315.

[30] Bortolotto Z. A., Bashir Z. I., Davies C. H., Collingridge G. L. (1994) A molecular switch activated by metabotropic glutamate receptors regulates induction of long-term potentiation. *Nature,* 368:740-743.

[31] Ben-Ari Y., Aniksztejn L., Bregestovski P. (1992) Protein kinase C modulation of NMDA currents: an important link for LTP induction. *Trends Neurosci.,* 15:333-339.

[32] Fernandez de Sevilla D., Buno W. (2010) The muscarinic long-term enhancement of NMDA and AMPA receptor-mediated transmission at Schaffer collateral synapses develop through different intracellular mechanisms. *J. Neurosci.,* 30:11032-11042.

[33] Fernandez de Sevilla D., Nunez A., Borde M., Malinow R., Buno W. (2008) Cholinergic-mediated IP_3-receptor activation induces long-lasting synaptic enhancement in CA1 pyramidal neurons. *J. Neurosci.*, 28:1469-1478.

[34] Holbro N., Grunditz A., Oertner T. G. (2009) Differential distribution of endoplasmic reticulum controls metabotropic signaling and plasticity at hippocampal synapses. *Proc. Natl. Acad. Sci. USA,* 106:15055-15060.

[35] Fujii S., Matsumoto M., Igarashi K., Kato H., Mikoshiba K. (2000) Synaptic plasticity in hippocampal CA1 neurons of mice lacking type 1 inositol-1,4,5-trisphosphate receptors. *Learn Mem.*, 7:312-320.

[36] Taufiq A. M., Fujii S., Yamazaki Y., Sasaki H., Kaneko K., Li J., Kato H., Mikoshiba K. (2005) Involvement of IP_3 receptors in LTP and LTD induction in guinea pig hippocampal CA1 neurons. *Learn Mem.*, 12:594-600.

[37] Fujii S., Saito K., Miyakawa H., Ito K., Kato H., (1991) Reversal of long-term potentiation (depotentiation) induced by tetanus stimulation of the input to CA1 neurons of guinea pig hippocampal slices. *Brain Res.,* 555:112-122.

[38] Bashir Z. I., Collingridge G. L. (1994) An investigation of depotentiation of long-term potentiation in the CA1 region of the hippocampus. *Exp. Brain Res.,* 100:437-443.

[39] O'Dell T. J., Kandel E. R. (1994) Low-frequency stimulation erases LTP through an NMDA receptor-mediated activation of protein phosphatases. *Learn Mem.,* 1:129-139.

[40] Yamazaki Y., Sugihara T., Goto J., Chida K., Fujiwara H., Kaneko K., Fujii S., Mikoshiba K. (2011) Role of inositol 1,4,5-trisphosphate receptors in the postsynaptic expression of guinea pig hippocampal mossy fiber depotentiation. *Brain Res.,* 1387: 19-28.

[41] Fujii S., Kuroda Y., Miura M., Furuse H., Sasaki H., Kaneko K., Ito K., Chen Z., Kato H. (1996) The long-term suppressive effect of prior activation of synaptic inputs by low-frequency stimulation on induction of long-term potentiation in CA1 neurons of guinea pig hippocampal slices. *Exp. Brain Res.,* 111:305-312.

[42] Yamazaki Y., Fujii S., Goto J., Sugihara T., Sugita M., Fujiwara H., Kaneko K., Aihara T, Mikoshiba K (2012) Suppressive effect of preconditioning low-frequency stimulation on subsequent induction of long-term potentiation by high frequency stimulation in hippocampal CA3 neurons. *Brain Res.,* 1449:15-23.

[43] Weisskopf M. G., Castillo P. E., Zalutsky R. A., Nicoll R. A. (1994) Mediation of hippocampal mossy fiber long-term potentiation by cyclic AMP. *Science,* 265:1878-1882.

[44] Nicoll R. A., Schmitz D. (2005) Synaptic plasticity at hippocampal mossy fibre synapses. *Nat. Rev. Neurosci.*, 6:863-876.

[45] Ito I., Sugiyama H. (1991) Roles of glutamate receptors in long-term potentiation at hippocampal mossy fiber synapses. *Neuroreport*, 2:333-336.

[46] Yeckel M. F., Kapur A., Johnston D. (1999) Multiple forms of LTP in hippocampal CA3 neurons use a common postsynaptic mechanism. *Nat. Neurosci.*, 2:625-633.

[47] Kapur A., Yeckel M., Johnston D. (2001) Hippocampal mossy fiber activity evokes Ca^{2+} release in CA3 pyramidal neurons via a metabotropic glutamate receptor pathway. *Neuroscience,* 107:59-69.

[48] Sokolov M. V., Rossokhin A. V., A MK., Gasparini S., Berretta N., Cherubini E., Voronin L. L. (2003) Associative mossy fibre LTP induced by pairing presynaptic stimulation with postsynaptic hyperpolarization of CA3 neurons in rat hippocampal slice. *Eur. J. Neurosci.,* 17:1425-1437.

[49] Fukunaga K., Stoppini L., Miyamoto E., Muller D. (1993) Long-term potentiation is associated with an increased activity of Ca^{2+}/calmodulin-dependent protein kinase II. *J. Biol. Chem.*, 268:7863-7867.

[50] Lee H. K., Choi S. S., Han E. J., Han K. J., Suh H. W. (2003) Role of glutamate receptors and an on-going protein synthesis in the regulation of phosphorylation of Ca^{2+}/calmodulin-dependent protein kinase II in the CA3 hippocampal region in mice administered with kainic acid intracerebroventricularly. *Neurosci. Lett.,* 348:93-96.

[51] Morioka M., Nagahiro S., Fukunaga K., Miyamoto E., Ushio Y. (1997) Calcineurin in the adult rat hippocampus: different distribution in CA1 and CA3 subfields. *Neuroscience,* 78:673-684.

[52] Bear M. F., Abraham W. C. (1996) Long-term depression in hippocampus. *Annu. Rev. Neurosci.*, 19:437-462.

[53] Alagarsamy S., Sorensen S. D., Conn P. J. (2001) Coordinate regulation of metabotropic glutamate receptors. *Curr. Opin. Neurobiol.*, 11:357-362.

[54] Mao L. M., Liu X. Y., Zhang G. C., Chu X. P., Fibuch E. E., Wang L. S., Liu Z., Wang J. Q. (2008) Phosphorylation of group I metabotropic glutamate receptors (mGluR1/5) in vitro and in vivo. *Neuropharmacology,* 55:403-408.

[55] Parekh A. B., Putney J. W. Jr. (2005) Store-operated calcium channels. *Physiol. Rev.,* 85:757-810.

[56] Roos J., DiGregorio P. J., Yeromin A. V., Ohlsen K., Lioudyno M., Zhang S., Safrina O., Kozak J. A., Wagner S. L., Cahalan M. D., Velicelebi G., Stauderman K. A. (2005) STIM1, an essential and conserved component of store-operated Ca^{2+} channel function. *J. Cell Biol.*, 169:435-445.

[57] Zhang S. L., Yu Y., Roos J., Kozak J. A., Deerinck T. J., Ellisman M. H., Stauderman K. A., Cahalan M. D. (2005) STIM1 is a Ca^{2+} sensor that activates CRAC channels and migrates from the Ca^{2+} store to the plasma membrane. *Nature,* 437:902-905.

[58] Liou J., Kim M. L., Heo W. D., Jones J. T., Myers J. W., Ferrell J. E. Jr., Meyer T. (2005) STIM is a Ca^{2+} sensor essential for Ca^{2+}-store-depletion-triggered Ca^{2+} influx. *Curr. Biol.,* 15:1235-1241.

[59] Zhang S. L., Yeromin A. V., Zhang X. H., Yu Y., Safrina O., Penna A., Roos J., Stauderman K. A., Cahalan M. D. (2006) Genome-wide RNAi screen of Ca^{2+} influx identifies genes that regulate Ca^{2+} release-activated Ca^{2+} channel activity. *Proc. Natl. Acad. Sci. USA,* 103:9357-9362.

[60] Feske S., Gwack Y., Prakriya M., Srikanth S., Puppel S. H., Tanasa B., Hogan P. G., Lewis R. S., Daly M., Rao A. (2006) A mutation in Orai1 causes immune deficiency by abrogating CRAC channel function. *Nature,* 441:179-185.

[61] Vig M., Peinelt C., Beck A., Koomoa D. L., Rabah D., Koblan-Huberson M., Kraft S., Turner H., Fleig A., Penner R., Kinet J. P. (2006) CRACM1 is a plasma membrane protein essential for store-operated Ca^{2+} entry. *Science,* 312:1220-1223.

[62] Hagenston A. M., Rudnick N. D., Boone C. E., Yeckel M. F. (2009) 2-Aminoethoxydiphenyl-borate (2-APB) increases excitability in pyramidal neurons. *Cell Calcium,* 45:310-317.

[63] Steinbeck J. A., Henke N., Opatz J., Gruszczynska-Biegala J., Schneider L., Theiss S., Hamacher N., Steinfarz B., Golz S., Brustle O., Kuznicki J., Methner A. (2011) Store-operated calcium entry modulates neuronal network activity in a model of chronic epilepsy. *Exp. Neurol.,* 232:185-194.

[64] Bienenstock E. L., Cooper L. N., Munro P. W. (1982) Theory for the development of neuron selectivity: orientation specificity and binocular interaction in visual cortex. *J. Neurosci.,* 2:32-48.

[65] Fujii S., Yamazaki Y., Kuroda Y., Mikoshiba K. (2010) Involvement of inositol-1,4,5-trisphosphate receptors in the bidirectional synaptic plasticity induced in hippocampal CA1 neurons by 1-10 Hz low-frequency stimulation. *Neuroscience,* 168:346-358.

[66] Pozo K., Goda Y. (2010) Unraveling mechanisms of homeostatic synaptic plasticity. *Neuron,* 66:337-351.

[67] Li L., Stefan M. I., Le Novere N. (2012) Calcium input frequency, duration and amplitude differentially modulate the relative activation of calcineurin and CaMKII. *PLoS One*, 7:e43810.

[68] Malinow R., Schulman H., Tsien R. W. (1989) Inhibition of postsynaptic PKC or CaMKII blocks induction but not expression of LTP. *Science,* 245:862-866.

[69] Silva A. J., Stevens C. F., Tonegawa S., Wang Y. (1992) Deficient hippocampal long-term potentiation in alpha-calcium-calmodulin kinase II mutant mice. *Science,* 257:201-206.

[70] Lee S. J., Escobedo-Lozoya Y., Szatmari E. M., Yasuda R. (2009) Activation of CaMKII in single dendritic spines during long-term potentiation. *Nature,* 458:299-304.

[71] Lisman J., Yasuda R., Raghavachari S. (2012) Mechanisms of CaMKII action in long-term potentiation. *Nat. Rev. Neurosci.,* 13:169-182.

[72] Lu Y. F., Hayashi Y., Moriwaki A., Tomizawa K., Matsui H. (1996) FK506, a Ca^{2+}/calmodulin-dependent phosphatase inhibitor, inhibits the induction of long-term potentiation in the rat hippocampus. *Neurosci. Lett.,* 205:103-106.

[73] van Woerden G. M., Hoebeek F. E., Gao Z., Nagaraja R. Y., Hoogenraad CC, Kushner SA, Hansel C, De Zeeuw CI, Elgersma Y (2009) betaCaMKII controls the direction of plasticity at parallel fiber-Purkinje cell synapses. *Nat. Neurosci.*, 12:823-825.

In: Metabotropic Glutamate Receptors
Editor: M. Foster Olive

ISBN: 978-1-63117-451-3

Chapter 3

Group I mGluR Receptors as Mediators of Learning and Memory

Lucas R. Watterson*[*] *and M. Foster Olive
Department of Psychology,
Arizona State University, Tempe, AZ, US

Abstract

Group I metabotropic glutamate receptors (mGluR1 and mGluR5) are predominantly expressed on postsynaptic elements and are intricately involved in various forms of neural plasticity, including long-term potentiation (LTP) and long-term depression (LTD). Given the well-established role of these and other forms of neural plasticity in learning and memory, it comes as no surprise that pharmacological or genetic deletion of Group I metabotropic glutamate receptors (mGluR1 and mGluR5) have deleterious effects on learning and memory in numerous behavioral paradigms. In contrast, positive allosteric modulation of these receptors, especially mGluR5, can enhance synaptic plasticity and improve various aspects of cognition, including spatial learning, behavioral flexibility, and the extinction of fear-related and drug-seeking behaviors. In this chapter we review the current literature regarding the role of Group I mGluR receptors in synaptic plasticity, learning and memory.

Keywords: Glutamate, cognition, learning, memory, synaptic plasticity, mGluR1, mGluR5

Introduction

Glutmate is the primary excitatory chemical messenger in the central nervous system (CNS) and is responsible for approximately 70% of all neurotransmission in the brain. When released into the synaptic cleft, glutamate can mediate rapid excitatory transmission via

[*] Department of Psychology, Arizona State University, PO Box 871104, Tempe, AZ 85287-1104, USA. Email: foster.olive@asu.edu.

interactions with ionotropic glutamate receptors (iGluRs), which are glutamate-gated cation channels that include N-methyl-D-aspartate (NMDA), α-amino-3-hydroxy-5-methylisoxazole-4-propionic acid (AMPA), and kainic acid (KA) receptors. iGluRs are widely distributed throughout the brain, are essential mediators of homeostatic processes throughout the CNS, and can often cause severe side effects when targeted pharmacologically. As a result, the therapeutic potential of iGluR ligands for the treatment neuropsychiatric disorders has thus far proved to be disappointingly limited [1-3]. However, glutamate also binds to metabotropic G-protein coupled receptors (mGluRs) that mediate slower modulatory transmission through second messenger system cascades. mGluRs are more diverse in their distribution throughout the CNS, and when targeted pharmacologically, are less prone to lead to severe side effects [1, 2]. As a result, mGluRs have received a great deal of attention in recent years as promising targets of drug action for many CNS disorders [1-4]. Furthermore, as reviewed below, the subset of mGluRs known as Group I have received the most attention to date and have shown promise as potential CNS pharmacotherapeutics.

To date, eight mGluR receptor subtypes (mGluR1–8) have been characterized and are subdivided into one of three categories based largely on their sequence homology, but also on their neuroanatomical (synaptic and regional) distribution, agonist selectivity, and intracellular signaling mechanisms. These three groups are referred to as Group I mGluRs (mGluR1 and mGluR5), Group II (mGluR2 and mGluR3), and Group III (mGluR4, mGluR6, mGluR7, and mGluR8) [5-7]. Furthermore, with the exception of mGluR2, all mGluRs can exist in different splice variants; for example, genes encoding mGluR1 receptors can be alternatively spliced to form mGluR1a-f subtypes [6, 8, 9]. In non-human mammals, the two prominent mGluR5 isoforms are mGluR5a and 5b, while 5c, 5d, and 5e have also been identified in humans [10, 11]. For a review of Group II and Group III splice variants, see [11].

Group I mGluRs are predominantly located post-synaptically on the perisynaptic annulus of dendritic spines [12] and are physically linked with NMDA receptors by various scaffolding proteins including PSD-95, Shank, the Homer family of proteins, and via direction interactions. However, a small number of Group I receptors are also found on presynaptic terminals where they play a role in modulating the release of glutamate [13]. mGluR5 receptors are widely distributed throughout the brain whereas mGluR1 is more limited, and with the exception of the hippocampus which contains high concentrations of both mGluR1 and mGluR5 receptors, the distribution of both of these receptors in the CNS is primarily complementary [14]. Specifically, the highest concentration of mGluR5 receptors are found in forebrain regions such as the cerebral cortex, dorsal and ventral striatum, olfactory bulb and tubercle, lateral septum, and hippocampus [15, 16], whereas mGluR1 receptors are highly expressed the cerebellum, thalamus, olfactory bulb, lateral septum, superior colliculus, and pallidal regions [17], where mGluR5 receptors are few or non-existent. Because the hippocampus is one of the few brain regions that contains high levels of both mGluR1 (excluding the CA1 region) and mGluR5 (CA1-CA3 and dentate gyrus) receptors, the majority of studies focusing on the role of Group I in synaptic plasticity, learning, and memory have largely focused on this region.

Signaling Mechanisms of Group I mGluRs

Group I mGluRs are coupled to $G_{q/11}$ G-proteins, activation of which increases phospholipase C (PLC) activity and calcium release from intracellular stores resulting in enhanced cell excitability and activation of numerous intracellular signaling molecules such as protein kinase C (PKC), protein kinase A (PKA), mitogen-activated protein kinase (MAPK), extracellular signal-related kinase (ERK), and cyclic adenosine monophosphate (cAMP) response element binding protein (CREB) [1, 9]. On the other hand, Group II and Group III mGluRs are coupled to $G_{i/o}$ G-proteins, are typically located presynaptically, and inhibit the activity of adenylyl cyclase and formation of intracellular cAMP. Thus, through their functioning as releasing-regulating autoreceptors, both Group II and Group III mGluRs comprise part of a negative feedback system that inhibits excess glutamate release. Finally, mGluRs, primarily mGluR3 and mGluR5, are localized on glial cells, with mGluR3 being the most common on astrocytes where it plays a role in regulating extracellular glutamate concentrations as part of the tripartite synapse, as well as neuron-to astrocyte transmission [11, 18, 19].

As mentioned above, mGluR5 receptors are physically coupled to NMDA receptors via various scaffolding proteins and biochemically interact through PKC. As a result of these molecular and biochemical interactions, activation of mGluR5 receptors results in enhanced functionality of the NMDA receptor. This mGluR5-NMDA interaction has been observed in numerous brain preparations, whereby activation of mGluR5 receptors with an orthosteric agonist (such as chlorohydroxyphenylglycine (CHPG) or dihydroxyphenylglycine (DHPG)) or positive allosteric modulators (PAMs, see below) potentiates NMDA receptor-mediated responses to exogenously applied glutamate or NMDA.

Group I mGluRs and Synaptic Plasticity

Numerous studies in the last fifteen years have established that Group I mGluR receptors are integral to both the initiation and maintenance of synaptic plasticity [20-22]. As mentioned, Group I mGluR5 receptors are primarily located post-synaptically and their role in intiatiating and maintaining synaptic plasticity (both LTP and LTD) occurs through a number of mechanisms. These include facilitation of NMDA receptor function (discussed below in further detail), de novo synthesis of AMPA GluR2 subunits, post-synaptic scaffolding proteins, release of calcium into the cytosol from the endoplasmic reticulum, and resulting activiation of downstream effector proteins such as MAP and ERK kinases which in turn activate transcription factors to modulate gene expression, as well as via changes in phospholipid signaling [20-22]. Group I mGluR receptors also regulate the synthesis of retrograde signaling molecules such endocannabinoids, which can presynaptically stimulate CB1 receptors to initiate plasticity processes in axon terminals. Not surprisingly, the most studied and understood mechanism for the induction and maintenance of LTP by Group I receptors is through their positive coupling to and potentiation of post-synaptic NMDA receptors [23-32]. As a result, Group I mGluR receptor activation induces PKC activity which can phosphorylate (among other substrates) specific subunits of the NMDA receptor, resulting in enhanced NMDA receptor function [33].

Inhibitory Effects of Group I mGluR Inactivation on Learning and Memory

The now well established linkage between Group I mGluR and NMDA receptors has been further strengthened by an number of studies showing that Group I mGluR receptors play an integral role in NMDA-dependent synaptic plasticity as well as learning and memory [34-36]. For instance, genetic or pharmacoloigical blockade of Group I mGluR receptors attenuates the expression of LTP in regions such as the hippocampus (dentate gyrus and CA1 region), amygdala, and dorsal and ventral striatum [23-32, 37-47]. These regions are essential in contextual/spatial learning, reward cue/emotional learning, and habit/incentive learning, respectively. In addition, microinjection studies have also shown that local infusions of Group I mGluR receptor antagonists into the hippocampus produce decrements in contextual [48] inhibitory avoidance learning [49], disruption of emotional learning when infused into the amygdala [47, 50, 51], and disruption of habit learning when infused into the dorsal striatum [52]. In contrast, local stimulation of Group I mGluR receptors in the amygdala has been shown to augment emotional learning [53].

The role of Group I mGluRs on spatial learning tasks has especially received much attention. Spatial learning is well established to be hippocampal dependent, and genetic deletion or pharmacological blockade of Group I mGluR receptors inhibits the induction and/or expression of LTP in various brain regions including the CA1 and dentate gyrus regions of the hippocampus [37-45]. Not surprisingly, both genetic knockout and pharmacological blockade studies have revealed Group I receptors are necessary for spatial navigational learning [36], although the studies assessing mGluR5 receptors far outnumbered those assessing mGluR1. In genetic knockout mice, deletion of both mGluR1 [54] and mGluR5 [41, 55] show impaired performance on the Morris water maze. However, some studies have shown that impairments are not global. For example, Naie et al. [44] showed that pharmacological blockade of mGluR1 impaired reference but not working memory on the radial arm maze. Numerous independent studies have shown that the mGluR5 antagonist 2-methyl-6-(phenylethynyle)pyridine (MPEP) leads to robust spatial learning deficits on the Morris water maze and other spatial tasks [2, 42, 43, 56-62]. However, it should be noted that other laboratories have not found spatial learning memory impairments following MPEP treatment [63-65]. In addition, the more selective mGluR5 antagonist 3-[2-methyl-1,3-thiazol-4-yl)ethynyl]pyridine (MTEP) has been shown to impair reference memory, but not working memory, in the radial arm maze [66], was shown by Naie et al. [44]. Finally, the mGluR5 antagonist fenobam, which binds to the same site as MPEP but with lower affinity, also impairs acquisition learning on the Morris water maze [58].

In fear-conditioning tasks, a type of context-specific associative learning dependent on both hippocampal and amygdaloid functioning, it has been shown that mGluR5 knockout mice show decreased fear conditioning to footshock-associated contexts as compared to wild-type controls as well as decrements in the extinction of conditioned fear-related behaviors [55]. In a study by Maciejak et al. [48], infusions of the mGluR1 receptor antagonist (RS)-1-aminoindan-1,5- dicarboxylic acid (AIDA) into the hippocampus increased freezing to a fear-conditioned context 24 hours following conditioning trials, whereas intrahippocampal injection of the Group I mGluR agonist DHPG decreased freezing. In this same study, however, intra-hippocampal infusions of the mGluR5 antagonist MPEP or mGluR5 agonist

CHPG did not have any effects, suggesting differential role for mGluR1 and mGluR5 receptors in fear-conditioned behaviors. Furthermore, others have shown that systemic injections of MPEP, MTEP or fenobam prior to fear-conditioning trials attenuates the acquisition and expression of conditioned fear [58, 66, 67]. In addition to impairing the acquisition of conditioned fear, the mGluR5 antagonist MPEP has also been shown to impair the consolidation of extinction learning following fear conditioning [68]. In a recent study, it was shown that spontaneous recovery of fear was attenuated following prolonged extinction training, an effect that was blocked by systemic injections of the mGluR5 antagonist MTEP, but not by systemic injections of the mGluR1 antagonist 7-(hydroxyimino)cyclopropa[b]chromen-1a-carboxylate ethyl ester (CPCCOEt), again suggesting differential roles for mGluR1 and mGluR5 receptors in fear-related behaviors [69].

Facilitatory Effects of Group I mGluR Activation on Learning and Memory

It is now well established that mGluR5 PAMs can facilitate the induction of synaptic plasticity. For example, the mGluR5 PAMs 3-cyano-N-(1,3-diphenyl-1H-pyrazol-5-yl)benzamide (CDPPB) and(S)-(4-fluoro-phenyl)-(3-[3-(4-fluorophenyl)-[1,2,4]-oxadiazol-5-yl]piperidin-1-yl)methanone (ADX47273) increase the expression of markers of synaptic plasticity in the PFC and hippocampus, such as LTP [70], increased phosphorylation of the NR1 and NR2B subunits of the NMDA receptor and the GluR1 subunit of the AMPA receptor, and activate α-calmodulin-dependent kinase II, CREB, and ERK [71-73]. 4-Nitro-N-(1,3-diphenyl-1H-pyrazol-5-yl)benzamide (VU-29), an mGluR5 PAM and CDPPB analogue, facilitated the induction of LTP in hippocampal CA1 slice preparations, an effect blocked by AP-5, confirming that increased NMDA receptor functioning is necessary for the expression of LTP with mGluR5 allosteric activation [74]. VU-29 also potentiated DHPG-induced LTD in this region. In this same report, it was shown that systemic administration of CDPPB or ADX47273 improved performance in the Morris water maze [74]. Consistent with these findings, systemic or intracerebroventricular infusion of mGluR5 PAMs been shown to improve spatial memory in various other spatial tasks [75, 76]. CDPPB and ADX47273 have also been reported to improve cognition as measured by novel object recognition [71, 77], conditioned avoidance [78, 79], reversal learning [80], and five-choice serial reaction time tests [72]. Support for mGluR5 receptor PAM-induced increases in prefrontal cortical functioning have been demonstrated using in vivo electrophysiological techniques, where CDPPB was shown to reverse changes in prefrontal cortex neuronal firing patterns induced by the psychotomimetic NMDA receptor antagonist MK-801 [2, 81, 82], effects that have been replicated in studies of reversal of deficits in cognitive set-shifting [83, 84].

Numerous studies have shown that Group I agonist can potentiate extinction learning following acquisition of fear-responding and drug-seeking in various tasks. The mGluR PAM ADX47273 enhanced extinction of conditioned fear after a single retrieval session [80]. CDPPB facilitated the extinction of a cocaine-induced conditioned place preference [85] and reduces extinction responding following intravenous cocaine and methamphetamine self-administration [86, 87] and oral alcohol self-administration [88]. However, CDPPB failed to facilitate extinction of methamphetamine-seeking when extinction sessions were conducted in a context different than the context during methamphetamine self-administration [89]. Importantly, while mGluR5 PAM activation of NMDA receptors can produce excitotoxicity at high doses [90], other have shown that repeated CDPPB administration does not produce evidence of neurotoxicity [85]. Interesting, it has been shown that the mGluR5 antagonist MPEP facilitates the conditioned rewarding effects of abused drugs such as cocaine, nicotine, heroin, and ketamine. Furthermore, co-treatment of MPEP with clonidine and buspirone, two drugs that alone do not produce CPP, led to a conditioned place preference when paired with MPEP during conditioning [91]. Thus, mGluR5 receptor PAMs may be a novel class of compounds by which to facilitate the extinction of drug-stimuli or drug-context associations. However, mGluR5 receptor antagonists may potentiate the rewarding effects of some non-glutamatergic compounds.

Conclusion

It is now established that Group I mGluR receptors play an integral role in synaptic plasticity and various learning and memory processes including the acquisition and maintenance of spatial and contextual learning, fear-related behaviors, and addiction-like behaviors following exposure to rewarding drugs. While Group I mGluR antagonists have largely been shown to impair various types of learning and memory, they have also shown promise in impairing the acquisition of conditioned fear or condition drug-seeking and in decreasing the expression of these behaviors following conditioning. In addition, Group I mGluR activation appear to facilitate the extinction of conditioned fear and drug-seeking behaviors, suggesting that compounds targets these receptors have potential as adjunct pharmacotherapeutics to exposure or other extinction-based therapies. Thus, Group I mGluR receptors remain potential novel pharmacological targets for the treatment of disorders of learning, memory, anxiety, or substance abuse.

References

[1] Niswender CM, Conn PJ (2010) Metabotropic glutamate receptors: physiology, pharmacology, and disease. *Annu. Rev. Pharmacol. Toxicol.* 50:295-322.

[2] Homayoun H, Moghaddam B (2010) Group 5 metabotropic glutamate receptors: role in modulating cortical activity and relevance to cognition. *Eur. J. Pharmacol.* 639:33-9.

[3] Gravius A, Pietraszek M, Dekundy A, Danysz W (2010) Metabotropic glutamate receptors as therapeutic targets for cognitive disorders. *Curr. Top Med. Chem.* 10:187-206.

[4] Nicoletti F, Bockaert J, Collingridge GL, Conn PJ, Ferraguti F, Schoepp DD, Wroblewski JT, Pin JP (2011) Metabotropic glutamate receptors: from the workbench to the bedside. *Neuropharmacology* 60:1017-41.

[5] Nakanishi S (1994) Metabotropic glutamate receptors: synaptic transmission, modulation, and plasticity. *Neuron* 13:1031-7.

[6] Pin JP, Acher F (2002) The metabotropic glutamate receptors: structure, activation mechanism and pharmacology. *Curr. Drug Targets CNS Neurol. Disord.* 1:297-317.

[7] Schoepp DD, Jane DE, Monn JA (1999) Pharmacological agents acting at subtypes of metabotropic glutamate receptors. *Neuropharmacology* 38:1431-76.

[8] Pin JP, Duvoisin R (1995) The metabotropic glutamate receptors: structure and functions. *Neuropharmacology* 34:1-26.

[9] Hermans E, Challiss RA (2001) Structural, signalling and regulatory properties of the group I metabotropic glutamate receptors: prototypic family C G-protein-coupled receptors. *Biochem. J.* 359:465-84.

[10] Minakami R, Katsuki F, Yamamoto T, Nakamura K, Sugiyama H (1994) Molecular cloning and the functional expression of two isoforms of human metabotropic glutamate receptor subtype 5. *Biochem. Biophys. Res. Commun.* 199:1136-43.

[11] Mukherjee S, Manahan-Vaughan D (2013) Role of metabotropic glutamate receptors in persistent forms of hippocampal plasticity and learning. *Neuropharmacology* 66:65-81.

[12] Lujan R, Nusser Z, Roberts JD, Shigemoto R, Somogyi P (1996) Perisynaptic location of metabotropic glutamate receptors mGluR1 and mGluR5 on dendrites and dendritic spines in the rat hippocampus. *Eur. J. Neurosci.* 8:1488-1500.

[13] Anwyl R (1999) Metabotropic glutamate receptors: electrophysiological properties and role in plasticity. *Brain Res. Rev.* 29:83-120.

[14] Shigemoto R, Mizuno N (2000). Metabotropic glutamate receptors - immunocytochemical and in situ hybridization analysis. In: Ottersen OP, Storm-Mathisen J (eds) *Handbook of Chemical Neuroanatomy: Metabotropic Glutamate Receptors: Immunocytochemical and In Situ Hybridization Analyses*. London: Elsevier, p. 63-98.

[15] Shigemoto R, Nomura S, Ohishi H, Sugihara H, Nakanishi S, Mizuno N (1993) Immunohistochemical localization of a metabotropic glutamate receptor, mGluR5, in the rat brain. *Neurosci. Lett.* 163:53-7.

[16] Romano C, Sesma MA, McDonald CT, O'Malley K, van den Pol AN, Olney JW (1995) Distribution of metabotropic glutamate receptor mGluR5 immunoreactivity in rat brain. *J. Comp. Neurol.* 355:455-69.

[17] Shigemoto R, Nakanishi S, Mizuno N (1992) Distribution of the mRNA for a metabotropic glutamate receptor (mGluR1) in the central nervous system: an in situ hybridization study in adult and developing rat. *J. Comp. Neurol.* 322:121-35.

[18] Aronica E, Gorter JA, Ijlst-Keizers H, Rozemuller AJ, Yankaya B, Leenstra S, Troost D (2003) Expression and functional role of mGluR3 and mGluR5 in human astrocytes and glioma cells: opposite regulation of glutamate transporter proteins. *Eur. J. Neurosci.* 17:2106-18.

[19] Perea G, Navarrete M, Araque A (2009) Tripartite synapses: astrocytes process and control synaptic information. *Trends Neurosci.* 32:421-31.

[20] Anwyl R (2009) Metabotropic glutamate receptor-dependent long-term potentiation. *Neuropharmacology* 56:735-40.

[21] Bellone C, Luscher C, Mameli M (2008) Mechanisms of synaptic depression triggered by metabotropic glutamate receptors. *Cell Mol. Life Sci.* 65:2913-23.

[22] Gladding CM, Fitzjohn SM, Molnar E (2009) Metabotropic glutamate receptor-mediated long-term depression: molecular mechanisms. *Pharmacol. Rev.* 61:395-412.

[23] Alagarsamy S, Marino MJ, Rouse ST, Gereau RWt, Heinemann SF, Conn PJ (1999) Activation of NMDA receptors reverses desensitization of mGluR5 in native and recombinant systems. *Nat. Neurosci.* 2:234-40.

[24] Alagarsamy S, Sorensen SD, Conn PJ (2001) Coordinate regulation of metabotropic glutamate receptors. *Curr. Opin. Neurobiol.* 11:357-62.

[25] Attucci S, Carla V, Mannaioni G, Moroni F (2001) Activation of type 5 metabotropic glutamate receptors enhances NMDA responses in mice cortical wedges. *Br. J. Pharmacol.* 132:799-806.

[26] Awad H, Hubert GW, Smith Y, Levey AI, Conn PJ (2000) Activation of metabotropic glutamate receptor 5 has direct excitatory effects and potentiates NMDA receptor currents in neurons of the subthalamic nucleus. *J. Neurosci.* 20:7871-9.

[27] Benquet P, Gee CE, Gerber U (2002) Two distinct signaling pathways upregulate NMDA receptor reponses via two distinct metabotropic glutamate receptor subtypes. *J. Neurosci.* 22:9679-9686.

[28] Doherty AJ, Palmer MJ, Henley JM, Collingridge GL, Jane DE (1997) (RS)-2-chloro-5-hydroxyphenylglycine (CHPG) activates mGlu5, but not mGlu1, receptors expressed in CHO cells and potentiates NMDA responses in the hippocampus. *Neuropharmacology* 36:265-7.

[29] Kotecha SA, Jackson MF, Al-Mahrouki A, Roder JC, Orser BA, MacDonald JF (2003) Co-stimulation of mGluR5 and N-methyl-D-aspartate receptors is required for potentiation of excitatory synaptic transmission in hippocampal neurons. *J. Biol. Chem.* 278:27742-9.

[30] Kotecha SA, MacDonald JF (2003) Signaling molecules and receptor transduction cascades that regulate NMDA receptor-mediated synaptic transmission. *Int. Rev. Neurobiol.* 54:51-106.

[31] Pisani A, Gubellini P, Bonsi P, Conquet F, Picconi B, Centonze D, Bernardi G, Calabresi P (2001) Metabotropic glutamate receptor 5 mediates the potentiation of N-methyl-D-aspartate responses in medium spiny striatal neurons. *Neuroscience* 106:579-587.

[32] Ugolini A, Corsi M, Bordi F (1999) Potentiation of NMDA and AMPA responses by the specific mGluR5 agonist CHPG in spinal cord motoneurons. *Neuropharmacology* 38:1569-76.

[33] Lu WY, Xiong ZG, Lei S, Orser BA, Dudek E, Browning MD, MacDonald JF (1999) G-protein-coupled receptors act via protein kinase C and Src to regulate NMDA receptors. *Nat. Neurosci.* 2:331-8.

[34] Riedel G, Platta B, Micheaub J (2003) Glutamate receptor function in learning and memory. *Behav. Brain Res.* 140:1-47.

[35] Simonyi A, Schachtman TR, Christoffersen GR (2005) The role of metabotropic glutamate receptor 5 in learning and memory processes. *Drug News Perspect.* 18:353-61.

[36] Simonyi A, Schachtman TR, Christoffersen GR (2010) Metabotropic glutamate receptor subtype 5 antagonism in learning and memory. *Eur. J. Pharmacol.* 639:17-25.

[37] Balschun D, Manahan-Vaughan D, Wagner T, Behnisch T, Reymann KG, Wetzel W (1999) A specific role for group I mGluRs in hippocampal LTP and hippocampus-dependent spatial learning. *Learn. Mem.* 6:138-52.

[38] Balschun D, Wetzel W (2002) Inhibition of mGluR5 blocks hippocampal LTP in vivo and spatial learning in rats. *Pharmacol. Biochem. Behav*. 73:375-380.

[39] Bikbaev A, Neyman S, Ngomba RT, Conn J, Nicoletti F, Manahan-Vaughan D (2008) MGluR5 mediates the interaction between late-LTP, network activity, and learning. *PLoS One* 3:e2155.

[40] Jia Z, Lu Y, Henderson J, Taverna F, Romano C, Abramow-Newerly W, Wojtowicz JM, Roder J (1998) Selective abolition of the NMDA component of long-term potentiation in mice lacking mGluR5. *Learn. Mem*. 5:331-43.

[41] Lu YM, Jia Z, Janus C, Henderson JT, Gerlai R, Wojtowicz JM, Roder JC (1997) Mice lacking metabotropic glutamate receptor 5 show impaired learning and reduced CA1 long-term potentiation (LTP) but normal CA3 LTP. *J. Neurosci*. 17:5196-205.

[42] Manahan-Vaughan D, Braunewell KH (2005) The metabotropic glutamate receptor, mGluR5, is a key determinant of good and bad spatial learning performance and hippocampal synaptic plasticity. *Cereb. Cortex* 15:1703-13.

[43] Naie K, Manahan-Vaughan D (2004) Regulation by metabotropic glutamate receptor 5 of LTP in the dentate gyrus of freely moving rats: relevance for learning and memory formation. *Cereb. Cortex* 14:189-98.

[44] Naie K, Manahan-Vaughan D (2005) Pharmacological antagonism of metabotropic glutamate receptor 1 regulates long-term potentiation and spatial reference memory in the dentate gyrus of freely moving rats via N-methyl-D-aspartate and metabotropic glutamate receptor-dependent mechanisms. *Eur. J. Neurosci*. 21:411-21.

[45] Neyman S, Manahan-Vaughan D (2008) Metabotropic glutamate receptor 1 (mGluR1) and 5 (mGluR5) regulate late phases of LTP and LTD in the hippocampal CA1 region in vitro. *Eur. J. Neurosci*. 27:1345-1352.

[46] Fendt M, Schmid S (2002) Metabotropic glutamate receptors are involved in amygdaloid plasticity. *Eur. J. Neurosci*. 15:1535-41.

[47] Rodrigues SM, Bauer EP, Farb CR, Schafe GE, LeDoux JE (2002) The group I metabotropic glutamate receptor mGluR5 is required for fear memory formation and long-term potentiation in the lateral amygdala. *J. Neurosci*. 22:5219-5229.

[48] Maciejak P, Taracha E, Lehner M, Szyndler J, Bidzinski A, Skorzewska A, Wislowska A, Zienowicz M, Plaznik A (2003) Hippocampal mGluR1 and consolidation of contextual fear conditioning. *Brain Res. Bull*. 62:39-45.

[49] Simonyi A, Serfozo P, Shelat PB, Dopheide MM, Coulibaly AP, Schachtman TR (2007) Differential roles of hippocampal metabotropic glutamate receptors 1 and 5 in inhibitory avoidance learning. *Neurobiol. Learn. Mem*. 88:305-311.

[50] Bonini JS, Rodrigues L, Kerr DS, Bevilaqua LR, Cammarota M, Izquierdo I (2003) AMPA/kainate and group-I metabotropic receptor antagonists infused into different brain areas impair memory formation of inhibitory avoidance in rats. *Behav. Pharmacol*. 14:161-6.

[51] Kim J, Lee S, Park H, Song B, Hong I, Geum D, Shin K, Choi S (2007) Blockade of amygdala metabotropic glutamate receptor subtype 1 impairs fear extinction. *Biochem. Biophys. Res. Commun*. 355:188-93.

[52] Packard MG, Vecchioli SF, Schroeder JP, Gasbarri A (2001) Task-dependent role for dorsal striatum metabotropic glutamate receptors in memory. *Learn. Mem*. 8:96-103.

[53] Rudy JW, Matus-Amat P (2009) DHPG activation of group 1 mGluRs in BLA enhances fear conditioning. *Learn. Mem*. 16:421-5.

[54] Aiba A, Chen C, Herrup K, Rosenmund C, Stevens CF, Tonegawa S (1994) Reduced hippocampal long-term potentiation and context-specific deficit in associative learning in mGluR1 mutant mice. *Cell* 79:365-75.

[55] Xu J, Zhu Y, Contractor A, Heinemann SF (2009) mGluR5 has a critical role in inhibitory learning. *J. Neurosci.* 29:3676-3684.

[56] Car H, Stefaniuk R, Wisniewska RJ (2007) Effect of MPEP in Morris water maze in adult and old rats. *Pharmacol. Rep.* 59:88-93.

[57] Christoffersen GR, Simonyi A, Schachtman TR, Clausen B, Clement D, Bjerre VK, Mark LT, Reinholdt M, Schmith-Rasmussen K, Zink LV (2008) MGlu5 antagonism impairs exploration and memory of spatial and non-spatial stimuli in rats. *Behav. Brain Res.* 191:235-245.

[58] Jacob W, Gravius A, Pietraszek M, Nagel J, Belozertseva I, Shekunova E, Malyshkin A, Greco S, Barberi C, Danysz W (2009) The anxiolytic and analgesic properties of fenobam, a potent mGlu5 receptor antagonist, in relation to the impairment of learning. *Neuropharmacology* 57:97-108.

[59] Homayoun H, Stefani MR, Adams BW, Tamagan GD, Moghaddam B (2004) Functional interaction between NMDA and mGlu5 receptors: effects on working memory, instrumental learning, motor behaviors, and dopamine release. *Neuropsychopharmacology* 29:1259-1269.

[60] Ballard TM, Woolley ML, Prinssen E, Huwyler J, Porter R, Spooren W (2005) The effect of the mGlu5 receptor antagonist MPEP in rodent tests of anxiety and cognition: a comparison. *Psychopharmacology* (Berl) 179:218-29.

[61] Steckler T, Oliveira AF, Van Dyck C, Van Craenendonck H, Mateus AM, Langlois X, Lesage AS, Prickaerts J (2005) Metabotropic glutamate receptor 1 blockade impairs acquisition and retention in a spatial Water maze task. *Behav. Brain Res.* 164:52-60.

[62] De Leonibus E, Manago F, Giordani F, Petrosino F, Lopez S, Oliverio A, Amalric M, Mele A (2009) Metabotropic glutamate receptors 5 blockade reverses spatial memory deficits in a mouse model of Parkinson's disease. *Neuropsychopharmacology* 34:729-38.

[63] Petersen S, Bomme C, Baastrup C, Kemp A, Christoffersen GRJ (2002) Differential effects of mGluR1 and mGluR5 antagonism on spatial learning in rats. *Pharmacol. Biochem. Behav.* 73:381-389.

[64] Campbell UC, Lalwani K, Hernandez L, Kinney GG, Conn PJ, Bristow LJ (2004) The mGluR5 antagonist 2-methyl-6-(phenylethynyl)-pyridine (MPEP) potentiates PCP-induced cognitive deficits in rats. *Psychopharmacology* (Berl) 175:310-8.

[65] Semenova S, Markou A (2007) The effects of the mGluR5 antagonist MPEP and the mGluR2/3 antagonist LY341495 on rats' performance in the 5-choice serial reaction time task. *Neuropharmacology* 52:863-872.

[66] Gravius A, Dekundy A, Nagel J, More L, Pietraszek M, Danysz W (2008) Investigation on tolerance development to subchronic blockade of mGluR5 in models of learning, anxiety, and levodopa-induced dyskinesia in rats. *J. Neural Transm.* 115:1609-19.

[67] Schulz B, Fendt M, Gasparini F, Lingenhohl K, Kuhn R, Koch M (2001) The metabotropic glutamate receptor antagonist 2-methyl-6-(phenylethynyl)-pyridine (MPEP) blocks fear conditioning in rats. *Neuropharmacology* 41:1-7.

[68] Fontanez-Nuin DE, Santini E, Quirk GJ, Porter JT (2011) Memory for fear extinction requires mGluR5-mediated activation of infralimbic neurons. *Cereb. Cortex* 21:727-735.

[69] Mao SC, Chang CH, Wu CC, Orejanera MJ, Manzoni OJ, Gean PW (2013) Inhibition of spontaneous recovery of fear by mGluR5 after prolonged extinction training. *PLoS One* 8:e59580.

[70] Kroker KS, Rast G, Rosenbrock H (2011) Differential effect of the mGlu5 receptor positive allosteric modulator ADX-47273 on early and late hippocampal LTP. *Neuropharmacology* 61:707-14.

[71] Uslaner JM, Parmentier-Batteur S, Flick RB, Surles NO, Lam JS, McNaughton CH, Jacobson MA, Hutson PH (2009) Dose-dependent effect of CDPPB, the mGluR5 positive allosteric modulator, on recognition memory is associated with GluR1 and CREB phosphorylation in the prefrontal cortex and hippocampus. *Neuropharmacology* 57:531-538.

[72] Liu F, Grauer S, Kelley C, Navarra R, Graf R, Zhang G, Atkinson PJ, Wantuch C, Popiolek M, Day M, Khawaja X, Smith D, Olsen M, Kouranova E, Gilbert A, Lai M, Pausch MH, Pruthi F, Pulicicchio C, Brandon NJ, Comery TA, Beyer CE, Logue S, Rosenzweig-Lipson S, Marquis KL (2008) ADX47273: a novel metabotropic glutamate receptor 5 selective positive allosteric modulator with preclinical antipsychotic-like and pro-cognitive activities. *J. Pharmacol. Exp. Ther.* 327:827-839.

[73] Parmentier-Batteur S, Obrien JA, Doran S, Nguyen SJ, Flick RB, Uslaner JM, Chen H, Finger EN, Williams TM, Jacobson MA, Hutson PH (2012) Differential effects of the mGluR5 positive allosteric modulator CDPPB in the cortex and striatum following repeated administration. *Neuropharmacology* 62:1453-60.

[74] Ayala JE, Chen Y, Banko JL, Sheffler DJ, Williams R, Telk AN, Watson NL, Xiang Z, Zhang Y, Jones PJ, Lindsley CW, Olive MF, Conn PJ (2009) mGluR5 positive allosteric modulators facilitate both hippocampal LTP and LTD and enhance spatial learning. *Neuropsychopharmacology* 34:2057-2071.

[75] Balschun D, Zuschratter W, Wetzel W (2006) Allosteric enhancement of metabotropic glutamate receptor 5 function promotes spatial memory. *Neuroscience* 142:691-702.

[76] Fowler SW, Walker JM, Klakotskaia D, Will MJ, Serfozo P, Simonyi A, Schachtman TR (2013) Effects of a metabotropic glutamate receptor 5 positive allosteric modulator, CDPPB, on spatial learning task performance in rodents. *Neurobiol. Learn. Mem.* 99:25-31.

[77] Reichel CM, Schwendt M, McGinty JF, Olive MF, See RE (2011) Loss of object recognition memory produced by exteded access to methamphetamine self-administration is reversed by positive allosteric modulation of metabotropic glutamate receptor 5. *Neuropsychopharmacology* 36:782-792.

[78] Schlumberger C, Pietraszek M, Gravius A, Danysz W (2010) Effects of a positive allosteric modulator of mGluR5 ADX47273 on conditioned avoidance response and PCP-induced hyperlocomotion in the rat as models for schizophrenia. *Pharmacol. Biochem. Behav.* 95:23-30.

[79] Spear N, Gadient RA, Wilkins DE, Do M, Smith JS, Zeller KL, Schroeder P, Zhang M, Arora J, Chhajlani V (2011) Preclinical profile of a novel metabotropic glutamate receptor 5 positive allosteric modulator. *Eur. J. Pharmacol.* 659:146-54.

[80] Xu J, Zhu Y, Kraniotis S, He Q, Marshall JJ, Nomura T, Stauffer SR, Lindsley CW, Conn PJ, Contractor A (2013) Potentiating mGluR5 function with a positive allosteric modulator enhances adaptive learning. *Learn. Mem.* 20:438-45.

[81] Homayoun H, Moghaddam B (2006) Bursting of prefrontal cortex neurons in awake rats is regulated by metabotropic glutamate 5 (mGlu5) receptors: rate-dependent influence and interaction with NMDA receptors. *Cereb. Cortex* 16:93-105.

[82] Homayoun H, Moghaddam B (2008) Orbitofrontal cortex neurons as a common target for classic and glutamatergic antipsychotic drugs. *Proc. Natl. Acad. Sci. U. S. A.* 105:18041-6.

[83] Stefani MR, Moghaddam B (2010) Activation of type 5 metabotropic glutamate receptors attenuates deficits in cognitive flexibility induced by NMDA receptor blockade. *Eur. J. Pharmacol.* 639:26-32.

[84] Darrah JM, Stefani MR, Moghaddam B (2008) Interaction of N-methyl-D-aspartate and group 5 metabotropic glutamate receptors on behavioral flexibility using a novel operant set-shift paradigm. *Behav. Pharmacol.* 19:225-234.

[85] Gass JT, Olive MF (2009) Positive allosteric modulation of mGluR5 receptors facilitates extinction of a cocaine contextual memory. *Biol. Psychiatry* 65:717-720.

[86] Cleva RM, Hicks MP, Gass JT, Wischerath KC, Plasters ET, Widholm JJ, Olive MF (2011) mGluR5 positive allosteric modulation enhances extinction learning following cocaine self-administration. *Behav. Neurosci.* 125:10-19.

[87] Kufahl PR, Hood LE, Nemirovsky NE, Barabas P, Halstengard C, Villa A, Moore E, Watterson LR, Olive MF (2012) Positive allosteric modulation of mGluR5 accelerates extinction learning but not relearning following methamphetamine self-administration. *Front Pharmacol.* 3:194.

[88] Gass JT, Chandler LJ (2013) The plasticity of extinction: contribution of the prefrontal cortex in treating addiction through inhibitory learning. *Front Psychiatry* 4:46.

[89] Widholm JJ, Gass JT, Cleva RM, Olive MF (2011) The mGluR5 positive allosteric modulator CDPPB does not alter extinction or contextual renewal of methamphetamine-seeking behavior in rats. *J. Addict. Res. Ther.* S1:004.

[90] Parmentier-Batteur S, Hutson PH, Menzel K, Uslaner JM, Mattson BA, O'Brien JA, Magliaro BC, Forest T, Stump CA, Tynebor RM, Anthony NJ, Tucker TJ, Zhang XF, Gomez R, Huszar SL, Lambeng N, Faure H, Le Poul E, Poli S, Rosahl TW, Rocher JP, Hargreaves R, Williams TM (2013) Mechanism based neurotoxicity of mGlu5 positive allosteric modulators - development challenges for a promising novel antipsychotic target. *Neuropharmacology*, in press.

[91] Rutten K, Van Der Kam EL, De Vry J, Bruckmann W, Tzschentke TM (2011) The mGluR5 antagonist 2-methyl-6-(phenylethynyl)-pyridine (MPEP) potentiates conditioned place preference induced by various addictive and non-addictive drugs in rats. *Addict. Biol.* 16:108-15.

In: Metabotropic Glutamate Receptors
Editor: M. Foster Olive

ISBN: 978-1-63117-451-3

Chapter 4

Group I Metabotropic Glutamate Receptors: A Potential Therapeutic Target for Amyloidogenic Disorders

Cara J. Westmark*
University of Wisconsin Madison, Madison, Wisconsin, US

Abstract

The Group I metabotropic glutamate receptors (mGluR1 and mGluR5) are glutamate-activated, G-protein-coupled receptors widely expressed in the central nervous system (CNS). We have identified amyloid-beta protein precursor (*App*) mRNA as a synaptic target that is translationally regulated through metabotropic glutamate receptor 5 (mGluR5) signaling. *App* mRNA codes for amyloid-beta protein precursor (AβPP), which is cleaved by β- and γ-secretases to produce amyloid-beta (Aβ), the predominant protein found in the senile plaques that are characteristic of Alzheimer's disease and Down syndrome. Recent evidence suggests aberrant accumulation of AβPP cleavage products in fragile X syndrome and autism. Thus, we hypothesize that inhibition of mGluR5 is a potential therapeutic strategy to modulate AβPP and catabolite levels in multiple neurological disorders characterized by abnormal AβPP production and processing. Antagonists of mGluR5 are currently under investigation for a range of indications including anxiety, epilepsy, pain, depression, Parkinson's disease, gastroesophageal reflux, fragile X syndrome, autism and addiction and could be repurposed for the treatment of Alzheimer's disease and Down syndrome.

Keywords: Alzheimer's disease, amyloid-beta, amyloid-beta protein precursor, autism, Down syndrome, fragile X mental retardation protein, fragile X syndrome, metabotropic glutamate receptor 5

* Corresponding Author address: University of Wisconsin, Department of Neurology, Medical Sciences Center, 1300 University Avenue, Madison, WI 53706. Email: westmark@wisc.edu.

Introduction

Neurological disorders as diverse as Alzheimer's disease, Down syndrome, fragile X syndrome, autism, epilepsy and Parkinson's disease may share a common pathological denominator in the overexpression of AβPP and its catabolites [1]. AβPP synthesis is regulated through a mGluR5 and fragile X mental retardation protein (FMRP) pathway [2]. This chapter reviews the state-of-the-art regarding the molecular mechanism underlying mGluR5/FMRP-mediated translation of AβPP and subsequent processing, the roles of AβPP catabolites and mGluR5 in neurological disorders and current and potential therapeutic approaches, including mGluR5 inhibitor therapy, targeted at normalizing AβPP and Aβ levels.

Molecular Mechanisms Underlying AβPP Synthesis and Processing

AβPP synthesis and processing involve an intricate combination of transcriptional, post-transcriptional, translational and post-translational events mediated by numerous nucleic acid binding proteins [3]. One of those RNA binding protein is FMRP, which has established roles in the transport, localization and/or translational repression of numerous dendritic mRNAs involved in synapse formation and synaptic plasticity [4]. Glutamate, or other Group I mGluR agonists such as (*S*)-3,5-dihydroxyphenylglycine (DHPG), activate mGluR5 signaling leading to the inactivation of FMRP and the translational de-repression of its mRNA targets. The translational regulation of *App* mRNA occurs via an mGluR5- and FMRP-dependent pathway. FMRP binds to a guanine-rich region in the coding region of *App* (mouse) and *APP* (human) mRNAs [2, 5]. The guanine-rich FMRP binding site (nucleotides 699-796) in *App* mRNA identified by ultraviolet crosslinking immunoprecipitation (CLIP) assays is supported by theoretical RNA/protein interaction studies, which used software developed by the Center for Genomic Regulation, Barcelona, Catalonia [fast predictions of RNA and protein interactions and domains algorithm (catRAPID)] [6]. Based on the physiochemical properties of the FMRP polypeptide and *App* mRNA nucleotide chains, the catRAPID software predicted that FMRP binds to nucleotides 650-751 and 751-852 of *App* mRNA, which are within the experimentally validated region [7]. Group I mGluR activation dislodges FMRP from the ribonucleoprotein (RNP) complex containing *App* mRNA thus increasing synaptic synthesis of AβPP [2]. FMRP expression is absent or greatly reduced in *Fmr1* knockout mice [8] and fragile X patients [9]. The *Fmr1* knockout mice exhibit elevated levels of AβPP and Aβ (2,10) and although few autopsy samples have been available for analysis, there is a trend for increased Aβ levels in fragile X patients [11]. Thus, *App* mRNA is a bona fide translational target of FMRP and mGluR5, and the loss or inactivation of FMRP results in elevated synthesis of AβPP.

AβPP can be cleaved by the non-amyloidogenic (α-secretase) or amyloidogenic (β- and γ-secretases) processing machinery to generate numerous AβPP catabolites [12]. The best studied is Aβ, an internal fragment of AβPP that begins approximately 99 residues from the carboxy-terminal end of the precursor protein and extends from the extracellular domain into the middle of the hydrophobic, membrane-spanning domain. Aβ is generated by amyloidogenic processing. Differential γ-secretase cleavage at the carboxy-terminal end of Aβ generates 37-49 amino acid Aβ peptides [13]. Aβ is found in normal biological fluids throughout life, but becomes toxic after over-expression and assembly into oligomeric aggregates [14]. N-terminal deletions of Aβ enhance aggregation of the peptide into neurotoxic β-sheet fibrils, which may nucleate senile plaques [15]. Alternatively, AβPP metabolism occurs via α-secretase cleavage within the Aβ region, which precludes formation of Aβ and releases an 82 kD soluble fragment (sAβPPα) into the extracellular fluid [16] where it has neuroprotective functions [17]. Both amyloidogenic and non-amyloidogenic processing release large soluble fragments of AβPP (sAβPPα and sAβPPβ) into the extracellular fluid and carboxy-terminal fragments into the cytosol. Group I mGluR stimulation accelerates sAβPP release into the extracellular fluid [18-21] and intraneuronal Aβ immunoreactivity [22] suggesting that Group I mGluRs can regulate AβPP biology at both the levels of synthesis and processing.

Research on Alzheimer's disease is shifting from a focus on Aβ plaques as the cause of the disease to the catalytic role of small soluble Aβ oligomers. Defective synaptic function occurs early in Alzheimer's disease prior to Aβ plaque accumulation and neuronal loss [23]. Aβ-derived diffusible ligands (ADDLs) can target and disrupt particular synapses and are associated with alterations in neurotransmitter release/uptake, changes in receptor [*N*-methyl-D-aspartate receptor (NMDAR), α-amino-3-hydroxy-5-methyl-4-isoxazolepropionic acid receptor (AMPAR), ephrin receptor B2 (EphB2) and insulin receptor] levels and localization, and impairment in synaptic plasticity [inhibition of long-term potentiation (LTP), facilitation/potentiation of long-term depression (LTD), cytoskeletal abnormalities and synapse deterioration/elimination] [24, 25]. Thus, excessive ADDL accumulation and activity at excitatory synapses is emerging as the causative agent in synaptic failure in Alzheimer's disease.

Accumulating evidence supports a model in which Aβ and mGluR5 function in a positive feedback loop that modulates AβPP synthesis and processing. As previously described, stimulation of mGluR5 alleviates FMRP-mediated translational repression of *App* mRNA leading to accumulation of AβPP. When AβPP is over-expressed or mutated, the β-secretase pathway is favored and liberates Aβ (26). Aβ clusters within excitatory synaptic membranes act as an extracellular scaffold for mGluR5 resulting in reduced mobility and increased activity of the receptor [27]. Thus, genetic mutations or environmental factors that exacerbate mGluR5 signaling or AβPP production could stimulate different points of a positive feedback loop that is perpetuated by Aβ (Figure 1). Additional supporting evidence for a mGluR5/Aβ-mediated positive feedback loop includes studies demonstrating: increased rodent Aβ secretion in response to treating rat cortical neurons with physiologically relevant concentrations of human Aβ dimer/trimer [28], stimulation of Group I mGluRs with glutamate resulted in dose- and time-dependent increases in the secretion of sAPP [29], and Aβ(1-42) activated the expression of β-site AβPP cleaving enzyme (BACE1) through a c-Jun N-terminal kinase (JNK)/c-jun signaling pathway that promotes *BACE1* transcription [30].

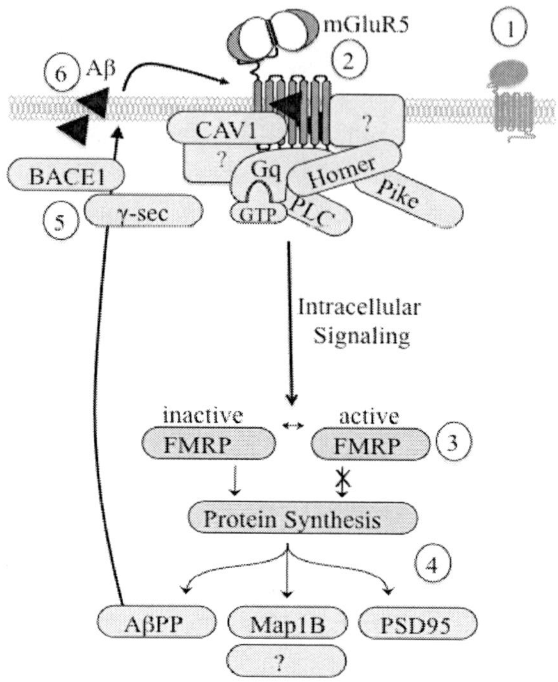

Figure 1. Model of mGluR5/Aβ-mediated positive feedback loop in neurons. ① mGluR5 is a primary target for drug development in numerous neurological and psychiatric disorders including fragile X syndrome. ② Cell stimulation causes clustering of mGluR5 receptors and altered interactions with scaffolding proteins such as caveolin and Homer, which likely alter downstream signaling events. ③ mGluR5 is known to signal through FMRP, a translational repressor. The absence of FMRP in *Fmr1* knockout cells results in constitutive, unregulated protein synthesis in response to mGluR5 stimulation. ④ Altered mGluR5/FMRP signaling modulates the synthesis of numerous synaptic proteins including AβPP, microtubule-associated protein 1B (Map1B) and postsynaptic density protein 95 (PSD95). Excessive production and accumulation of these synaptic proteins contributes to the pathology of fragile X syndrome. ⑤ Excessive AβPP production stimulates amyloidogenic processing in which BACE1 and γ-secretase cleave AβPP to generate Aβ. ⑥ Aβ laterally diffuses through cell membranes forming clusters that impede mGluR5 mobility and stimulate receptor activity, resulting in a positive feedback loop.

Roles of AβPP Catabolites and mGluR5 in Neurological Disorders

AβPP and its catabolites function in neurite outgrowth, dendritic spine formation, synapse formation, cell signaling, protein synthesis, synaptic transmission, learning and memory, apoptosis, and neuronal migration, motility and development [31-40]. Much remains to be learned regarding overlapping and sometimes opposing functions of specific AβPP catabolites. As Aβ exhibits hormetic effects (biphasic dose-response curves) with low concentrations enhancing synaptic plasticity and memory and high concentrations inducing cognitive impairment [41], it will be necessary to develop therapeutics that attenuate excessive catabolite activities while retaining normal physiological activities. Altered mGluR5 expression or activity is associated with several neurological and psychiatric disorders suggesting that mGluR5 inhibitors may be a viable therapeutic strategy to reduce AβPP production, processing and ensuing phenotypes (Figure 2).

The best characterized amyloidogenic disorder is Alzheimer's disease, which is characterized histologically by Aβ plaques, neurofibrillary tangles and neuronal cell death [42]. Patients suffer from cognitive dysfunction and behavioral problems. Alzheimer's disease currently afflicts 4.7 million individuals in the U.S. and is projected to increase to 13.8 million by 2050 [43]. In a small percentage of cases, Alzheimer's disease is directly associated with specific genetic mutations in the *APP* or *Presenilin* 1 genes; however, in most cases the etiology is unknown. The Aβ Hypothesis is the leading theory regarding the cause of Alzeimer's disease, and is supported by human studies associating early-onset in families with mutations in the *APP* and *Presenilin* genes [44] and rodent studies demonstrating Aβ neurotoxicity and rescue of disease phenotypes by reducing Aβ [45]. As previously described, AβPP synthesis is regulated through a mGluR5-dependent signaling pathway. There is a paucity studies examining mGluR5 expression in Alzheimer's disease with one study showing no change in mGluR5 expression in Alzheimer's disease cerebral cortex [46] and another study showing upregulated mGluR5 expression in the caudate nucleus in both normal aging and Alzheimer's disease [47]. The mGluR5 antagonist 2-methyl-6-(phenylethynyl)-pyridine (MPEP) blocks the translation of AβPP *in vitro* [2, 11], and chronic dosing with the mGluR5 antagonist fenobam in mice reduces Aβ levels *in vivo* [48]. Several antagonists of mGluR5 are in clinical trials for the treatment of fragile X syndrome [49] and could be repurposed for Alzheimer's disease. Thus, inhibition of mGluR5 signaling could be a viable therapeutic strategy to modulate AβPP and catabolite production in Alzheimer's disease.

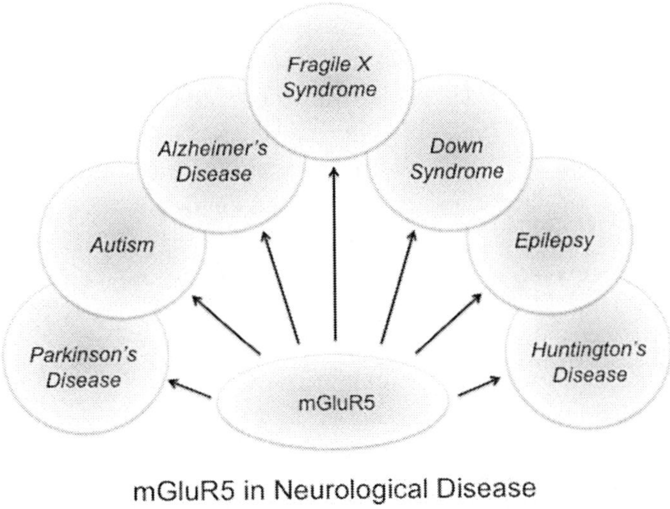

Figure 2. Altered mGluR5 expression or activity is associated with several neurological disorders including fragile X syndrome, Alzheimer's disease, Down syndrome, autism, epilepsy, Parkinson's disease and Huntington disease. In a mouse model of fragile X syndrome, mGluR5 signaling mediates the translation of AβPP, which is associated with elevated Aβ. In many other cases, the consequences of altered mGluR5 activity have not yet been determined, but the overexpression of AβPP and catabolites is predicted to be a common denominator in disease pathology that may be treatable with mGluR5 inhibitor therapy.

Down syndrome is the most common genetic cause of cognitive impairment occurring at a frequency of 1 in 690 live births [50]. The 50% overexpression of more than 300 genes due to trisomy of chromosome 21 results in intellectual disability, hypotonia, speech and language impairment, congenital heart defects, seizures, obstructive sleep apnea and craniofacial dysmorphology phenotypes [51]. The *APP* gene is one of the trisomic genes, and Down syndrome individuals over-express *APP* mRNA by four- to five-fold thus depositing extracellular Aβ at a greatly accelerated rate [52]. Down syndromc is highly comorbid with Alzheimer's disease with essentially all Down syndrome adults over the age of 40 years developing sufficient neuropathology to meet the diagnostic criteria of Alzheimer's disease [53-57]. mGluR5 expression is upregulated in Down syndrome brain [58] further supporting our theory that blockade of this pathway would be therapeutically efficacious in this amyloidogenic disorder.

Fragile X syndrome is the most common form of inherited intellectual disability and the leading known genetic cause of autism [59, 60]. Clinically, the disorder is characterized by moderate to severe mental retardation, autistic-like behaviors, seizures, macroorchidism, and facial characteristics including large, prominent ears and a long, narrow face [61]. At the neuroanatomical level, fragile X syndrome is characterized by an overabundance of long, thin tortuous postsynaptic spines with prominent heads and irregular dilations resembling the spines observed during normal, early neocortical development [62]. Typically, fragile X syndrome is caused by a trinucleotide repeat expansion in the promoter region of the fragile X mental retardation (*FMR1*) gene. The CGG repeat expansion leads to methylation of the promoter and lack of translation of FMRP. Loss of FMRP expression causes excessive protein synthesis of numerous dendritic mRNAs, including *APP* mRNA, that are normally translationally repressed by FMRP. Elevated levels of AβPP and/or a metabolite such as Aβ likely contribute to multiple fragile X phenotypes as evidenced by the rescue of seizures, anxiety, the ratio of immature/mature dendritic spine and mGluR-LTD by genetic knockdown of *App* in *Fmr1* knockout mice [11]. The mGluR theory of fragile X syndrome proposes that overactive signaling by Group I mGluRs contributes to many of the symptoms of fragile X [63]. In support of the theory, mutant *Fmr1* knockout mice expressing 50% fewer mGluR5 receptors exhibit rescue of phenotypic and behavioral abnormalities associated with fragile X syndrome [64]. In a preclinical drug study, the Roche mGluR5 inhibitor CTEP [65] demonstrated comprehensive phenotype correction in *Fmr1* knockout mice [66]. Thus, drug development efforts for fragile X syndrome have predominantly focused on mGluR5 inhibitors [67].

Autism is a cluster of complex neurobiological disorders known as autism spectrum disorders that normally present in the second or third years of life. Austin spectrum disorders are estimated to occur in 1 in 88 children with prevalence 4.7-fold higher in males [68]. The core features include impairments in social interaction and communication, and repetitive stereotyped behavior. Many autistic children are mentally retarded and half exhibit marked delay in motor milestones. The etiology of autism is not known but genetic as well as environmental factors likely affect the severity of symptoms [69-71]. Autism is highly comorbid with other developmental disorders such as fragile X syndrome and epilepsy [72, 73]. Subjects with severe autism exhibit elevated sAβPPα and reduced Aβ(1-40), Aβ(1-42) and sAβPPβ in blood plasma [74-76]. Postmorten analyses of brain tissue from subjects with idiopathic and duplication 15q11.2-q13 [dup(15)] autism reveal increased intraneuronal

Aβ(17-40/42) in almost all examined cortical and subcortical structures in children and adults. Aβ(17-40/42) is a product of non-amyloidogenic processing. In some older subjects (ages 39-52), diffuse plaques containing Aβ(1-40/42) were present [77]. AβPP levels are increased in the superior frontal cortex (Brodmann Area 9) of children with autism and decreased in the cerebellar vermis of adult subjects with autism [78]. These data suggest that age-associated alterations in AβPP processing occur in autism. FXS and autism have not been associated with an increased risk of Alzheimer's disease or dementia, albeit there have been few studies to date and it is difficult to measure cognitive decline in individuals with intellectual disability. Recent work strongly supports roles for AβPP and Aβ in defective synaptic function that occurs prior to Aβ plaque formation and neuronal loss in Alzheimer's disease [23] suggesting that these proteins can negatively impact neurological health regardless of the presence of Alzheimer's pathology.

There is significant overlap in the molecular pathways underlying fragile X syndrome and autism [79]. For example, FMRP and mGluR5 are also dysregulated in autism. FMRP is reduced in the cerebellar vermis and superior frontal cortex of adults with autism while mGluR5 expression is increased in the vermis and superior frontal cortex of autistic children [80, 81]. Treating mouse models of autism with mGluR5 inhibitors attenuates repetitive self-grooming, stereotypic behavior and some measures of sociability [82-84]. These data suggest that a common molecular pathway involving mGluR5 and FMRP contributes to aberrant synaptic plasticity in fragile X syndrome and autism.

Epilepsy is a family of disorders characterized by abnormal electrical discharges in the brain and is typically manifested by brief episodes of altered or diminished consciousness, involuntary movements or convulsions. In the majority of cases, the etiology of epilepsy in not known; however, brain injury, infections, tumors, stroke and high fevers can all cause seizures. The cumulative lifetime risk of unprovoked seizures is 4.1% in the general population with a higher incidence in the elderly [85]. Epilepsy is highly comorbid with Alzheimer's disease, Down syndrome, fragile X syndrome and autism [73, 86-92]. Support for an underlying common mechanism between these disorders centers on Aβ and is supported by: elevated AβPP levels [93] and Aβ-immunoreactive senile plaques [94] in temporal lobe epilepsy, seizure history in many pedigrees with autosomal dominant early onset Alzheimer's disease [95-105], and greater intraneuronal Aβ accumulation and a higher risk of sudden unexpected death in epilepsy in individuals with dup(15) autism compared with subjects with idiopathic autism [77].

The aberrant expression of mGluR5 in epilepsy may be the cause or a consequence of seizure activity. The expression of mGluR5 is elevated in temporal lobe epilepsy [106-108] as well as in glioneuronal tumors from patients with intractable epilepsy [109]. Seizures induce upregulation of mGluR5 in reactive astrocytes in a rat model of mesial temporal lobe epilepsy [110]. Synaptic activation of Group I mGluRs induces and maintains epileptiform discharges in hippocampal slices [111-115]. In rodent models of complex partial seizures and status epilepticus, MPEP alone or in combination with antieptileptic drugs attenuates seizures [116-118]. Transgenic mouse models that overexpress AβPP and Aβ are more susceptible to chemically- [119] and audiogenic-induced seizures [120]. Seizures can be attenuated in *Fmr1* knockout mice by mGluR5 blockade [121] or genetic reduction of AβPP/Aβ levels [11] and in mouse models of Alzheimer's disease by passive immunotherapy against Aβ (120) or mGluR5 blockade [120, 122]. Thus, mGluR5 antagonists are a viable therapeutic strategy to reduce Aβ levels and associated seizures in epilepsy.

In addition to the aforementioned disorders, abnormal AβPP or catabolite levels have been reported in Parkinson's disease. Parkinson's disease is a motor system disorder resulting from the loss of dopamine-producing brain cells. The primary symptoms of Parkinson's disease are tremor, stiffness, bradykinesia and impaired balance and coordination. Newly diagnosed, unmedicated, non-demented Parkinson's disease patients exhibit reduced levels of Aβ(1-40), Aβ(1-42) and Aβ(1-38) in cerebrospinal fluid [123]. Lower levels of Aβ(1-42) in cerebrospinal fluid correlate with Pittsburgh Compound B-stained Aβ deposits in Alzheimer's disease brain [124] suggesting that the senile plaques in the brain act as a sink for Aβ(1-42) during disease progression resulting in lower levels in the cerebrospinal fluid and blood plasma. A reduced Aβ(1-42)/Aβ(1-40) plasma ratio is an independent risk factor for Alzheimer's disease irrespective of the total Aβ load [125-130]. Lower Aβ(1-42)/Aβ(1-40) blood plasma ratios are also observed in Down syndrome [131] and fragile X syndrome [11]. There is a significant upregulation of brain mGluR5 in human Parkinson's disease [132] as well as in a primate model of the disorder [133] suggesting that mGluR5 is a viable therapeutic target in Parkinson's disease [134]. mGluR5 inhibitors improve motor, anxiety and cognitive deficits in animal models of Parkinson's disease [135-139], and reduce levodopa-induced dyskinesias in human Parkinson's disease without changing the antiparkinsonian effects of dopaminergic therapy [140, 141]. Since Alzheimer's disease pathology is highly comorbid in Parkinson's disease cases, particularly those with early dementia, treatments that modulate AβPP production and processing are expected to benefit both disorders.

Huntington's disease is an autosomal-dominant disorder that leads to movement, cognitive and psychiatric problems. The disorder is caused by a polyglutamine expansion in the huntingtin protein, which is associated with excitotoxic death of striatal neurons. Huntington's and Alzheimer's diseases are both progressive neurodegenerative disorders associated with insoluble aggregated proteins, huntingtin in the former and Aβ in the latter case. The two disorders rarely coexist although the initial stages of Alzheimer's-like lesions develop early in Huntington patients [142]. Mutated huntingtin protein functions in the attenuation of mGluR1a signaling [143]; however, mGluR5 activation causes higher levels of extracellular-regulated kinase (ERK) activation in the striatum of a Huntington mouse model compared to controls [144]. MPEP rescues motor coordination and hyperactivity and increases survival in the Huntington disease mice [145]. Thus, mGluR5 inhibitor therapy is also a viable therapeutic strategy to retard disease progression in Huntington's disease.

Psychiatric disorders such as schizophrenia, bipolar disorder and major depression are also associated with reduced brain FMRP [146] and altered Aβ levels [147-150]. Another chapter in this book by Suzuki, Kawamoto and Ohta discusses the therapeutic potential of mGluR1 for the treatment of schizophrenia. Major depression is a risk factor for developing Alzheimer's disease [151]. Thus, numerous neurological and psychiatric disorders are associated with skewed levels of AβPP, Aβ and/or FMRP suggesting that AβPP metabolism plays an important role in neurological health. Although many questions remain regarding the expression levels and role(s) of AβPP and Aβ, particularly in those disorders not definitively associated with an increased risk for Alzheimer's disease, accumulating evidence suggests that mGluR5 may be a viable therapeutic target to modulate AβPP and catabolite production and ensuing phenotypes in numerous neurological and psychiatric disorders.

Therapeutic Approaches to Modulate AβPP and Aβ

Therapeutic approaches to treat Alzheimer's disease have focused on cholinesterase inhibitors, Aβ immunotherapy and secretase inhibitors. Two classes of drugs, acetylcholinesterase inhibitors [(donepezil (Aricept®), rivastigmine (Exelon®), galantamine (Reminyl®) and tacrine (Cognex®] and NMDAR antagonists [memantine (Namenda®)] are currently approved by the Food and Drug Administration (FDA) to treat cognitive symptoms of Alzheimer's disease. These drugs act on healthy neurons to compensate for lost acetylcholine activity or modulate NMDAR activity, respectively. They improve cognitive ability for a year or less, but do not reduce Aβ or neurofibrillary tangle accumulation and subsequent disease progression. Aβ immunotherapy has proved very effective in Aβ clearance as well as associated cognitive decline; however, there are questions about its safety [152]. Human clinical trials utilizing active immunization against Aβ were suspended because 6% of the patients developed meningoencephalitis [153]. In Alzheimer's disease mouse models characterized by prominent vascular Aβ deposition (APP23 and PDAPP mice), Aβ immunization caused hemorrhages [154, 155]. β-secretase (aka: memapsin 2; BACE1) is a type I transmembrane protein containing an aspartic protease catalytic domain. β-secretase cleaves AβPP within endosomes of neurons in the brain to generate the amino-terminal end of Aβ [156]). Continuous administration of a β-secretase inhibitor reduces brain and plasma Aβ by 50% and inhibits age-related cognitive decline in an Alzheimer mouse model [157]. γ-secretase is a membrane-embedded aspartyl protease complex that contains presenilin as the catalytic component. This complex cleaves the 99-residue carboxy-terminal fragment of AβPP to release Aβ and the AβPP intracellular domain. β-secretase inhibitors are currently in early phase clinical trials, but their inability to penetrate the blood-brain barrier impacts their effectiveness. γ-secretase is an essential component of the Notch signaling pathway, and utility of γ-secretase inhibitors is limited by their toxicity [158]. Thus, novel approaches to modulate AβPP synthesis and processing are required.

Group I mGluRs regulate the synthesis [2] and metabolism [159] of AβPP suggesting that inhibition of this pathway could normalize AβPP and Aβ production as well as seizure, behavioral and pathological phenotypes associated with amyloid-related disorders. Indeed, mGluR5 blockade reduces Aβ levels [48] as well as pentylenetetrazol (PTZ)-induced seizure severity [122], audiogenic seizure incidence [120, 121], anxiety phenotypes and dendritic spine protrusion morphology [121, 160] in mouse models of fragile X syndrome, Alzheimer's disease or Down syndrome.

There are eight subtypes of mGluR that have been classified into groups based on their sequence homology and signaling properties [161]. Group I mGluRs (mGluR1 and mGluR5) couple to Gq and activate phospholipase C. Group II mGluRs (mGluR2 and mGluR3) and Group III mGluRs (mGluR4, mGluR6, mGluR7 and mGluR8) couple to Gi/Go and inhibit adenylyl cyclase. Group I receptors are generally postsynaptic in location whereas groups 2 and 3 are presynaptic. mGluRs contain a heptahelical domain in the membrane region, a large extracellular amino terminal domain where the glutamate binding site is found and an intracellular carboxy terminal domain. The amino terminal domain has a bilobate Venus Flytrap domain where glutamate binds and the closed conformation of this domain is required for receptor activation. Competitive antagonists of mGluR prevent complete closing of the bilobular Venus Flytrap domain. Allosteric modulators are non-competitive ligands that bind to the transmembrane heptahelical domain. Negative modulators inhibit receptor activation without affecting orthosteric agonist binding and positive modulators enhance agonist activation but cannot activate receptors alone. The well-studied drugs MPEP and fenobam are potent, non-competitive, selective and systemically active allosteric antagonists of mGluR5 [162, 163].

Inhibitors of mGluR5 are under study for a range of indications including anxiety [164, 165], epilepsy [166], pain [167], depression [168], Parkinson's disease [169, 170], gastroesophageal reflux disease [171], fragile X syndrome, autism [63, 172, 173] and addiction [174-176]. Several current studies testing mGluR5 imaging drugs and inhibitors are registered with ClinicalTrials.gov (Table 1) and discussed below.

Radioactive compounds are under clinical evaluation for the optimization of positron emission tomography (PET) as an imaging tool to detect mGluR5. A study sponsored by the National Institute of Mental Health (NIMH) entitled, "Evaluation of Two Radioactive Chemicals to Image mGluR5 Receptors in Brain," will test two radioactive chemicals, [^{11}C]SP203 and [^{11}C]FPEB, and a research drug STX107 that can attach to mGluR5 to optimize PET conditions for mGluR5 imaging in healthy individuals. The selected PET ligands have high affinity for mGluR5, and the negative allosteric modulator STX107 will be used to block radioligand brain uptake. Two serial PET scans with the radioligands will be performed under baseline conditions on one day and another PET scan under receptor blockade on a different day. The radioligands will be evaluated for peak brain uptake (higher uptake is better for a more robust PET signal), time-stability of distribution volume and the ratio of specific to non-displaceable uptake in the brain. A second NIMH-sponsored study entitled, "PET Imaging of Brain mGluR5 Receptors Using [^{18}F]SP203," will also quantify mGluR5 binding parameters in the brains of healthy individuals. A study sponsored by the Institute for Neurodegenerative Disorders, "A PET Brain Imaging Study of mGluR5 in Subjects with Neuropsychiatric Conditions (FPEB)," will test [^{18}F]FPEB in a Phase I clinical trial in individuals with Parkinson's disease, Huntington's disease, autism spectrum disorder, fragile X syndrome, Alzheimer's disease or mild cognitive impairment. A recently terminated PET study, "Brain Glutamate Receptors and Cocaine Dependence" sponsored by the National Institute on Drug Abuse (NIDA) had proposed to determine whether the density of mGluR5 is altered in participants with cocaine addiction compared to healthy controls. Thus, the development of mGluR5 radioligands for PET analysis may benefit multiple neurological, psychiatric and addiction disorders. PET analysis has been used to assess pharmacokinetics of mGluR5 inhibitors. Recently completed studies sponsored by AstraZeneca included, "An Open Label Positron Emission Tomography (PET) Study to Determine Central mGluR5 Receptor Occupancy of AZD2516," and "Positron Emission Tomography (PET) Study with [^{11}C]AZ12713580 to Determine Central mGluR5 Receptor Occupancy of AZD2066". These pharmacokinetic studies will inform efficacy and dosing predictions for future drug development [177].

Table 1. mGluR5 Compounds in Clinical Trials

Compound	Sponsor	Study Population
mGluR5 Radioligands		
[^{11}C]FPEB [^{11}C]SP203	NIMH	healthy subjects
[^{18}F]SP203 [^{11}C]SP203	NIMH	healthy subjects
[^{18}F]FPEB	Institute for Neurodegenerative Disorders	Parkinson's disease Huntington disease autism spectrum disorders fragile X syndrome Alzheimer's disease mild cognitive impairment
[^{18}F]SP203	NIDA	cocaine addiction
mGluR5 Antagonists		
Acamprosate	Indiana University	fragile X syndrome
Acamprosate	Mayo Clinic	alcoholism
Acamprosate	Autism Speaks	autism
ADX10059	Addex Pharma S.A.	migraine
ADX10059	Addex Pharma S.A.	gastroesophageal reflux
AFQ056	Novartis	obsessive-compulsive disorder
AFQ056	Novartis	Parkinson's disease
AFQ056	Novartis	fragile X syndrome
AZD2516	AstraZeneca	healthy subjects
AZD2066	AstraZeneca	healthy subjects
RO4917523	Hoffmann-La Roche	fragile X syndrome
RO4917523	Hoffmann-La Roche	major depression disorder
STX107	Seaside Therapeutics	healthy subjects fragile X syndrome

Clinical trials to evaluate behavioral endpoints are also on-going. Acamprosate was approved in the U.S. for abstinence maintenance in 2004, and has been or is currently under study in 20 clinical trials for disorders ranging from alcohol, gambling and cocaine dependence to anxiety, autism, fragile X syndrome, schizophrenia and tinnitus. Acamprosate likely possesses mGluR5 inhibitor activity [178], and has been investigated in fragile X syndrome where it is associated with improved linguistic communication and global clinical benefit [179, 180]. A completed Phase 4 clinical trial of acamprosate in alcoholism sponsored by the Mayo Clinic is investigating the relationship between genetically determined variation in the NMDAR and treatment response to acamprosate. Autism Speaks is currently recruiting for a Phase 2 trial to assess improvement in social deficits in response to acamprosate. Addex Pharma S.A. is evaluating their mGluR5 inhibitor ADX10059 in Phase 2 trials for the prevention of migraine (study terminated) and gastroesophageal reflux (study completed). The ADX10059 trial for migraine prevention was terminated because there was a higher than expected incidence of abnormalities in liver function tests in this population. Novartis Pharmaceutics is testing their mGluR5 inhibitor AFQ056 in Phase 2 clinical trials for obsessive compulsive disorder, Parkinson's disease and fragile X syndrome. Previous studies in human and monkey Parkinson's disease models indicate that AFQ056 attenuates levodopa-induced dyskinesias without changing the antiparkinsonian effects of dopaminergic therapy [140, 141, 181]. AFQ056 is associated with improvement in Aberrant Behavior Checklist-Community Edition scores in fragile X syndrome patients with full methylation of the FMR1 promoter [182]. Preclinical studies with *Fmr1* knockout mice or hippocampal neurons, respectively, indicate that AFQ056 rescues inhibition of the startle response, social behavior and elongated spines [183-185]. Hoffmann-La Roche is testing their mGluR5 inhibitor, RO4917523, in Phase 2 clinical trials for fragile X syndrome and major depression disorder. The mGluR5 inhibitor STX107 developed by Merck and licensed by Seaside Therapeutics, Inc. passed Phase 1 clinical testing in fragile X patients. Thus, successful outcomes from these current and recently completed clinical trials will provide critical tools and pharmacokinetic and safety data regarding mGluR5 inhibitors.

In addition to the mGluR5 antagonist and radiotracer drugs currently described at Clinical Trials.gov, the mGluR5 inhibitor fenobam has completed a clinical trial in fragile X syndrome with promising results. A pilot open label, single dose trial of fenobam in adults with fragile X syndrome was conducted to provide an initial evaluation of safety and pharmacokinetics [172]. Fenobam was developed by McNeil Laboratories in the late 1970s as an anxiolytic agent and was tested in phase I clinical trials (oral administration) and displayed no harmful side effects [186-188] while one phase II clinical trial reported psychostimulating properties [189]. Fenobam was later discovered to be a potent and highly selective noncompetitive antagonist of mGluR5 [163, 190]. In 2006, fenobam was designated as an orphan drug for the treatment of fragile X syndrome. In the first clinical trial of fenobam in fragile X syndrome [172], there were no significant adverse reactions and a single dose elicited clinical improvement in a subset of patients in terms of calmed behavior and prepulse inhibition.

Conclusion

In conclusion, AβPP is translationally regulated through an FMRP/mGluR5-dependent pathway suggesting that mGluR5 inhibitors are a viable therapeutic strategy to decrease the production of AβPP and subsequent processing to Aβ. Inhibitors of mGluR5 are currently under study for the treatment of numerous neurological, psychiatric and addiction disorders, and if shown efficacious and safe in reducing Aβ and attenuating cognitive deficits, could be rapidly repurposed for the treatment of Alzheimer's disease and Down syndrome.

References

[1] Westmark CJ (2013) What's hAPPening at synapses? The role of amyloid beta-protein precursor and beta-amyloid in neurological disorders. *Mol. Psychiatry* 18:425-434.

[2] Westmark CJ, Malter JS (2007) FMRP mediates mGluR5-dependent translation of amyloid precursor protein. *PLoS Biol.* 5:e52.

[3] Westmark CJ, Malter JS (2012) The regulation of AβPP expression by RNA-binding proteins. *Aging Res. Rev.* 11:450-459.

[4] Santoro MR, Bray SM, Warren ST (2012) Molecular mechanisms of Fragile X Syndrome: a twenty-year perspective. *Annu. Rev. Pathol.* 7:219-245.

[5] Lee EK, Kim HH, Kuwano Y, Abdelmohsen K, Srikantan S, Subaran SS et al. (2010) hnRNP C promotes APP translation by competing with FMRP for APP mRNA recruitment to P bodies. *Nat. Struct. Mol. Biol.* 17:732-739.

[6] Bellucci M, Agostini F, Masin M, Tartaglia GG (2011) Predicting protein associations with long noncoding RNAs. *Nat. Methods* 8:444-445.

[7] Cirillo D, Agostini F, Klus P, Marchese D, Rodriguez S, Bolognesi B et al. (2013) Neurodegenerative diseases: quantitative predictions of protein-RNA interactions. *RNA* 19:129-140.

[8] The Dutch-Belgian Fragile X Consortium (1994) Fmr1 knockout mice: a model to study fragile X mental retardation. *Cell* 78:23-33.

[9] Siomi H, Siomi MC, Nussbaum RL, Dreyfuss G (1993) The protein product of the fragile X gene, FMR1, has characteristics of an RNA-binding protein. *Cell* 74:291-298.

[10] Liao L, Park SK, Xu T, Vanderklish P, Yates JR 3rd (2008) Quantitative proteomic analysis of primary neurons reveals diverse changes in synaptic protein content in fmr1 knockout mice. *Proc. Natl. Acad. Sci. USA* 105:15281-15286.

[11] Westmark CJ, Westmark PR, O'Riordan KJ, Ray BC, Hervey CM, Salamat MS et al. (2011) Reversal of fragile X phenotypes by manipulation of AβPP/Abeta levels in fmr1 mice. *PLoS One* 6:e26549.

[12] Haass C, Kaether C, Thinakaran G, Sisodia S (2012) Trafficking and proteolytic processing of APP. *Cold Spring Harb. Perspect. Med.* 2:a006270.

[13] Steiner H, Fluhrer R, Haass C (2008) Intramembrane proteolysis by gamma-secretase. *J. Biol. Chem.* 283:29627-29631.

[14] Klein WL, Stine WB Jr, Teplow DB (2004) Small assemblies of unmodified amyloid beta-protein are the proximate neurotoxin in Alzheimer's disease. *Neurobiol. Aging* 25:569-580.

[15] Pike CJ, Overman MJ, Cotman CW (1995) Amino-terminal deletions enhance aggregation of beta-amyloid peptides in vitro. *J. Biol. Chem.* 270:23895-23898.

[16] Gandy SE, Caporaso GL, Buxbaum JD, de Cruz Silva O, Iverfeldt K, Nordstedt C et al. (1993) Protein phosphorylation regulates relative utilization of processing pathways for Alzheimer beta/A4 amyloid precursor protein. *Ann. N. Y. Acad. Sci.* 695:117-121.

[17] Mucke L, Abraham CR, Masliah E (1996) Neurotrophic and neuroprotective effects of hAPP in transgenic mice. *Ann. N. Y. Acad. Sci.* 777:82-88.

[18] Lee RK, Jimenez J, Cox AJ, Wurtman RJ (1996) Metabotropic glutamate receptors regulate APP processing in hippocampal neurons and cortical astrocytes derived from fetal rats. *Ann. N. Y. Acad. Sci.* 777:338-343.

[19] Lee RK, Wurtman RJ, Cox AJ, Nitsch RM (1995) Amyloid precursor protein processing is stimulated by metabotropic glutamate receptors. *Proc. Natl. Acad. Sci. USA* 92:8083-8087.

[20] Nitsch RM, Deng A, Wurtman RJ, Growdon JH (1997) Metabotropic glutamate receptor subtype mGluR1α stimulates the secretion of the amyloid beta-protein precursor ectodomain. *J. Neurochem.* 69(2):704-712.

[21] Ulus IH, Wurtman RJ (1997) Metabotropic glutamate receptor agonists increase release of soluble amyloid precursor protein derivatives from rat brain cortical and hippocampal slices. *J. Pharmacol. Exp. Ther.* 281:149-154.

[22] Stephenson DT, Clemens JA (1998) Metabotropic glutamate receptor activation in vivo induces intraneuronal amyloid immunoreactivity in guinea pig hippocampus. *Neurochem. Int.* 33:83-93.

[23] Coleman P, Federoff H, Kurlan R (2004) A focus on the synapse for neuroprotection in Alzheimer disease and other dementias. *Neurology* 63:1155-1162.

[24] Lacor PN, Buniel MC, Chang L, Fernandez SJ, Gong Y, Viola KL et al. (2004) Synaptic targeting by Alzheimer's-related amyloid beta oligomers. *J. Neurosci.* 24: 10191-10200.

[25] Ferreira ST, Klein WL (2011) The Aβ oligomer hypothesis for synapse failure and memory loss in Alzheimer's disease. *Neurobiol. Learn. Mem.* 96:529-543.

[26] Mattson MP (1997) Cellular actions of beta-amyloid precursor protein and its soluble and fibrillogenic derivatives. *Physiol. Rev.* 77:1081-1132.

[27] Renner M, Lacor PN, Velasco PT, Xu J, Contractor A, Klein WL et al. (2010) Deleterious effects of amyloid beta oligomers acting as an extracellular scaffold for mGluR5. *Neuron* 66:739-754.

[28] Marsden IT, Minamide LS, Bamburg JR (2011) Amyloid-β-induced amyloid-β secretion: a possible feed-forward mechanism in Alzheimer's disease. *J. Alzheimers Dis.* 24:681-691.

[29] Jolly-Tornetta C, Gao ZY, Lee VM, Wolf BA (1998) Regulation of amyloid precursor protein secretion by glutamate receptors in human Ntera 2 neurons. *J. Biol. Chem.* 273:14015-14021.

[30] Guglielmotto M, Monteleone D, Giliberto L, Fornaro M, Borghi R, Tamagno E et al. (2011) Amyloid-β_{42} activates the expression of BACE1 through the JNK pathway. *J. Alzheimers Dis.* 27:871-883.

[31] Hoe HS, Lee HK, Pak DT (2012) The upside of APP at synapses. *CNS Neurosci. Ther.* 18:47-56.

[32] Cao Q, Huang YS, Kan MC, Richter JD (2005) Amyloid precursor proteins anchor CPEB to membranes and promote polyadenylation-induced translation. *Mol. Cell Biol.* 25:10930-10939.

[33] Claasen AM, Guevremont D, Mason-Parker SE, Bourne K, Tate WP, Abraham WC et al. (2009) Secreted amyloid precursor protein-alpha upregulates synaptic protein synthesis by a protein kinase G-dependent mechanism. *Neurosci. Lett.* 460:92-96.

[34] Taylor CJ, Ireland DR, Ballagh I, Bourne K, Marechal NM, Turner PR et al. (2008) Endogenous secreted amyloid precursor protein-alpha regulates hippocampal NMDA receptor function, long-term potentiation and spatial memory. *Neurobiol. Dis.* 31:250-260.

[35] Ishida A, Furukawa K, Keller JN, Mattson MP (1997) Secreted form of beta-amyloid precursor protein shifts the frequency dependency for induction of LTD, and enhances LTP in hippocampal slices. *Neuroreport* 8:2133-2137.

[36] Zheng H, Koo EH (2011) Biology and pathophysiology of the amyloid precursor protein. *Mol. Neurodegener.* 6:27.

[37] Nikolaev A, McLaughlin T, O'Leary DD, Tessier-Lavigne M (2009) APP binds DR6 to trigger axon pruning and neuron death via distinct caspases. *Nature* 457:981-989.

[38] Nalbantoglu J, Tirado-Santiago G, Lahsaini A, Poirier J, Goncalves O, Verge G et al. (1997) Impaired learning and LTP in mice expressing the carboxy terminus of the Alzheimer amyloid precursor protein. *Nature* 387:500-505.

[39] Koffie RM, Hyman BT, Spires-Jones TL (2011) Alzheimer's disease: synapses gone cold. *Mol. Neurodegener*. 6:63.

[40] Blanchard BJ, Thomas VL, Ingram VM (2002) Mechanism of membrane depolarization caused by the Alzheimer Aβ1-42 peptide. *Biochem. Biophys. Res. Commun.* 293:1197-1203.

[41] Puzzo D, Privitera L, Palmeri A (2012) Hormetic effect of amyloid-beta peptide in synaptic plasticity and memory. *Neurobiol. Aging* 33:1484.

[42] Mann DM (1985) The neuropathology of Alzheimer's disease: a review with pathogenetic, aetiological and therapeutic considerations. *Mech. Ageing Dev.* 31:213-255.

[43] Hebert LE, Weuve J, Scherr PA, Evans DA (2013) Alzheimer disease in the United States (2010-2050) estimated using the 2010 census. *Neurology* 80:1778-1783.

[44] Bertram L, Tanzi RE (2012) The genetics of Alzheimer's disease. *Prog. Mol. Biol. Transl. Sci.* 107:79-100.

[45] LaFerla FM, Green KN (2012) Animal models of Alzheimer disease. *Cold Spring Harb. Perspect. Med.* 2:11.

[46] Albasanz JL, Dalfo E, Ferrer I, Martin M (2005) Impaired metabotropic glutamate receptor/phospholipase C signaling pathway in the cerebral cortex in Alzheimer's disease and dementia with Lewy bodies correlates with stage of Alzheimer's-disease-related changes. *Neurobiol. Dis*. 20:685-693.

[47] Tsamis KI, Mytilinaios DG, Njau SN, Baloyannis SJ (2013) Glutamate receptors in human caudate nucleus in normal aging and Alzheimer's disease. *Curr. Alzheimer Res.* 10:469-475.

[48] Malter JS, Ray BC, Westmark PR, Westmark CJ (2010) Fragile X Syndrome and Alzheimer's Disease: another story about APP and beta-amyloid. *Curr. Alzheimer Res.* 7:200-206.

[49] Krueger DD, Bear MF (2011) Toward fulfilling the promise of molecular medicine in fragile X syndrome. *Annu. Rev. Med.* 62:411-429.

[50] Parker SE, Mai CT, Canfield MA, Rickard R, Wang Y, Meyer RE et al. (2010) Updated national birth prevalence estimates for selected birth defects in the United States, 2004-2006. *Birth Defects Res. A Clin. Mol. Teratol.* 88:1008-1016.

[51] Lott IT (2012) Neurological phenotypes for Down syndrome across the life span. *Prog. Brain Res.* 197:101-121.

[52] Beyreuther K, Pollwein P, Multhaup G, Monning U, Konig G, Dyrks T et al. (1993) Regulation and expression of the Alzheimer's beta/A4 amyloid protein precursor in health, disease, and Down's syndrome. *Ann. N. Y. Acad. Sci.* 695:91-102.

[53] Wisniewski K, Howe J, Williams DG, Wisniewski HM (1978) Precocious aging and dementia in patients with Down's syndrome. *Biol. Psychiatry* 13:619-627.

[54] Mann DM, Esiri MM (1989) The pattern of acquisition of plaques and tangles in the brains of patients under 50 years of age with Down's syndrome. *J. Neurol. Sci.* 89:169-179.

[55] Wisniewski KE, Wisniewski HM, Wen GY (1985) Occurrence of neuropathological changes and dementia of Alzheimer's disease in Down's syndrome. *Ann. Neurol.* 17:278-282.

[56] Lott IT, Head E (2005) Alzheimer disease and Down syndrome: factors in pathogenesis. *Neurobiol. Aging* 26:383-389.

[57] Head E, Silverman W, Patterson D, Lott IT (2012) Aging and Down syndrome. *Curr. Gerontol. Geriatr. Res.* 2012:412536.

[58] Oka A, Takashima S (1999) The up-regulation of metabotropic glutamate receptor 5 (mGluR5) in Down's syndrome brains. *Acta Neuropathol.* (Berl) 97:275-278.

[59] Wang LW, Berry-Kravis E, Hagerman RJ (2010) Fragile X: leading the way for targeted treatments in autism. Neurother*a*peutics 7:264-274.

[60] Heulens I, Kooy F (2011) Fragile X syndrome: from gene discovery to therapy. *Front Biosci.* 16:1211-1232.

[61] Hagerman RJ, Hagerman PJ (2002) Physical and behavioral phenotype. In: Hagerman RJ, Hagerman PJ (eds) *Fragile X Syndrome: Diagnosis, Treatment, amd Research.* 3rd Ed. Baltimore: John Hopkins University Press, pp. 3-109.

[62] Beckel-Mitchener A, Greenough WT (2004) Correlates across the structural, functional, and molecular phenotypes of fragile X syndrome. *Ment. Retard. Dev. Disabil. Res. Rev.* 10:53-59.

[63] Bear MF, Huber KM, Warren ST (2004) The mGluR theory of fragile X mental retardation. *Trends Neurosci.* 27:370-377.

[64] Dolen G, Osterweil E, Rao BS, Smith GB, Auerbach BD, Chattarji S et al. (2007) Correction of fragile X syndrome in mice. *Neuron* 56:955-962.

[65] Lindemann L, Jaeschke G, Michalon A, Vieira E, Honer M, Spooren W et al. (2011) CTEP: a novel, potent, long-acting, and orally bioavailable metabotropic glutamate receptor 5 inhibitor. *J. Pharmacol. Exp. Ther.* 339:474-486.

[66] Michalon A, Sidorov M, Ballard TM, Ozmen L, Spooren W, Wettstein JG et al. (2012) Chronic pharmacological mGlu5 inhibition corrects Fragile X in adult mice. *Neuron* 74:49-56.

[67] Dolen G, Carpenter RL, Ocain TD, Bear MF (2010) Mechanism-based approaches to treating fragile X. *Pharmacol. Ther.* 127:78-93.

[68] Autism and Developmental Disabilities Monitoring Network Surveillance Year 2008 Principal Investigators, Centers for Disease Control and Prevention (2012) Prevalence of autism spectrum disorders--Autism and Developmental Disabilities Monitoring Network, 14 sites, United States, 2008. *MMWR Surveill Summ.* 61:1-19.

[69] Newschaffer CJ, Croen LA, Daniels J, Giarelli E, Grether JK, Levy SE et al. (2007) The epidemiology of autism spectrum disorders. *Annu. Rev. Public Health* 28:235-258.

[70] Hessl D, Dyer-Friedman J, Glaser B, Wisbeck J, Barajas RG, Taylor A et al. (2001) The influence of environmental and genetic factors on behavior problems and autistic symptoms in boys and girls with fragile X syndrome. *Pediatrics* 108:E88.

[71] Spencer CM, Alekseyenko O, Hamilton SM, Thomas AM, Serysheva E, Yuva-Paylor LA et al. (2011) Modifying behavioral phenotypes in Fmr1KO mice: genetic background differences reveal autistic-like responses. *Autism. Res.* 4:40-56.

[72] Clifford S, Dissanayake C, Bui QM, Huggins R, Taylor AK, Loesch DZ (2007) Autism spectrum phenotype in males and females with fragile X full mutation and premutation. *J. Autism. Dev. Disord.* 37:738-747.

[73] Amiet C, Gourfinkel-An I, Bouzamondo A, Tordjman S, Baulac M, Lechat P et al. (2008) Epilepsy in autism is associated with intellectual disability and gender: evidence from a meta-analysis. *Biol. Psychiatry* 64:577-582.

[74] Sokol DK, Chen D, Farlow MR, Dunn DW, Maloney B, Zimmer JA et al. (2006) High levels of Alzheimer beta-amyloid precursor protein (APP) in children with severely autistic behavior and aggression. *J. Child Neurol.* 21:444-449.

[75] Ray B, Long JM, Sokol DK, Lahiri DK (2011) Increased secreted amyloid precursor protein-alpha (sAPPα) in severe autism: proposal of a specific, anabolic pathway and putative biomarker. *PLoS One* 6:e20405.

[76] Bailey AR, Giunta BN, Obregon D, Nikolic WV, Tian J, Sanberg CD et al. (2008) Peripheral biomarkers in autism: secreted amyloid precursor protein-alpha as a probable key player in early diagnosis. *Int. J. Clin. Exp. Med.* 1:338-344.

[77] Wegiel J, Frackowiak J, Mazur-Kolecka B, Schanen NC, Cook EH,Jr, Sigman M et al. (2012) Abnormal intracellular accumulation and extracellular Aβ deposition in idiopathic and Dup15q11.2-q13 autism spectrum disorders. *PLoS One* 7:e35414.

[78] Fatemi SH, Folsom TD, Kneeland RE, Yousefi MK, Liesch SB, Thuras PD (2013) Impairment of fragile X mental retardation protein-metabotropic glutamate receptor 5 signaling and its downstream cognates ras-related C3 botulinum toxin substrate 1, amyloid beta A4 precursor protein, striatal-enriched protein tyrosine phosphatase, and homer 1, in autism: a postmortem study in cerebellar vermis and superior frontal cortex. *Mol. Autism* 4:21.

[79] Berry-Kravis E, Knox A, Hervey C (2011) Targeted treatments for fragile X syndrome. *J. Neurodev. Disord.* 3:193-210.

[80] Fatemi SH, Folsom TD, Kneeland RE, Liesch SB (2011) Metabotropic glutamate receptor 5 upregulation in children with autism is associated with underexpression of both Fragile X mental retardation protein and GABAA receptor beta 3 in adults with autism. *Anat. Rec.* (Hoboken) 294:1635-1645.

[81] Fatemi SH, Folsom TD (2011) Dysregulation of fragile x mental retardation protein and metabotropic glutamate receptor 5 in superior frontal cortex of individuals with autism: a postmortem brain study. *Mol. Autism* 2:6.

[82] Silverman JL, Tolu SS, Barkan CL, Crawley JN (2010) Repetitive self-grooming behavior in the BTBR mouse model of autism is blocked by the mGluR5 antagonist MPEP. *Neuropsychopharmacology* 35:976-989.

[83] Burket JA, Herndon AL, Winebarger EE, Jacome LF, Deutsch S (2011). Complex effects of mGluR5 antagonism on sociability and stereotypic behaviors in mice: possible implications for the pharmacotherapy of autism spectrum disorders. *Brain Res. Bull.* 86:152-158.

[84] Silverman JL, Smith DG, Rizzo SJ, Karras MN, Turner SM, Tolu SS et al. (2012) Negative allosteric modulation of the mGluR5 receptor reduces repetitive behaviors and rescues social deficits in mouse models of autism. *Sci. Transl. Med.* 4:131ra51.

[85] Hauser WA, Annegers JF, Kurland LT (1993) Incidence of epilepsy and unprovoked seizures in Rochester, Minnesota: 1935-1984. *Epilepsia* 34:453-468.

[86] Hauser WA, Morris ML, Heston LL, Anderson VE (1986) Seizures and myoclonus in patients with Alzheimer's disease. *Neurology* 36:1226-1230.

[87] Mendez M, Lim G (2003) Seizures in elderly patients with dementia: epidemiology and management. *Drugs Aging* 20:791-803.

[88] Pueschel SM, Louis S, McKnight P (1991) Seizure disorders in Down syndrome. *Arch. Neurol.* 48:318-320.

[89] Wisniewski KE, Segan SM, Miezejeski CM, Sersen EA, Rudelli RD (1991) The Fra(X) syndrome: neurological, electrophysiological, and neuropathological abnormalities. *Am. J. Med. Genet.* 38:476-480.

[90] Musumeci SA, Hagerman RJ, Ferri R, Bosco P, Dalla Bernardina B, Tassinari CA et al. (1999) Epilepsy and EEG findings in males with fragile X syndrome. *Epilepsia* 40:1092-1099.

[91] Volkmar FR, Nelson DS (1990) Seizure disorders in autism. *J. Am. Acad. Child Adolesc. Psychiatry* 29:127-129.

[92] Giovanardi Rossi P, Posar A, Parmeggiani A (2000) Epilepsy in adolescents and young adults with autistic disorder. *Brain Dev.* 22:102-106.

[93] Sheng JG, Boop FA, Mrak RE, Griffin WS (1994) Increased neuronal beta-amyloid precursor protein expression in human temporal lobe epilepsy: association with interleukin-1 alpha immunoreactivity. *J. Neurochem.* 63:1872-1879.

[94] Mackenzie IR, Miller LA (1994) Senile plaques in temporal lobe epilepsy. *Acta Neuropathol.* 87:504-510.

[95] Abe M, Sonobe N, Fukuhara R, Mori Y, Ochi S, Matsumoto T et al. (2012) Phenotypical difference of amyloid precursor protein (APP) V717L mutation in Japanese family. *BMC Neurol.* 12:38-2377-12-38.

[96] Cabrejo L, Guyant-Marechal L, Laquerriere A, Vercelletto M, De la Fourniere F, Thomas-Anterion C et al. (2006) Phenotype associated with APP duplication in five families. *Brain* 129:2966-2976.

[97] Edwards-Lee T, Ringman JM, Chung J, Werner J, Morgan A, St. George Hyslop P et al. (2005) An African American family with early-onset Alzheimer disease and an APP (T714I) mutation. *Neurology* 64:377-379.

[98] Ezquerra M, Carnero C, Blesa R, Gelpi JL, Ballesta F, Oliva R (1999) A presenilin 1 mutation (Ser169Pro) associated with early-onset AD and myoclonic seizures. *Neurology* 52:566-570.

[99] Furuya H, Yasuda M, Terasawa KJ, Tanaka K, Murai H, Kira J et al. (2003) A novel mutation (L250V) in the presenilin 1 gene in a Japanese familial Alzheimer's disease with myoclonus and generalized convulsion. *J. Neurol. Sci.* 209:75-77.

[100] Gomez-Tortosa E, Barquero S, Baron M, Gil-Neciga E, Castellanos F, Zurdo M et al. (2010) Clinical-genetic correlations in familial Alzheimer's disease caused by presenilin 1 mutations. *J. Alzheimers Dis.* 19:873-884.

[101] Jayadev S, Leverenz JB, Steinbart E, Stahl J, Klunk W, Yu CE et al. (2010) Alzheimer's disease phenotypes and genotypes associated with mutations in presenilin 2. *Brain* 133:1143-1154.

[102] Rudzinski LA, Fletcher RM, Dickson DW, Crook R, Hutton ML, Adamson J et al (2008) Early onset familial Alzheimer Disease with spastic paraparesis, dysarthria, and seizures and N135S mutation in PSEN1. *Alzheimer Dis. Assoc. Disord.* 22:299-307.

[103] Shrimpton AE, Schelper RL, Linke RP, Hardy J, Crook R, Dickson DW et al. (2007) A presenilin 1 mutation (L420R) in a family with early onset Alzheimer disease, seizures and cotton wool plaques, but not spastic paraparesis. *Neuropathology* 27:228-232.

[104] Snider BJ, Norton J, Coats MA, Chakraverty S, Hou CE, Jervis R et al. (2005) Novel presenilin 1 mutation (S170F) causing Alzheimer disease with Lewy bodies in the third decade of life. *Arch. Neurol.* 62:1821-1830.

[105] Wu YY, Cheng IH, Lee CC, Chiu MJ, Lee MJ, Chen TF et al. (2011) Clinical phenotype of G206D mutation in the presenilin 1 gene in pathologically confirmed familial Alzheimer's disease. *J. Alzheimers Dis.* 25:145-150.

[106] Tang FR, Lee WL, Yeo TT (2001) Expression of the group I metabotropic glutamate receptor in the hippocampus of patients with mesial temporal lobe epilepsy. *J. Neurocytol.* 30:403-411.

[107] Notenboom RG, Hampson DR, Jansen GH, van Rijen PC, van Veelen CW, van Nieuwenhuizen O et al. (2006) Up-regulation of hippocampal metabotropic glutamate receptor 5 in temporal lobe epilepsy patients. *Brain* 129:96-107.

[108] Kandratavicius L, Rosa-Neto P, Monteiro MR, Guiot MC, Assirati JA Jr, Carlotti CG Jr et al. (2013) Distinct increased metabotropic glutamate receptor type 5 (mGluR5) in temporal lobe epilepsy with and without hippocampal sclerosis. *Hippocampus* 23:1212-30.

[109] Aronica E, Yankaya B, Jansen GH, Leenstra S, van Veelen CW, Gorter JA et al. (2001) Ionotropic and metabotropic glutamate receptor protein expression in glioneuronal tumours from patients with intractable epilepsy. *Neuropathol. Appl. Neurobiol.* 27:223-237.

[110] Aronica E, van Vliet EA, Mayboroda OA, Troost D, da Silva FH, Gorter JA (2000) Upregulation of metabotropic glutamate receptor subtype mGluR3 and mGluR5 in reactive astrocytes in a rat model of mesial temporal lobe epilepsy. *Eur. J. Neurosci.* 12:2333-2344.

[111] Lee AC, Wong RK, Chuang SC, Shin HS, Bianchi R (2002) Role of synaptic metabotropic glutamate receptors in epileptiform discharges in hippocampal slices. *J. Neurophysiol.* 88:1625-1633.

[112] Merlin LR (2002) Differential roles for mGluR1 and mGluR5 in the persistent prolongation of epileptiform bursts. *J. Neurophysiol.* 87:621-625.

[113] Chuang SC, Zhao W, Bauchwitz R, Yan Q, Bianchi R, Wong RK (2005) Prolonged epileptiform discharges induced by altered group I metabotropic glutamate receptor-mediated synaptic responses in hippocampal slices of a fragile X mouse model. *J. Neurosci.* 25:8048-8055.

[114] Wong RK, Bianchi R, Chuang SC, Merlin LR (2005) Group I mGluR-induced epileptogenesis: distinct and overlapping roles of mGluR1 and mGluR5 and implications for antiepileptic drug design. *Epilepsy Curr.* 5:63-68.

[115] Bianchi R, Wong RKS, Merlin LR (2012) Glutamate Receptors in Epilepsy: Group I mGluR-Mediated Epileptogenesis. In Noebels JL, Avoli M, Rogawski MA, Olsen RW, Delgado-Escueta AV (Eds). *Jasper's Basic Mechanisms of the Epilepsies.* 4th ed. Bethesda, MD.

[116] Borowicz KK, Luszczki JJ, Czuczwar SJ (2009) 2-methyl-6-phenylethynyl-pyridine (MPEP), a non-competitive mGluR5 antagonist, differentially affects the anticonvulsant activity of four conventional antiepileptic drugs against amygdala-kindled seizures in rats. *Pharmacol. Rep.* 61:621-630.

[117] Tang FR, Chen PM, Tang YC, Tsai MC, Lee WL (2007) 2-methyl-6-phenylethynyl-pyridine (MPEP), a metabotropic glutamate receptor 5 antagonist, with low doses of

MK801 and diazepam: a novel approach for controlling status epilepticus. *Neuropharmacology* 53:821-831.

[118] Jesse CR, Savegnago L, Rocha JB, Nogueira CW (2008) Neuroprotective effect caused by MPEP, an antagonist of metabotropic glutamate receptor mGluR5, on seizures induced by pilocarpine in 21-day-old rats. *Brain Res.* 1198:197-203.

[119] Westmark CJ, Westmark P, Beard A, Hildebrandt S, Malter JS (2008) Seizure susceptibility and mortality in mice that over-express amyloid precursor protein. *Int. J. Clin. Exp. Pathol.* 1:157-157-168.

[120] Westmark CJ, Westmark PR, Malter JS (2010) Alzheimer's disease and Down syndrome rodent models exhibit audiogenic seizures. *J. Alzheimers Dis*. 20:1009-1013.

[121] Yan QJ, Rammal M, Tranfaglia M, Bauchwitz RP (2005) Suppression of two major Fragile X Syndrome mouse model phenotypes by the mGluR5 antagonist MPEP. *Neuropharmacology* 49:1053-1066.

[122] Westmark CJ, Westmark PR, Malter JS (2009) MPEP reduces seizure severity in Fmr-1 KO mice over expressing human Abeta. *Int. J. Clin. Exp. Pathol.* 3:56-68.

[123] Alves G, Bronnick K, Aarsland D, Blennow K, Zetterberg H, Ballard C et al. (2010) CSF amyloid-beta and tau proteins, and cognitive performance, in early and untreated Parkinson's disease: the Norwegian ParkWest study. *J. Neurol. Neurosurg. Psychiatry* 81:1080-1086.

[124] Fagan AM, Shaw LM, Xiong C, Vanderstichele H, Mintun MA, Trojanowski JQ et al. (2011) Comparison of Analytical Platforms for Cerebrospinal Fluid Measures of β-amyloid 1-42, total tau, and p-tau181 for identifying Alzheimer disease amyloid plaque pathology. *Arch. Neurol.* 68:1137-1144.

[125] Graff-Radford NR, Crook JE, Lucas J, Boeve BF, Knopman DS, Ivnik RJ et al. (2007) Association of low plasma Aβ42/Aβ40 ratios with increased imminent risk for mild cognitive impairment and Alzheimer disease. *Arch. Neurol.* 64:354-362.

[126] van Oijen M, Hofman A, Soares HD, Koudstaal PJ, Breteler MM (2006) Plasma Aβ(1-40) and Aβ(1-42) and the risk of dementia: a prospective case-cohort study. *Lancet Neurol.* 5:655-660.

[127] Sobow T, Flirski M, Kloszewska I, Liberski PP (2005) Plasma levels of alpha beta peptides are altered in amnestic mild cognitive impairment but not in sporadic Alzheimer's disease. *Acta Neurobiol. Exp.* (Wars) 65:117-124.

[128] Schupf N, Tang MX, Fukuyama H, Manly J, Andrews H, Mehta P et al. (2008) Peripheral Abeta subspecies as risk biomarkers of Alzheimer's disease. *Proc. Natl. Acad. Sci. USA* 105:14052-14057.

[129] Ringman JM, Younkin SG, Pratico D, Seltzer W, Cole GM, Geschwind DH et al. (2008) Biochemical markers in persons with preclinical familial Alzheimer disease. *Neurology* 71:85-92.

[130] Mehta PD, Pirttila T, Mehta SP, Sersen EA, Aisen PS, Wisniewski HM (2000) Plasma and cerebrospinal fluid levels of amyloid beta proteins 1-40 and 1-42 in Alzheimer disease. *Arch. Neurol.* 57:100-105.

[131] Mehta PD, Capone G, Jewell A, Freedland RL (2007) Increased amyloid beta protein levels in children and adolescents with Down syndrome. *J. Neurol. Sci.* 254:22-27.

[132] Price DL, Rockenstein E, Ubhi K, Phung V, MacLean-Lewis N, Askay D et al. (2010) Alterations in mGluR5 expression and signaling in Lewy body disease and in

transgenic models of alpha-synucleinopathy--implications for excitotoxicity. *PLoS One* 5:e14020.

[133] Sanchez-Pernaute R, Wang JQ, Kuruppu D, Cao L, Tueckmantel W, Kozikowski A et al. (2008) Enhanced binding of metabotropic glutamate receptor type 5 (mGluR5) PET tracers in the brain of parkinsonian primates. *Neuroimage* 42:248-251.

[134] Johnson KA, Conn PJ, Niswender CM (2009) Glutamate receptors as therapeutic targets for Parkinson's disease. *CNS Neurol. Disord. Drug Targets* 8:475-491.

[135] De Leonibus E, Manago F, Giordani F, Petrosino F, Lopez S, Oliverio A et al. (2009) Metabotropic glutamate receptors 5 blockade reverses spatial memory deficits in a mouse model of Parkinson's disease. *Neuropsychopharmacology* 34:729-738.

[136] Johnston TH, Fox SH, McIldowie MJ, Piggott MJ, Brotchie JM (2010) Reduction of L-DOPA-induced dyskinesia by the selective metabotropic glutamate receptor 5 antagonist 3-[(2-methyl-1,3-thiazol-4-yl)ethynyl]pyridine in the 1-methyl-4-phenyl-1,2,3,6-tetrahydropyridine-lesioned macaque model of Parkinson's disease. *J. Pharmacol. Exp. Ther.* 333:865-873.

[137] Rylander D, Iderberg H, Li Q, Dekundy A, Zhang J, Li H et al. (2010) A mGluR5 antagonist under clinical development improves L-DOPA-induced dyskinesia in parkinsonian rats and monkeys. *Neurobiol. Dis.* 39:352-361.

[138] Chen L, Liu J, Ali U, Gui ZH, Hou C, Fan LL et al. (2011) Chronic, systemic treatment with a metabotropic glutamate receptor 5 antagonist produces anxiolytic-like effects and reverses abnormal firing activity of projection neurons in the basolateral nucleus of the amygdala in rats with bilateral 6-OHDA lesions. *Brain Res. Bull.* 84:215-223.

[139] Morin N, Gregoire L, Morissette M, Desrayaud S, Gomez-Mancilla B, Gasparini F et al. (2013) MPEP, an mGlu5 receptor antagonist, reduces the development of L-DOPA-induced motor complications in de novo parkinsonian monkeys: biochemical correlates. *Neuropharmacology* 66:355-364.

[140] Berg D, Godau J, Trenkwalder C, Eggert K, Csoti I, Storch A et al. (2011) AFQ056 treatment of levodopa-induced dyskinesias: results of 2 randomized controlled trials. *Mov. Disord.* 26:1243-1250.

[141] Gregoire L, Morin N, Ouattara B, Gasparini F, Bilbe G, Johns D et al. (2011) The acute antiparkinsonian and antidyskinetic effect of AFQ056, a novel metabotropic glutamate receptor type 5 antagonist, in L-Dopa-treated parkinsonian monkeys. *Parkinsonism Relat. Disord.* 17:270-276.

[142] Jellinger KA (1998) Alzheimer-type lesions in Huntington's disease. *J. Neural Transm.* 105:787-799.

[143] Anborgh PH, Godin C, Pampillo M, Dhami GK, Dale LB, Cregan SP et al. (2005) Inhibition of metabotropic glutamate receptor signaling by the huntingtin-binding protein optineurin. *J. Biol. Chem.* 280:34840-34848.

[144] Ribeiro FM, Paquet M, Ferreira LT, Cregan T, Swan P, Cregan SP et al. (2010) Metabotropic glutamate receptor-mediated cell signaling pathways are altered in a mouse model of Huntington's disease. *J. Neurosci.* 30:316-324.

[145] Schiefer J, Sprunken A, Puls C, Luesse HG, Milkereit A, Milkereit E et al. (2004) The metabotropic glutamate receptor 5 antagonist MPEP and the mGluR2 agonist LY379268 modify disease progression in a transgenic mouse model of Huntington's disease. *Brain Res.* 1019:246-254.

[146] Fatemi SH, Kneeland RE, Liesch SB, Folsom TD (2010) Fragile X mental retardation protein levels are decreased in major psychiatric disorders. *Schizophr. Res.* 124:246-247.

[147] Albertini V, Benussi L, Paterlini A, Glionna M, Prestia A, Bocchio-Chiavetto L et al. (2012) Distinct cerebrospinal fluid amyloid-beta peptide signatures in cognitive decline associated with Alzheimer's disease and schizophrenia. *Electrophoresis* 33:3738-3744.

[148] Jakobsson J, Zetterberg H, Blennow K, Johan Ekman C, Johansson AG, Landen M (2013) Altered concentrations of amyloid precursor protein metabolites in the cerebrospinal fluid of patients with bipolar disorder. *Neuropsychopharmacology* 38:664-672.

[149] Pomara N, Bruno D, Sarreal AS, Hernando RT, Nierenberg J, Petkova E et al. (2012) Lower CSF amyloid beta peptides and higher F2-isoprostanes in cognitively intact elderly individuals with major depressive disorder. *Am. J. Psychiatry* 169:523-530.

[150] Baba H, Nakano Y, Maeshima H, Satomura E, Kita Y, Suzuki T et al. (2012) Metabolism of amyloid-beta protein may be affected in depression. *J. Clin. Psychiatry* 73:115-120.

[151] Gracia-Garcia P, de la Camara C, Santabarbara J, Lopez-Anton R, Quintanilla MA, Ventura T et al. (2013) Depression and incident Alzheimer disease: the impact of disease severity. *Am. J. Geriatr. Psychiatry*, in press.

[152] Morgan D (2011) Immunotherapy for Alzheimer's disease. *J. Intern. Med.* 269:54-63.

[153] Orgogozo JM, Gilman S, Dartigues JF, Laurent B, Puel M, Kirby LC et al. (2003) Subacute meningoencephalitis in a subset of patients with AD after Aβ42 immunization. *Neurology* 61:46-54.

[154] Pfeifer M, Boncristiano S, Bondolfi L, Stalder A, Deller T, Staufenbiel M et al. (2002) Cerebral hemorrhage after passive anti-Aβ immunotherapy. *Science* 298:1379.

[155] Racke MM, Boone LI, Hepburn DL, Parsadainian M, Bryan MT, Ness DK et al. (2005) Exacerbation of cerebral amyloid angiopathy-associated microhemorrhage in amyloid precursor protein transgenic mice by immunotherapy is dependent on antibody recognition of deposited forms of amyloid beta. *J. Neurosci.* 25:629-636.

[156] AK, Brindisi M, Tang J (2012) Developing beta-secretase inhibitors for treatment of Alzheimer's disease. *J. Neurochem.* 120 Suppl 1:71-83.

[157] Chang WP, Huang X, Downs D, Cirrito JR, Koelsch G, Holtzman DM et al. (2011) Beta-secretase inhibitor GRL-8234 rescues age-related cognitive decline in APP transgenic mice. *FASEB J.* 25:775-784.

[158] Wolfe MS (2012) γ-Secretase inhibitors and modulators for Alzheimer's disease. *J. Neurochem.* 120 Suppl 1:89-98.

[159] Ulus IH, Wurtman RJ (1997) Metabotropic glutamate receptor agonists increase release of soluble amyloid precursor protein derivatives from rat brain cortical and hippocampal slices. *J. Pharmacol. Exp. Ther.* 281:149-154.

[160] de Vrij FM, Levenga J, van der Linde HC, Koekkoek SK, De Zeeuw CI, Nelson DL et al. (2008) Rescue of behavioral phenotype and neuronal protrusion morphology in Fmr1 KO mice. *Neurobiol. Dis.* 31:127-132.

[161] Pin JP, Acher F (2002) The metabotropic glutamate receptors: structure, activation mechanism and pharmacology. *Curr. Drug Targets CNS Neurol. Disord.* 1:297-317.

[162] Malherbe P, Kratochwil N, Zenner MT, Piussi J, Diener C, Kratzeisen C et al. (2003) Mutational analysis and molecular modeling of the binding pocket of the metabotropic

glutamate 5 receptor negative modulator 2-methyl-6-(phenylethynyl)-pyridine. *Mol. Pharmacol.* 64:823-832.

[163] Porter RH, Jaeschke G, Spooren W, Ballard TM, Buttelmann B, Kolczewski S et al. (2005) Fenobam: a clinically validated nonbenzodiazepine anxiolytic is a potent, selective, and noncompetitive mGlu5 receptor antagonist with inverse agonist activity. *J. Pharmacol. Exp. Ther.* 315:711-721.

[164] Nordquist RE, Steckler T, Wettstein JG, Mackie C, Spooren W (2008) Metabotropic glutamate receptor modulation, translational methods, and biomarkers: relationships with anxiety. *Psychopharmacology* (Berl) 199:389-402.

[165] Brodkin J, Bradbury M, Busse C, Warren N, Bristow LJ, Varney MA (2002) Reduced stress-induced hyperthermia in mGluR5 knockout mice. *Eur. J. Neurosci.* 16:2241-2244.

[166] Moldrich RX, Chapman AG, De Sarro G, Meldrum BS (2003) Glutamate metabotropic receptors as targets for drug therapy in epilepsy. *Eur. J. Pharmacol.* 476:3-16.

[167] Walker K, Bowes M, Panesar M, Davis A, Gentry C, Kesingland A et al. (2001) Metabotropic glutamate receptor subtype 5 (mGlu5) and nociceptive function. I. Selective blockade of mGlu5 receptors in models of acute, persistent and chronic pain. *Neuropharmacology* 40:1-9.

[168] Hughes ZA, Neal SJ, Smith DL, Sukoff Rizzo SJ, Pulicicchio CM, Lotarski S et al. (2013) Negative allosteric modulation of metabotropic glutamate receptor 5 results in broad spectrum activity relevant to treatment resistant depression. *Neuropharmacology* 66:202-214.

[169] Breysse N, Baunez C, Spooren W, Gasparini F, Amalric M (2002) Chronic but not acute treatment with a metabotropic glutamate 5 receptor antagonist reverses the akinetic deficits in a rat model of parkinsonism. *J. Neurosci.* 22:5669-5678.

[170] Caraci F, Battaglia G, Sortino MA, Spampinato S, Molinaro G, Copani A et al. (2012) Metabotropic glutamate receptors in neurodegeneration/neuroprotection: Still a hot topic? *Neurochem. Int.* 61:559-65.

[171] Frisby CL, Mattsson JP, Jensen JM, Lehmann A, Dent J, Blackshaw LA (2005) Inhibition of transient lower esophageal sphincter relaxation and gastroesophageal reflux by metabotropic glutamate receptor ligands. *Gastroenterology* 129:995-1004.

[172] Berry-Kravis E, Hessl D, Coffey S, Hervey C, Schneider A, Yuhas J et al. (2009) A pilot open label, single dose trial of fenobam in adults with fragile X syndrome. *J. Med. Genet.* 46:266-271.

[173] Dolen G, Osterweil E, Rao BS, Smith GB, Auerbach BD, Chattarji S et al. (2007) Correction of Fragile X syndrome in mice. *Neuron* 56:955-962.

[174] Chiamulera C, Epping-Jordan MP, Zocchi A, Marcon C, Cottiny C, Tacconi S et al. (2001) Reinforcing and locomotor stimulant effects of cocaine are absent in mGluR5 null mutant mice. *Nat. Neurosci.* 4:873-874.

[175] Stoker AK, Olivier B, Markou A (2012) Involvement of metabotropic glutamate receptor 5 in brain reward deficits associated with cocaine and nicotine withdrawal and somatic signs of nicotine withdrawal. *Psychopharmacology* (Berl) 221:317-327.

[176] Sinclair CM, Cleva RM, Hood LE, Olive MF, Gass JT (2012) mGluR5 receptors in the basolateral amygdala and nucleus accumbens regulate cue-induced reinstatement of ethanol-seeking behavior. *Pharmacol. Biochem. Behav.* 101:329-335.

[177] Kagedal M, Cselenyi Z, Nyberg S, Raboisson P, Stahle L, Stenkrona P, et al (2013) A positron emission tomography study in healthy volunteers to estimate mGluR5 receptor occupancy of AZD2066 - estimating occupancy in the absence of a reference region. *Neuroimage* 82C:160-169.

[178] Harris BR, Prendergast MA, Gibson DA, Rogers DT, Blanchard JA, Holley RC et al. (2002) Acamprosate inhibits the binding and neurotoxic effects of trans-ACPD, suggesting a novel site of action at metabotropic glutamate receptors. *Alcohol. Clin. Exp. Res.* 26:1779-1793.

[179] Erickson CA, Early M, Stigler KA, Wink LK, Mullett JE, McDougle CJ (2011) An open-label naturalistic pilot study of acamprosate in youth with autistic disorder. *J. Child. Adolesc. Psychopharmacol.* 21:565-569.

[180] Erickson CA, Mullett JE, McDougle CJ (2010) Brief report: acamprosate in fragile X syndrome. *J. Autism Dev. Disord.* 40:1412-1416.

[181] Stocchi F, Rascol O, Destee A, Hattori N, Hauser RA, Lang AE, et al (2013) AFQ056 in Parkinson patients with levodopa-induced dyskinesia: 13-week, randomized, dose-finding study. *Mov. Disord.* 28:1838-46.

[182] Jacquemont S, Curie A, des Portes V, Torrioli MG, Berry-Kravis E, Hagerman RJ et al. (2011) Epigenetic modification of the FMR1 gene in fragile X syndrome is associated with differential response to the mGluR5 antagonist AFQ056. *Sci. Transl. Med.* 3:64ra1.

[183] Levenga J, Hayashi S, de Vrij FM, Koekkoek SK, van der Linde HC, Nieuwenhuizen I et al. (2011) AFQ056, a new mGluR5 antagonist for treatment of fragile X syndrome. *Neurobiol. Dis.* 42:311-317.

[184] Gantois I, Pop AS, de Esch CE, Buijsen RA, Pooters T, Gomez-Mancilla B et al. (2013) Chronic administration of AFQ056/Mavoglurant restores social behaviour in Fmr1 knockout mice. *Behav. Brain Res.* 239:72-79.

[185] Pop AS, Levenga J, de Esch CE, Buijsen RA, Nieuwenhuizen IM, Li T et al. (2013) Rescue of dendritic spine phenotype in Fmr1 KO mice with the mGluR5 antagonist AFQ056/Mavoglurant. *Psychopharmacology* (Berl), in press.

[186] Pecknold J, McClure D, Appeltauer L (1980) Fenobam in anxious outpatients. *Curr. Ther. Res.* 274:119-123.

[187] Pecknold JC, McClure DJ, Appeltauer L, Wrzesinski L, Allan T (1982) Treatment of anxiety using fenobam (a nonbenzodiazepine) in a double-blind standard (diazepam) placebo-controlled study. *J. Clin. Psychopharmacol.* 2:129-133.

[188] Lapierre Y, Oyewumi L (1982) Fenobam: another anxiolytic? *Curr. Ther. Res.* 31:95-101.

[189] Friedmann C, Davis L, Ciccone P, and Rubin R (1980) Phase II double blind controlled study of a new anxiolytic, fenobam (McN-3377) vs placebo. *Curr. Ther. Res.* 274:144-151.

[190] Wallberg A, Nilsson K, Osterlund K, Peterson A, Elg S, Raboisson P et al. (2006) Phenyl ureas of creatinine as mGluR5 antagonists. A structure-activity relationship study of fenobam analogues. *Bioorg. Med. Chem. Lett.* 16:1142-1145.

In: Metabotropic Glutamate Receptors
Editor: M. Foster Olive

ISBN: 978-1-63117-451-3

Chapter 5

Metabotropic Glutamate Receptors in the Pathophysiology and Treatment of Schizophrenia and Major Depression

Kelly A. Newell**, *Natalie Matosin and Jeremy S. Lum

[1]Faculty of Science, Medicine and Health,
and Illawarra Health and Medical Research Institute,
University of Wollongong, NSW, Australia
[2]Schizophrenia Research Institute, NSW, Australia

Abstract

Glutamatergic signaling is implicated in the pathophysiology of several psychiatric disorders, including schizophrenia and major depression. While the ionotropic glutamate receptors in particular are believed to play an important role in these disorders, more recent interest has focused on the metabotropic class of glutamate receptors. Metabotropic glutamate receptors (mGluRs) have been identified as exciting novel therapeutic targets for the treatment of schizophrenia and major depression, showing promising evidence in preclinical models. However it is important to consider the status of mGluRs in the pathophysiology of these disorders as this may influence the therapeutic ability of this novel class of mGluR-targeting drugs. Furthermore, it is necessary to determine whether current therapeutics indirectly influence mGluRs as this may provide some understanding of the effects of the use of these novel drugs as adjuncts with current therapeutics. In this chapter we review the literature on the role of mGluRs in schizophrenia and major depression. We also review the studies assessing whether current antipsychotic and antidepressant therapeutics mediate their effects through mGluRs and we examine the potential of novel therapeutics targeting these receptors.

Keywords: Metabotropic glutamate receptor, mGluR, mGlu, GRM, schizophrenia, depression, pathophysiology, antipsychotic, antidepressant, novel therapeutics

* Corresponding Author. Mailing address: Faculty of Science, Medicine and Health, University of Wollongong, Northfields Avenue Wollongong NSW Australia, 2522. Email: knewell@uow.edu.au.

Introduction

Schizophrenia (SZ) and major depression (MD) are extremely debilitating mental disorders that affect approximately 1% and 10% of the worldwide population respectively. Monoaminergic systems (dopamine, serotonin and norepinephrine) are strongly implicated in the symptomology associated with these disorders. More recently, the glutamatergic system, particularly the *N*-methyl-D-aspartate receptor (NMDAR), has been implicated in the underlying pathophysiology [1, 2]. For many years therapeutics for SZ and MD have targeted the monoaminergic systems, despite their limited efficacy. With the implication of glutamatergic signaling, therapeutics that target the glutamatergic system have the potential to more effectively treat these disorders, thus reducing the burden on patients, their families and the economy. Correcting the glutamatergic dysfunction in SZ and MD, particularly through targeting the NMDAR, is a major direction of novel therapeutic research. However, direct pharmacological NMDAR modulation to correct these dysfunctions is problematic. In SZ, NMDAR agonists such as D-serine have shown therapeutic potential, but are confounded by large dose requirements and the associated risk of excitotoxicity [3]. In MD, there is strong evidence that although a single dose of the NMDAR antagonist ketamine provides a great anti-depressant effect in treatment-resistant patients, NMDAR antagonism induces psychotomimetic effects (similar to SZ symptoms) [3]. Furthermore, because NMDAR expression is widespread throughout the central nervous system, targeting the NMDAR directly can cause adverse effects due to binding in brain regions not implicated in psychiatric pathology.

As a result of the many issues associated with targeting the NMDAR directly, attention has shifted to the metabotropic family of glutamate receptors. Glutamate signals through two families of receptors: ionotropic (NMDAR, α-amino-3-hydroxy-5-methyl-4-isoxazolepropionic acid receptor [AMPAR] and kainate) and metabotropic. Metabotropic glutamate receptors (mGluRs) belong to the family III/class C G-protein coupled receptors and consists of: Group I (mGluR1 and mGluR5) which primarily couple to Gαq and activate phospholipase C to produce diacylglycerol and/or inositol triphosphate and; Group II (mGluR2 and 3) and Group III (mGluR4,6,7,8) which primarily couple to Gαi/o and inhibit adenylyl cyclase activity and cyclic AMP production [4] (Figure 1).

mGluRs in the Pathophysiology of Schizophrenia and Major Depression

Although the mGluR systems present exciting therapeutic targets for the treatment of SZ and MD, little is known about their role in the pathophysiology of these disorders. This is an important avenue to pursue as it will not only contribute to our understanding of the glutamatergic dysfunction associated with these disorders, but will also improve the quality and development of novel therapeutics which target the glutamatergic system.

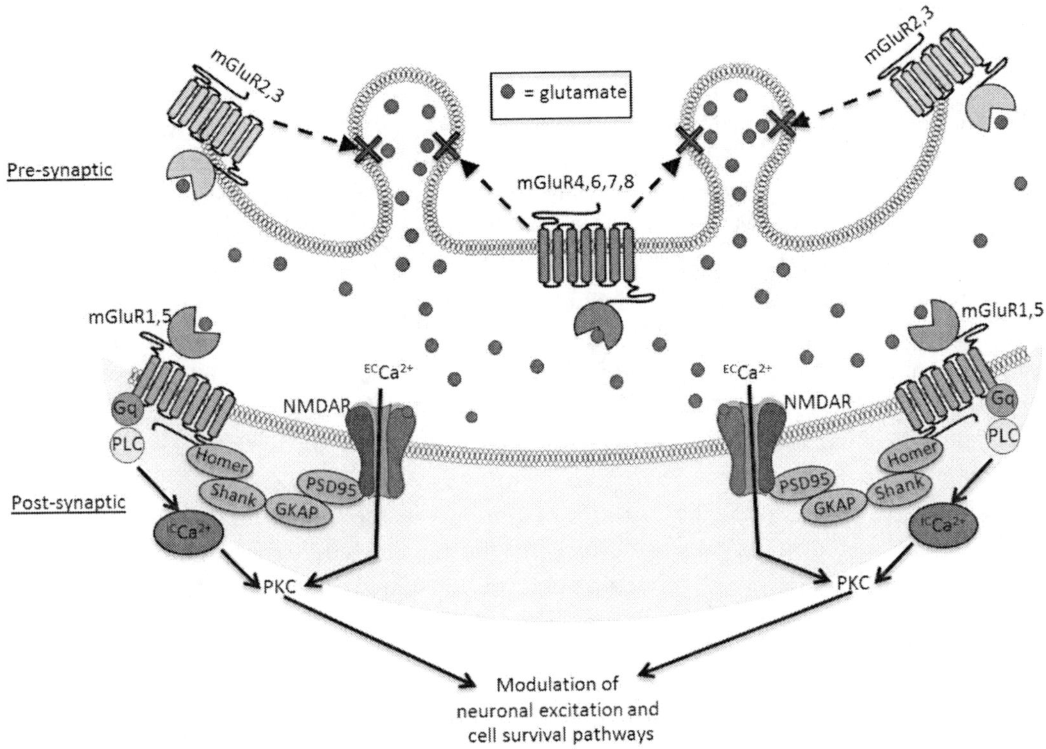

Figure 1. Schematic of metabotropic glutamate receptor (mGluR) locations at the neuronal synapse. Group 1 (mGluR1,5) and Group II (mGluR2,3) mGluRs are located peri-synaptically on the post-synaptic and pre-synaptic membranes respectively. Group III (mGluR4,6,7,8) are located pre-synaptically at the active zone. Group I mGluRs share a structural and functional link with the N-methyl-D-aspartate receptor (NMDAR) via the scaffold proteins Homer, Shank, GKAP, PSD-95, and act to regulate postsynaptic signaling. Group II and III mGluRs regulate glutamate (and GABA) release from the presynaptic terminal. Figure adapted from Matosin and Newell [4].

Group I mGluRs in Schizophrenia and Major Depression

Group I mGluRs, mGluR1 and mGluR5, are primarily located peri-synaptically on the post-synaptic membrane, where they slowly modulate glutamatergic signaling. mGluR1 and mGluR5 both form a physical and functional link to the NMDAR. The scaffold proteins Homer, SH_3 and multiple ankyrin repeat domains (SHANK), guanylate-kinase-associated protein (GKAP) and post-synaptic density 95 (PSD-95) form a physical link between the NMDAR and mGluR1/5 [4] (Figure 1). Therefore it is possible that NMDAR dysfunction in SZ or MD could lead to consequential alterations in mGluR1 and/or mGluR5. Alternatively, alterations in group I mGluRs may contribute to NMDAR dysfunction in SZ or MD. Although both mGluR1 and mGluR5 are linked to the NMDAR, it is primarily mGluR5 that shows promise as a therapeutic target to treat glutamatergic dysfunction in SZ and MD (discussed in detail below). This may be, in part, due to regional colocalisation differences with the NMDAR. mGluR5 and NMDAR are expressed together in brain regions including the hippocampus, striatum and neocortex, while mGluR1 is predominantly expressed in the cerebellum, thalamus, hippocampus, hypothalamus, amygdala, basal ganglia, brainstem and spinal cord [5–7].

mGluR1 and mGluR5 in the Pathophysiology of Schizophrenia

Although pharmacological manipulation of mGluR1 shows less therapeutic potential than mGluR5, it is evident that mGluR1 is altered in SZ (Table 1). A 12.4% increase in mGluR1 mRNA has been reported in the post-mortem prefrontal cortex of SZ subjects [8]. Similarly, in an independent population, increased mGluR1 protein (~10%) was reported in the prefrontal cortex of a small elderly cohort of SZ subjects [9]. mGluR1 has also been implicated at the genetic level, with two studies to date reporting associations between *GRM1* and SZ [10, 11]. Furthermore, mGluR1 knockout mice show deficits in prepulse inhibition, a behavioural phenotype that equates to the well-established phenomenon of sensorimotor gating deficits observed in SZ patients [12]. Evidence is therefore mounting that mGluR1 could indeed be involved in the pathophysiology of SZ. Even so, further research in larger and independent cohorts is needed to define the role of mGluR1 in SZ.

With regards to mGluR5, the majority of findings in the human brain suggest that this receptor is unchanged in SZ. We were recently the first to report that mGluR5 density, as measured by [^{3}H]MPEP radioligand binding, was unaltered in the dorsolateral prefrontal cortex of SZ subjects, accompanied by no change in mGluR5 monomer protein levels [13]. This finding is supported by several earlier studies reporting no change in mGluR5 mRNA (although one study did report an increase specifically in cortical layer III of the PFC) and no change in mGluR5 monomer protein in the SZ post-mortem brain (Table 1). These findings, particularly in the MPEP allosteric binding site that we reported [13], may be advantageous as they suggest that mGluR5-based therapeutics can bind to their target. Nonetheless, it is unknown whether mGluR5 function, regulation or interactions with other key proteins, such as the NMDAR, are affected in SZ, which may affect the benefits of mGluR5-based therapeutics. The scaffold proteins, Homer, SHANK, GKAP, and PSD-95 play an important role in linking mGluR5 (and mGluR1) to the NMDAR, and alterations in these scaffold proteins could affect the mGluR5/NMDAR association or coupling to downstream effectors. Both the *Homer1* gene and protein are reported to be altered in the prefrontal cortex and hippocampus in SZ [14, 15], with Homer1 knockout mice showing a SZ-like behavioural phenotype [16]. As Homer1 is critical for mGluR5/NMDAR coupling [17, 18], mGluR5 function may be altered, despite that mGluR5 protein levels appear undisturbed in SZ. It is unknown whether NMDAR/mGluR5 coupling is altered in SZ, although this could negatively impact on the effectiveness of mGluR5-based therapeutics to treat NMDAR dysfunctions.

While the majority of human studies suggest no involvement of mGluR5 in the pathophysiology of SZ (Table 1), these studies do not discriminate between cellular compartments, neuronal populations, cortical layers or receptor isoforms. There has been no investigation of specific mGluR5 isoforms, which can have different functions and cellular interactions [19]. mGluR5a/b, the most common splice variants, have a long intracellular C-terminus, allowing these receptor subtypes to closely interact with scaffold proteins and endogenous modulators to control downstream effectors [20]. mGluR5d, a lesser-known splice variant, has a shorter C-terminus and therefore fewer interactions with scaffold and intracellular proteins compared to mGluR5a/b. There is limited knowledge about the differing functions of these isoforms, or whether the ratio of mGluR5a/b:mGluR5d is altered across highly implicated brain regions [21].

Several endogenous modulators, including Norbin (neurochondrin), Tamalin (GRASP1), Preso1 (FRMPD4), and calcineurin (protein phosphatase 3), regulate mGluR5 function, notably phosphorylation as well as localisation within the cell. Although protein phosphorylation has been established as an important mechanism for regulating mGluR5 function, including receptor activation/deactivation and coupling to trafficking and/or scaffolding molecules [22, 23], mGluR5 phosphorylation has not been examined in the brains of patients with SZ. The mGluR5 regulator calcineurin has been shown to be unaltered in the SZ brain [24], however other endogenous regulators of mGluR5, including norbin, tamalin and preso1, have not been investigated in SZ. Despite this, both genetic and animal studies suggest their involvement. For example, Norbin knockout mice and calcineurin knockout mice generate a strong SZ-like behavioural phenotype [25, 26]. Furthermore, exome sequencing of multiplex pedigrees has shown the mGluR5/Tamalin association to be disrupted in SZ [16]. These studies suggest that mGluR5 itself or its regulation could be disrupted in SZ. Consistent with this: (1) mGluR5 has recently been reported as a top candidate gene for SZ [27, 28]; (2) mGluR5 knockout mice demonstrate a SZ-like behavioural phenotype [29]; and (3) rats treated with mGluR5 antagonists show SZ-like behaviours [4].

mGluR1 and mGluR5 in the Pathophysiology of Major Depression

There has been limited investigation of Group I mGluRs in MD, although studies to date suggest that they may be altered in this disorder (Table 1). One recent in vivo study found reduced mGluR5 binding in the prefrontal cortex, insula, thalamus, and hippocampus, of 11 unmedicated MD subjects compared to 11 matched controls, as measured via positron emission tomography [30]. Furthermore, mGluR5 binding in the hippocampus was negatively associated with the severity of depressive symptoms. It should be noted however that the authors found an association between a prior history of antidepressant use and mGluR5 binding in the left precentral gyrus, suggesting a possible interaction between mGluR5 and antidepressants. The same study also found reduced (-32%) mGluR5 (but not mGluR1) protein levels in the prefrontal cortex (BA10) of a cohort of 15 post-mortem MD subjects compared to matched controls. In contrast, in two post-mortem brain cohorts, we recently found no change in [^{3}H]MPEP binding to mGluR5 in the anterior cingulate cortex of subjects with MD [31]. This suggests that brain region or cohort differences may exist. Alternatively, the inconsistent results may be due to the different experimental procedures employed.

To our knowledge, no other studies have directly measured Group I mGluRs or scaffold proteins, such as Homer, in MD. However PSD-95, another major scaffold protein, has been shown to be downregulated in the prefrontal cortex in MD [32]. In addition, a single nucleotide polymorphism (SNP) in a putative regulatory region of the *Homer1* gene has been associated with MD in a Caucasian genome-wide analysis [33], suggesting that mGluR1 and/or mGluR5 and their anchoring to the cell membrane and/or interactions with the NMDAR could be disrupted in MD. The involvement of Group I mGluRs in MD is further supported by animal studies. Kovačević and colleagues have revealed that mGluR5 binding was reduced in the striatum of the Flinders Sensitive Line of rats (a rat model of MD), compared to Sprague-Dawley control rats, providing support for altered mGluR5 in MD [34]. In addition, increased mGluR1a protein was reported in the hippocampus of the olfactory bulbectomy rat model of depression [35].

It is not known whether Group I mGluR gene mutations are involved in MD. One study reported that SNPs in *GRM1* were associated with unipolar depression [36]. This same study also reported that *GRM1* alleles that were associated with a lower risk of unipolar depression were associated with lower hippocampal glutamate levels, a more pronounced normalisation of the hypothalamic-pituitary-adrenal (HPA) axis dysregulation and an increased likelihood for patients to respond to antidepressant medication. While further replication is needed, these findings suggest that *GRM1* polymorphisms could act as a biomarker for a subset of depressed patients, allowing a more tailored approach to antidepressant treatment [36].

Group II mGluRs in Schizophrenia and Major Depression

mGluR2 and 3 are primarily located pre-synaptically away from the active zone, where they regulate glutamate and GABA release via coupling to the G-protein Gi/o, inhibiting adenylyl cyclase activity and decreasing cAMP transduction [37] (Figure 1). mGluR2/3 based drugs therefore have the potential to treat glutamate dysfunction in SZ and MD [3, 38]. In addition to modulating glutamate and GABA release, mGluR2/3 also regulate dopamine release and antagonise serotonin 5HT2A receptors [39]. Considering that the dopaminergic and serotonergic systems are largely implicated in symptoms associated with SZ and MD [40, 41], it is possible that mGluR2/3 play a role in mediating changes in the dopaminergic and serotonergic systems in these disorders, or that changes in these systems could cause downstream alterations in mGluR2/3. mGluR2/3 are largely expressed in the hippocampus, cingulate cortex, amygdala, striatum, olfactory system, and cerebellum, with low expression in other brain regions. Notably, mGluR2 shows a stronger distribution in these regions than mGluR3 [38].

mGluR2 and mGluR3 in the Pathophysiology of Schizophrenia

As can be seen from Table 1, there have been inconsistent findings with regards to mGluR2 and 3 expression in the SZ brain. We recently reported no change in mGluR2/3 binding in the dorsolateral prefrontal cortex in SZ subjects [42]. This has been confirmed in the anterior cingulate cortex in a separate cohort [31]. One study, using a relatively small cohort (n=16/9), reported increased mGluR2/3 immunoreactivity in the prefrontal cortex in SZ [9]. In contrast, Corti et al [43] reported reduced mGluR3 dimers (but not monomers) in the frontal cortex of SZ patients, while another study in an independent cohort reported reduced mGluR3 protein (monomers), specifically in the dorsolateral prefrontal cortex in SZ [44]. Using qPCR, mGluR2 mRNA, but not mGluR3 mRNA, was reported to be reduced in the frontal cortex of a small cohort of untreated SZ subjects [45], suggesting differential involvement of these two receptors in SZ. Taken together, there is inconsistency in the post-mortem human brain findings with regards to mGluR2/3 in SZ. Some of this inconsistency has arisen from the difficulty of obtaining mGluR2 and mGluR3 specific antibodies and ligands, and the different methodologies used across the studies. However, these differences across cohorts also raise the notion that alterations in mGluR2/3 may be specific to subsets of patients. Defining patient subsets with these alterations, possibly with the use of peripheral biomarkers that are known to associate with brain pathological changes, and determining the consequences of such alterations will be an important way to move forward with post-mortem molecular analyses.

Several studies have consistently reported a genetic association of *GRM3* with SZ [46–49], although others in independent populations have reported no association [49–51], further supporting that alterations in Group II mGluRs may be specific to certain patient groups. While SNPs of various genes have been reported in SZ, finding the functional significance of such SNPs is imperative. One study reported associations between a *GRM3* SNP (rs6465084) and cognitive ability in healthy subjects [52], suggesting that the *GRM3* associations in SZ could be related to cognitive dysfunction. Supporting this, Egan et al [47] reported an association with a *GRM3* SNP (hCV11245618) and SZ, and also associated this with measures of prefrontal and hippocampal cognition. In addition, the high-risk *GRM3* allele was associated with excitatory amino acid transporter (EAAT) mRNA reductions in brain samples, but only weakly associated with mGluR3 mRNA levels and mGluR2/3 immunoreactivity [47]. Interestingly the presence of *GRM3* polymorphisms has been associated with olanzapine-induced improvements in negative symptoms, suggesting that it could be a useful predictor of olanzapine responsiveness [53]. While the mechanism underlying the *GRM3* influence on olanzapine response is unknown, the authors suggest that it may be related to olanzapine's effects on mGluR3 expression in the brain [53, 54], highlighting the need to characterise how current therapeutics influence the mGluR system.

mGluR2 and mGluR3 in the Pathophysiology of Major Depression

To our knowledge, only two published studies have assessed mGluR2/3 in post-mortem brain from subjects with MD (Table 1). While one study reported no change in mRNA expression in the anterior cingulate cortex and dorsolateral prefrontal cortex [55], another reported a large 67% increase in mGluR2/3 protein in the prefrontal cortex (BA10) in depressed subjects [56]. We have since found no change in mGluR2/3 binding in the anterior cingulate cortex in MD [31]. In addition, a preliminary study reported an association between one marker (rs6465084) in *GRM3* and MD in a Japanese population, with no association between *GRM2* and MD [57]. As this specific SNP was previously associated with cognitive ability, its association with MD may be related to cognitive alterations in this population of patients [52]. Minimal data is available directly from patients supporting or refuting the role of mGluR2/3 in MD, although several animal models of depression continue to suggest that Group II mGluRs are involved in its pathophysiology. Mice reared in isolation show increased mGluR2/3 binding density in the prefrontal cortex and hippocampus, suggesting mGluR2/3 is involved in the pathogenesis of depressive behaviour [58]. In contrast however, the olfactory bulbectomy and Flinders Sensitive Line rat models of depression both show a reduced expression of mGluR2/3 receptors in the hippocampus [35, 59]. This raises the question about animal model validity and relevance to psychiatric research [60]. Depression is a heterogeneous disorder, and while these animal models provide valuable information that we cannot obtain directly from patients, it is important that we understand exactly what is being modelled and to which aspect of this disease it is relevant. According to Kaffman and Krystal [60], one drawback of many animal models of depression is that they only examine adult animals and do not consider the neurodevelopmental component of these models. Considering that MD is thought to arise due to abnormalities during brain development, this developmental approach will be an important avenue to pursue in future studies.

Group III mGluRs in Schizophrenia and Major Depression

Little is known about group III mGluRs in relation to the pathophysiology of SZ and MD (Table 1). mGluR6 is restricted to the retina and therefore is not involved in these disorders [61]. mGluR4,7,8 are primarily located pre-synaptically, but in contrast to mGluR2/3, these receptors are located at the active zone, close to the site of synaptic vesicle/membrane fusion [62]. Group III mGluRs are believed to function as glutamatergic autoreceptors and GABAergic heteroreceptors. Like the Group II mGluRs, they couple to inhibitory cyclic AMP signal transduction and modulate calcium homeostasis and presynaptic excitability [63] (Figure 1).

mGluR4 is primarily located in the thalamus, hypothalamus, striatum and cerebellum [64]. One study found no change in mGluR4 mRNA in the thalamus of SZ subjects [65]. This is supported by a report of no change in mGluR4a protein levels in the prefrontal cortex (BA9,11,32,46) and striatum from SZ subjects compared to controls [9]. Furthermore, two studies have reported that *GRM4* is not associated with SZ [66, 67]. However, a recent study has reported that mGluR4 knockout mice show improved spatial short- and long-term memory and enhanced long-term potentiation in the CA1 of the hippocampus [68]. This suggests that an opposing mGluR mechanism (i.e. mGluR4 upregulation) could be involved in the deterioration of these behaviours observed in psychiatric illness, and that targeting mGluR4 pharmacologically could thus be an avenue to treat cognitive dysfunction.

While mGluR4 has not been directly examined in MD subjects, mGluR4 protein has been shown to be reduced in the hippocampus of the olfactory bulbectomy rat model of depression [35]. In addition, mGluR4 mRNA was recently shown to be reduced in the hippocampus of the maternal separation rat model of depression [69]. This preliminary data provides impetus to conduct more research on mGluR4 in MD.

Of the Group III receptors, mGluR7 is particularly gaining interest. mGluR7 is primarily located in the hippocampus, cerebral cortex and cerebellum [70–72]. Several *GRM7* gene polymorphisms have been associated with SZ [66, 73, 74]. While these studies have generally reported weak associations between *GRM7* SNPs and SZ, these findings are strengthened by the fact that they have been replicated in several independent cohorts. As with all genetic susceptibility, there have been populations that show no evidence of an association [75], and the only study to investigate mGluR7 in the brains of SZ subjects found no change in mGluR7 mRNA in the thalamus [65]. However, mGluR7 knockout mice show deficits in working memory and fear extinction [76]. Similarly, negative modulation of mGluR7 in rodents has been reported to impair cognition [77]. Furthermore, mGluR7 knockout and knockdown mice show reductions in glutamic acid decarboxylase (GAD) mRNA and protein expression in the hippocampus [78], a feature of SZ [79]. Collectively, these studies suggest that mGluR7 could be involved in SZ, or more specifically, the cognitive deficits present in SZ. mGluR7 could therefore be a valuable therapeutic target to treat the largely untreated cognitive deficits associated with this disorder.

With regards to MD, *GRM7* was recently implicated as a strong MD candidate gene [80], and mGluR7 knockout mice show antidepressant and anxiolytic behaviours [81], coupled with dysregulated HPA measures [82], suggesting a possible role of mGluR7 in depression. In contradiction, upregulation of mGluR7 with an mGluR7 agonist attenuates depressive-like behaviours in a mouse model [83]. While it is unclear why an mGluR7 agonist produces similar behavioural effects as a knockout of the mGluR7 gene, these studies do provide justification to continue research into the role of mGluR7 in MD, which poses an interesting novel therapeutic target. Recent data on the Wistar-Kyoto rat model of depression showed a selective increase in mGluR7 mRNA in the hippocampus, although this effect was not observed in the maternal separation rat model of depression [84]. Meanwhile, hippocampal mGluR7 immunoreactivity has been reported to be reduced in the olfactory bulbectomy model of depression [35]. While these findings again question the validity of animal models for researching psychiatric disorders, these models remain valuable in that we can measure aspects that are unobtainable in humans.

mGluR8 is located in key brain regions related to the pathophysiology of SZ, including the hippocampus, striatum and cerebral cortex [72, 85, 86]. *GRM8* has been associated with SZ in one population [87] but not in another [75]. A post-mortem study reported no change in mGluR8 mRNA in the thalamus of SZ subjects [65]. mGluR8 knockout mice show no SZ-like phenotype, as measured by prepulse inhibition and locomotion; however they do show anxiety-like behaviours [88, 89], suggesting that mGluR8 may be more relevant to anxiogenic pathophysiology rather than psychotic. In MD, only one recent study, on the maternal separation rat model of depression, reported that mGluR8 mRNA expression in the hippocampus was unaltered [90]. While a genome-wide scan identified an association between *GRM8* and trait depression [91], no genome-wide studies have identified an association between *GRM8* and MD, and to our knowledge there are no published studies examining specific SNPs of *GRM8* in MD. Based on the evidence to date, it appears that mGluR8 does not play a major role in either SZ or MD.

The Effects of Current Antipsychotic Therapeutics on mGluRs

Antipsychotics are the primary form of treatment for SZ patients. First (typical), second (atypical) and third generation (atypical) antipsychotics all competitively antagonise dopamine D_2 receptors. In addition, second and third generation antipsychotics also display affinities for other dopaminergic, serotonergic, muscarinic, adrenergic and histamatergic receptors, which are key to their therapeutic efficacy [98]. These antipsychotics are efficacious for positive SZ symptoms; however they display minimal benefits for negative and cognitive symptom profiles for many patients. Little is known about the interactions between current antipsychotics and the glutamatergic system, in particular the mGluRs, (Table 2). This is an important avenue of research for several reasons: (1) it provides a comparison with human brain data to discriminate between true pathological effects and effects caused by pre-mortem antipsychotic treatment; (2) it may indicate the benefits of specific antipsychotics in subgroups of patients that have alterations in the glutamate system (e.g. see [53]); and (3) it may inform the use of mGluR novel therapeutics as adjuncts with the current antipsychotics.

Table 1. Summary of mGluR investigations in schizophrenia and major depression post-mortem human brain samples

Study	Cohort	Brain region	Analysis	Result
[31]	15 SZ 15 MD 15 BP 15 CT	Anterior cingulate cortex	Receptor autoradiography: $[^3H]$MPEP labelling of mGluR5; $[^3H]$LY341495 labelling of mGluR2/3.	No significant differences in mGluR5 or mGluR2/3 binding in SZ, MD or BP, but differential age associations were present.
	12 CT 12 MDNP 12 MDP			No significant differences in mGluR5 or mGluR2/3 binding in MDNP or MDP, but differential age associations were present.
[92]	20 SZ 19 BP 29 CT	Prefrontal cortex (BA9)	qRT-PCR and western blot analysis of mGluR5.	Reduction in mGluR5 protein in SZ (32-53%) and BP (39-63%) compared to CT. Reduction in mGluR5 mRNA in BP (23%) but not SZ.
	15 SZ 14 MD 15 BP 14 CT	Lateral cerebella		Reduction in mGluR5 protein in SZ (60-75%), BP (54-79%) and MD (49-64%) compared to CT. Reduction in mGluR5 mRNA in SZ (51%) and MD (47%) compared to CT.
[13]	37 SZ 37 CT	Dorsolateral prefrontal cortex (BA46)	Receptor autoradiography: $[^3H]$MPEP labelling of mGluR5. Western blot analysis of mGluR5.	No change in mGluR5 binding or protein expression in SZ.
[55]	12 DP 12CT	Anterior cingulate cortex	qRT-PCR analysis of mGluR1,2,3	No change in mGluR1, 2 or 3 in DP.
	14 DP 14 CT	Dorsolateral prefrontal cortex		
[93]	21 SZ 35 CT	Prefrontal cortex (BA10)	Western blot analysis of mGluR5.	No change in mGluR5 protein expression in SZ.
[30]	11 MD 11 CT	Whole brain	In vivo $[^{11}C]$ABP688 binding to mGluR5 using PET.	Reductions (9.6%-22.6%) in mGluR5 binding in PFC, insula, thalamus, temporal cortex, parietal cortex, hippocampus in MD.
	15 MD 15 CT	Prefrontal cortex (BA10)	Western blot analysis of mGluR5 and mGluR1.	32% reduction in mGluR5 protein levels in MD. No change in mGluR1 protein levels.
[42]	37 SZ 37 CT	Dorsolateral prefrontal cortex (BA46)	Receptor autoradiography: $[^3H]$LY341495 labelling of mGluR2/3.	No change in mGluR2/3 binding, but a negative association with age in CT and SZ subjects.
[8]	28 SZ 14 SZA 42 CT	Prefrontal cortex (BA9)	qRT-PCR of mGluR1 and mGluR5.	12.4% increase in mGluR1a levels in SZ. mGluR5 mRNA levels were not altered in SZ.

Study	Cohort	Brain region	Analysis	Result
[56]	11 MD 11 CT	Prefrontal cortex (BA10)	Western blot analysis of mGluR2/3.	67% increase in mGluR2/3 protein levels in MD.
[44]	15 SZ 15 CT	Dorsolateral prefrontal cortex (BA9), cortical regions (BA 38,24, 11,7,19,4), nucleus accumbens, caudate, thalamus, hippocampus, cerebellum	Western blot analysis of mGluR2 and 3.	Reduction (~20%) in mGluR3 protein in the DLPFC only. No change in other regions or in mGluR2.
[45]	7 SZ (untreated) 7 CT	Frontal cortex	Receptor autoradiography: [^{3}H]LY341495 labelling of mGluR2/3; qPCR analysis of mGluR2 and mGluR3.	Decrease (~60%) in mGluR2 mRNA, but not mGluR3, in SZ. No change in mGluR2/3 binding.
[94]	34 SZ 70 CT	Dorsolateral prefrontal cortex	qRT-PCR analysis of mGluR3	Increased (~15%) mGluR3 mRNA in the dorsolateral prefrontal cortex; no change in the prefrontal cortex or hippocampus.
	30 SZ 71 CT	Hippocampus		
	33 SZ 33 BP 35 CT	Prefrontal cortex		
[43]	20 SZ 35 CT	Prefrontal cortex (BA10)	Western blot analysis of mGluR2/3 and mGluR3	49% decrease in mGluR3 dimer protein, but not monomer, in SZ. No change in mGluR2/3 protein.
[9]	16 SZ 9 CT	Prefrontal cortex: BA9,11,32,46; caudate, putamen, nucleus accumbens	Western blot analysis of mGluR1a, mGluR2/3, mGluR4a, and mGluR5.	Increase in mGluR1a (~10%) and mGluR2/3 immunoreactivity in prefrontal cortex in SZ. No change in mGluR4a or mGluR5 in the prefrontal cortex. No change in any mGluRs in the striatum.
[95]	20 SZ 20 CT	Dorsolateral prefrontal cortex (BA46)	Immunocytochemistry and western blot analysis of mGluR2/3.	No change in cellular localization or laminar distribution of mGluR2/3 in SZ. No change in mGluR2/3 protein expression in SZ.
[96]	5 SZ 6 CT	Dentate gyrus, cornu ammonis, parahippocampal gyrus	In situ hybridisation of transcripts encoding for mGluR5.	No change in mGluR5 mRNA in SZ subjects.
[65]	12 SZ 8 CT	Thalamus	In situ hybridisation of transcripts encoding for mGluR1,2,3,4,5,7,8	No change in any mGluR mRNA in SZ.
[97]	7 SZ 10 CT	Prefrontal cortex (BA9, 10, 11)	In situ hybridisation of transcripts encoding for mGluR5.	Increased mGluR5 mRNA (27.9%) in layer III of BA 11 in SZ. No change in BA9 or BA10

Abbreviations: BA, Brodmann's area; BP, bipolar disorder; CT, control; DP, depression; MD, major depression; MDNP, major depression with no psychosis; MDP, major depression with psychosis; mGluR, metabotropic glutamate receptor; SZA schizoaffective.

Group I mGluRs

Volk et al [8] reported that chronic haloperidol or olanzapine treatment did not alter mGluR1a mRNA in the monkey prefrontal cortex. In addition they found that prefontal cortex mGluR1a mRNA levels did not differ between SZ patients that had received antipsychotics at the time of death, and those that did not. We have recently shown in the human brain that lifetime antipsychotic dose (as measured by chlorpromazine equivalents), is not associated with mGluR5 binding or protein levels, at least in the dorsolateral prefrontal cortex [13]. Furthermore, we have shown that treatments with typical (haloperidol) and atypical (olanzapine) antipsychotics do not influence mGluR5 binding density in the rat prefrontal cortex, hippocampus or striatum. However, another study has shown that mGluR5 (pan)mRNA expression in the cortex and striatum was increased by chronic typical (haloperidol) and atypical (sertindole) antipsychotic treatment [99].

In addition, acute treatment with haloperidol and the atypical antipsychotics quetiapine, clozapine and amisulpride, but not sertindole, have been shown to increase Homer1a mRNA expression in the striatum, but not the hippocampus [99]. As Homer1a can reduce mGluR5 coupling to postsynaptic effectors and therefore alter mGluR5 signaling [100], antipsychotic effects on Homer1 have the potential to affect mGluR5 function.

Group II mGluRs

While minimal research has examined the association between current antipsychotics and Group II mGluRs, our findings [42] and those of Crook et al [95], both demonstrate no association between antipsychotic dose (converted to chlorpromazine equivalents) and mGluR2/3 binding or immunodensity in the post-mortem human brain [42, 95]. While these human studies are limited by the collective examination of mGluR2/3, they are supported by the report of no change to mGluR3 protein levels specifically, following haloperidol or risperidone treatment in rats [44]. Furthermore, Tascedda et al [54] reported that chronic olanzapine or haloperidol treatment produced no change in mGluR2 or mGluR3 mRNA levels, while chronic clozapine treatment increased mGluR3, but not mGluR2, mRNA expression in the rat frontal cortex [54]. In contrast however, Sartorius et al [94] reported that mGluR3 mRNA was not changed in the rat frontal cortex following chronic clozapine or haloperidol treatment. While Sartorius et al [94] and Tascedda et al [54] used similar dosing regimens, they differed slightly in the method used to detect mRNA expression, which could account for the differences observed between the studies.

Group III mGluRs

To our knowledge there has only been one published study that reported on the effects of current antipsychotic treatment on mGluR expression. Tascedda et al [54] showed that antipsychotic treatment (haloperidol, olanzapine, clozapine) caused no significant change in mGluR7 mRNA expression in the rat hippocampus or frontal cortex. To date, no studies have examined the effects of current antipsychotics on mGluR4 or 8.

The Effects of Current Antidepressant Therapeutics on mGluRs

Current antidepressant drugs are primarily classed into one of three categories: (1) Selective Serotonin Reuptake Inhibitors (SSRIs); (2) Serotonin-Norepinephrine Reuptake Inhibitors (SNRIs); or (3) Tricyclic antidepressants (TCAs). As can be seen from these classifications, most antidepressants exert their effects on monoaminergic systems. Antidepressants primarily work by increasing extracellular monoamine levels in the synaptic cleft, however there is strong evidence that patients need to consistently dose with these drugs for 3-4 weeks before beneficial effects are seen [101]. This suggests that downstream intracellular pathways might be recruited before the onset of therapeutic effects. As with antipsychotics, there has been limited research into the area of antidepressant and mGluR interactions, although current knowledge suggests that existing antidepressants might mediate some effects through mGluRs (Table 2).

Group I mGluRs

Smialowska et al [102] found that chronic (21 days) imipramine (a TCA) treatment significantly increased the level of mGluR5a immunoreactivity in the CA1 field of the hippocampus, to 148% of the control, without changing mGluR1 immunoreactivity [102]. However, another study found that total mGluR5 protein levels were unaltered following 21 days of imipramine treatment [103]. It is unknown if SNRIs affect Group I mGluRs, however 14-days of venlafaxine treatment (an SNRI) resulted in increased Homer1a mRNA expression in the rat cingulate and orbital cortex [103], suggesting a possible avenue to affect Group I mGluR function. In contrast, rats treated for 21 days with imipramine showed no change in Homer1a protein levels in the hippocampus, cortex or striatum [103]. Furthermore, numerous studies have found that antidepressant treatment alters NMDARs, with increases, decreases and altered stoichiometry being reported, depending on the brain region examined and the NMDAR subunit examined [104]. Due to the close signaling partnership between NMDAR and mGluR1/5, it is possible that mGluR1/5 and its signaling partners will be altered following antidepressant treatment, which might indicate a more (or less) powerful effect when combined with Group I mGluR-based novel therapeutics. Interestingly, it was shown in rats that chronic imipramine treatment (10mg/kg, orally, 7, 14 or 21 days) decreased the responsiveness of CA1 hippocampal neurons to Group I mGluR activation, and that this effect was still present 7 days after the imipramine treatment ceased [105]. While this suggests that changes in mGluR expression and/or function could influence the therapeutic benefits of imipramine, it also has important implications for the co-treatment of, or transition from, current antidepressants to novel mGluR-based therapies. mGluR5 negative allosteric modulators are in preclinical research for the treatment of depression [106]. These early findings of Zahoronda and Bijak [105] therefore suggest that patients already on antidepressant treatments may demonstrate differential responsiveness to drugs targeting mGluR. More studies are needed to determine if patients transitioning from imipramine treatment to mGluR-based treatments will benefit from an immediate transition, or will require a "wash-out" period in which the lasting effects of imipramine on mGluRs "wear-off" before mGluR-based therapies can commence.

Group II mGluRs

Group II mGluRs interact with the serotonergic system [45, 107], therefore it is probable that antidepressants influence Group II mGluRs via this interaction. Chronic fluoxetine treatment (an SSRI) however, resulted in no change to mGluR2/3 protein expression in the monkey prefrontal cortex [56], similar to chronic imipramine treatment (a TCA), which caused no change in mGluR2/3 binding in the rat hippocampus and frontal cortex [108]. In contrast, Matrisciano et al [103] showed that chronic (21days) and acute imipramine treatments cause opposing effects on mGluR2/3 protein, with increases reported in the hippocampus, cortex and striatum following chronic treatment, and reductions in the same regions after a single dose. Furthermore, chronic (14 days) treatment with the TCA amitriptyline reduced mGluR2/3 protein in the mouse hippocampus [35]. Although preliminary, these studies highlight that differences arise not only between antidepressant classes (e.g. SSRI, SNRI, TCA), but also within classes. Furthermore, the technique, animal species and dosing duration used may all influence these measures, which raises the question of which protocols are the most relevant to the clinical situation.

Group III mGluRs

Matrisciano et al [103] showed that imipramine treatment in rats caused no change to mGluR4 protein levels, although their small sample size was very small. More recently, O'Connor et al [90], found evidence that antidepressant therapies have differential effects depending on the underlying biology of the animal model. Antidepressant treatment in standard naïve Sprague-Dawley rats produced differential effects on Group III mGluRs compared to the same treatment in Sprague-Dawley rats that had experienced maternal separation (a model of depression). The O'Connor study demonstrates that using animal models to study the effects of these drugs has its limitations [90]. There are clear differences in how these drugs work in naïve animals, compared to animal models of psychopathology.

While naïve animals may be useful in advising the effects of these drugs in healthy individuals (who in fact have no psychopathology), the interaction of these drugs with the disease state, can cause different effects. Until we have a model that truly represents the disease state, our knowledge of the molecular effects of these drugs will be limited. Perhaps testing in a combination of models that represent different aspects of the disease state will help to overcome this limitation.

Table 2. Summary of mGluR investigations in animal models of antipsychotic or antidepressant treatment

Animals	Treatment	Analysis	Outcome	Study
• Sprague-Dawley rats • Male • Adult • n=6/group	Haloperidol (0.1mg/kg) or olanzapine (1mg/kg), 3 times per day, p.o. for 8, 16 or 36 days. Sacrificed 48 hours after last treatment.	Receptor autoradiography: [^{3}H]MPEP labelling of mGluR5 in prefrontal cortex, hippocampus, striatum.	No effect of haloperidol or olanzapine treatment on mGluR5 binding.	[13]
• Sprague-Dawley rats (naive or maternally separated) • Male • Adult • n=7-10/group	Fluoxetine (10mg/kg), i.p. for 21 days; ketamine (10mg/kg) for one day. Sacrificed 24 hours after the last treatment	qRT-PCR for mGluR4,7,8 in the hippocampus	Fluoxetine increased mGluR7 (~30%) and mGluR8 (~50%) mRNA in naive rats, but not in maternally separated rats. It increased mGluR4 mRNA in maternally separated rats (~50%), but not naive rats. Ketamine treatment reduced mGluR4 mRNA (~-30%) in naive rats, but not in maternally separated rats. It had no effect on mGluR7 or mGluR8.	[90]
• Wistar rats • Male • Adolescent/ • early adult • n=4-9/group	Imipramine (10mg/kg) i.p. twice a day for 21 days. Sacrificed 24 hours after last treatment.	Western blot analysis of mGluR5 and immunohistochemistry of mGluR1a and mGluR5a in the hippocampus.	mGluR5 protein increased (~35%) in response to imipramine in CA1 but not CA3 or dentate gyrus. No significant effect on mGluR1a.	[102]
• Macaque monkeys • Male • Young adult • n=6/group	Haloperidol (1.0-2.0mg/190mg pellet until plasma levels reached 1.5ng/ml), olanzapine (0.55-1.1mg/190mg pellet until plasma levels reached 15ng/ml), or placebo, p.o. twice daily for 17-27 months. Sacrificed on the same day as last treatment.	qRT-PCR for mGluR1a in prefrontal cortex.	No significant change in mGluR1a mRNA in response to haloperidol or olanzapine treatment.	[8]
• Rhesus monkeys • Male • Adult • n=6-7/group	Fluoxetine (dose escalated to 3mg/kg) p.o. for 39 weeks. Sacrificed 20-24 hours after last treatment.	Western blot analysis of mGluR2/3 protein in prefrontal cortex (BA 10).	No change in mGluR2/3 in response to fluoxetine.	[56]
• Sprague-Dawley rats • Male • Adult • n=7	Haloperidol (0.8mg/kg), or sertindole (2mg/kg) s.c. for 21 days. Sacrificed 90 mins after the last treatment.	In situ hybridisation of transcripts encoding for mGluR5 in cortex, striatum and hippocampus.	Increased mGluR5 mRNA in medial striatum (17-21%), anterior cingulate (8-9%) and motor cortices (11-12%) in response to both treatments. No difference in hippocampus. Increased mGluR5 mRNA in lateral striatum after sertindole treatment only (14%).	[99]

Table 2. (Continued)

Animals	Treatment	Analysis	Outcome	Study
• Sprague-Dawley rats • Male • Adult • n=10	Haloperidol (1.5 mg/kg/day) or risperidone (6 mg/kg/day) p.o. daily for 6 months.	Western blot analysis of mGluR3 in frontal cortex.	No effect of haloperidol or risperidone treatment on mGluR3 in the frontal cortex.	[44]
• Sprague-Dawley rats • Male • Adult • n=8-10	Haloperidol (0.08, 0.6, and 2 mg/kg) or clozapine (0.5, 5, and 10mg/kg) i.p., once daily for 28 days. Sacrificed 7 hours after last treatment.	qRT-PCR analysis of mGluR3 in frontal cortex.	mGluR3 mRNA expression did not change in the frontal cortex of rats treated with haloperidol or clozapine.	[94]
• C57Bl/Han mice • Male • Early adult • n=6	Amitriptyline (10mg/kg) i.p. daily for 14 days. Sacrificed 24 hours after last treatment.	Western blot analysis of mGluR1a, mGluR2/3, mGluR4 and mGluR7 in the hippocampus.	Reduced mGluR2/3 (-50%), mGluR4 (-39%) and mGluR7 (-64%), but not mGluR1a, in response to amitriptyline treatment.	[35]
• Wistar rats • Male • Adult • n=8	Imipramine (10mg/kg) p.o. for 21 days or 1 day. Sacrificed 48 hours after last treatment.	Membrane binding to mGluR2/3 using [^{3}H]LY341495, in the hippocampus and frontal cortex.	No change in mGluR2/3 binding density or affinity.	[108]
• Wistar rats • Male • Adult • n=6	Chronic (21 days) imipramine (10 mg/kg) or citalopram (10 mg/kg) i.p. Sacrificed 24 hours after final injection.	Western blot analysis of mGluR4 and mGluR7 in the hippocampus and frontal cortex.	mGluR7-immunoreactivity was decreased after citalopram in both the hippocampus and cerebral cortex (~-55%), but not following imipramine treatment. No changes were observed in mGluR4-immunoreactivity.	[109]
• Sprague-Dawley rats • Male • Adolescent/young adult • n=4/group	Imipramine (10mg/kg) i.p. for 1 or 21 days. Sacrificed 24 hours after last treatment.	Western blot analysis of mGluR1a, mGluR2/3, mGluR4, and qRT-PCR analysis of mGluR2 and mGluR3, in hippocampus, cerebral cortex, nucleus accumbens and striatum.	mGluR2/3 protein was increased in the hippocampus (~71%), cortex (~38%), striatum (~75%), nucleus accumbens (~63%) following 21 day treatment. mGluR2/3 protein was reduced in the hippocampus (~25%), striatum (~21%), and cortex (~27%) following 1 imipramine injection. mGluR1a was increased in the hippocampus (~45%) following 21 day treatment only. mGluR4 protein remained unaltered. mGluR2 and mGluR3 mRNA was unaltered.	[103]
• Sprague-Dawley rats • Male • Adult • n=6-8/group	Haloperidol (1mg/kg), clozapine (10mg/kg), or olanzapine (2mg/kg) s.c., daily for 1 (acute) or 21 (chronic) days. Sacrificed 6 hours after last treatment.	RNase Protection Assay.	Increase in mGluR3 mRNA expression in frontal cortex (43%) (but not striatum or hippocampus) following chronic clozapine treatment (but not olanzapine or haloperidol). No changes to mGluR1, mGluR4-8 mRNA after any treatment.	[54]

Abbreviations: i.p., intraperitoneal; mGluR, metabotropic glutamate receptor; p.o., peroral (by mouth); s.c., subcutaneous.

mGluRs as Novel Therapeutic Targets for Schizophrenia and Major Depression

mGluRs, in particular, Group I and II, have received growing attention over the past decade as potential novel therapeutic targets for treating SZ and MD. Treatments for SZ to date have typically focused on the dopaminergic system. While second generation antipsychotics have a more complex mechanism of action, these drugs are confounded by their limited efficacy for negative and cognitive SZ symptoms, their debilitating side effects including obesity and diabetes, and associated non-compliance [110]. Many avenues suggest that NMDAR dysfunction is a core biological alteration in SZ patients, at least in a subset of patients [1, 111]. Furthermore, NMDAR dysfunction in the hippocampus and frontal cortex in particular are believed to contribute to the cognitive symptoms observed in patients [1]. Targeting the NMDAR, and glutamatergic system more broadly, is therefore a possible therapeutic avenue to treat these aspects of SZ [4, 112, 113]. Directly targeting the NMDAR attracts issues such as excitotoxicity, as well as the possibility of unwanted side effects due to the wide distribution of NMDARs in the central nervous system. Targeting mGluRs is an attractive alternative to restore glutamatergic dysfunction and/or NMDAR dysfunction in SZ patients. To date Group II agonists have progressed to clinical trials, while mGluR5 positive allosteric modulators (PAMs) are of more recent interest, showing potential in preclinical studies. Minimal research has been conducted on Group III mGluRs, although they could be promising.

Traditional treatments for MD have typically targeted the serotonin and noradrenaline systems, and are well known to take several weeks to achieve therapeutic effects. In contrast, a single injection of the NMDAR antagonist ketamine has been shown to produce rapid antidepressant effects [114], suggesting that the glutamatergic system may be a more efficient target for depression treatment. NMDAR antagonists such as ketamine do not provide a long-term solution as they have the potential to cause SZ-like psychotomimetic effects [115]. Therefore, current research proposes that targeting the mGluR system may provide a way to generate rapid anti-depressant effects, without the psychotomimetic side effects. To date, Group I and II antagonists and negative allosteric modulators (NAMs) have received the most attention with regards to their potential as novel antidepressants.

Group I Novel Therapeutics

Group I mGluRs (mGluR1 and mGluR5) share a physical and functional link with the NMDAR and thus they offer an approach for the treatment of NMDAR dysfunction in SZ. mGluR5 PAMs have shown promise preclinically, purportedly due to their ability to upregulate NMDAR activity. However, mGluR5 are also highly expressed on microglia [116]. Microglia induced inflammation is believed to be involved in SZ [117]. Activating mGluR5 can inhibit microglial-associated inflammation and toxicity [118, 119], which may contribute to the therapeutic effects of mGluR5 PAMs seen in animal models. mGluR1 has been shown to modulate NMDAR currents by inducing phosphorylation of the NR2B subunit of the NMDAR [5], suggesting that mGluR1 PAMs might offer a similar therapeutic effect as targeting mGluR5 [5]. While this has not been directly investigated, it should be noted that mGluR1 and mGluR5 antagonists produce different behavioural responses which suggest that these receptors interact differently with the NMDAR; this may in part be related to regional differences in group I mGluR expression [5]. Therefore it is unclear whether mGluR1 PAMs will offer the same therapeutic value as mGluR5 PAMs. While it has been suggested that mGluR1 antagonists or NAMs could offer an alternative approach for the treatment of SZ [120], overall they show little promise so far. The mGluR1 NAM, BAY36-7620 attenuates selected NMDAR antagonist mediated behaviours only [38,121]. In addition, the mGluR1 antagonist EMQMCM, has been shown to have no effect on MK-801-induced prepulse inhibition or locomotor activity [122]. Furthermore, due to the high concentration of mGluR1 in the cerebellum, these mGluR1 NAMs also produce ataxic side effects in rodents [122]. mGluR1 antagonists have shown antidepressant properties in rodents, as measured by reduced immobility time in the forced swim and tail suspension tests [123], howver further research is needed to confirm their antidepressant potential.

We have recently reviewed the evidence of mGluR5 PAMs for the treatment of SZ [4]. mGluR5 PAMs including ADX47273, CDPPB, and various VU compounds have shown the ability to reverse positive, negative and cognitive SZ-like behaviours in rodent models suggesting mGluR5 could be a worthwhile target for SZ treatment [4]. In the rat hippocampus it was shown that ADX47273 induced long-term potentiation [124], suggesting that this could underlie its ability to enhance cognitive function in these rodents. Despite the potential of mGluR5 PAMs, they have been plagued by solubility issues and a lack of oral bioavailability. Recent developments of orally available mGluRs show they do have potential [125–127], however they need further investigation especially regarding their ability to modulate dopamine-mediated behaviours [21]. While there is a large body of animal behavioural studies, research thus far lacks a detailed understanding of the molecular and cellular alterations that these novel drugs produce in the brain.

A majority of the mGluR5 PAMs bind to the same region as the NAM, MPEP, within the mGluR5 transmembrane region (e.g. CDPPB, ADX-47273, VU-29). However, a small number of mGluR5 PAMs bind to a separate locus from the MPEP binding region (e.g. CPPHA) [128]. Regardless of differences in binding sites, these mGluR5 PAMs all increase intracellular calcium levels when co-activated with an mGluR5 agonist in various cell lines. However, CPPHA has been shown to inhibit the ERK1/2 signaling pathway during high agonist concentrations, something not observed with MPEP site PAMs [129]. This difference in activation of differing signaling systems, known as signaling bias, is hypothesised to be influenced not only by drug type, but also neuronal population [130]. mGluR5 PAMs which bind to the MPEP site, increase long-term potentiation and depression, which is strongly correlated to their efficacy in improving cognitive deficits involved in SZ [131]. In contrast, mGluR5 PAMs which bind to a locus separate from the MPEP site, have been shown to lack the ability to increase long-term potentiation and depression [130]. It is important that we understand the molecular and physiological consequences of the different PAMs, as it may provide an opportunity to preferentially target the activation of specific pathways that result in the desired behavioral response while minimizing unwanted effects.

In contrast to SZ, mGluR5 NAMs, including MPEP, MTEP, CTEP, GRN-52 and GRN-529, have been investigated for their antidepressant and anxiolytic potential. These drugs have been modelled from the original mGluR5 NAM, Fenobam, which has shown anxiolytic properties in the clinic [132]. These novel mGluR5 NAMs show promise, especially with their ability to reduce immobility time in the forced swim and tail suspension tests, two characteristic models of depressive behaviour [106, 133]. There is also evidence in rodents that some mGluR5 NAMs can cause SZ-like behaviours [122]. This raises serious concerns that mGluR5 NAMS could induce psychotomimetic effects, similar to ketamine. The Vanderbilt Center for Neuroscience Drug Discovery, have begun to address this issue. Using the premise of signaling bias, they have shown that it is possible to target specific signaling pathways to achieve the desired therapeutic response while having minimal effects on other mGluR5-mediated signaling pathways, thereby reducing adverse effects [130]. Specifically, the mGluR5 NAM MTEP has been shown to enhance the psychotomimetic effects of PCP [122], while newer analogues developed by the Vanderbilt Center (e.g. VU0285683) lack this property [134], indicating that the psychotomimetic limitations of mGluR5 NAMs can be overcome by specialist drug design [21]. With mGluR5 NAMs currently in clinical trials for the treatment of Fragile X Syndrome, we are eagerly awaiting the commencement and results from trials in MD patients.

Group II Novel Therapeutics

mGluR2/3 are primarily pre-synaptically located at the pre-terminal region of the axon, where they regulate glutamate and GABA release in response to excess synaptic glutamate [62] (Figure 1). Thus it is believed that the therapeutic potential of mGluR2/3 drugs lies in their ability to regulate glutamatergic tone. In addition to modulating glutamate and GABA release, mGluR2/3 also modulate dopamine release and antagonise serotonin 5HT2A receptors [39, 45], adding to the possible mechanism of the therapeutic efficacy of mGluR2/3 based drugs for SZ and MD. The glutamate binding sites on mGluR2 and mGluR3 are highly conserved, making it difficult to design orthosteric agonists/antagonists specific to mGluR2 or mGluR3; thus the majority of drugs have targeted these receptors together. Numerous studies have examined the therapeutic potential of mGluR2/3 agonists for the treatment of SZ. These studies, including the most prominent Lily compounds, LY379268, LY354740, LY2140023, have shown promise in preclinical studies demonstrating the ability to reverse positive, negative and cognitive SZ-like behaviours in rodents [135]. These preclinical findings have not translated quite so easily to human clinical trials, highlighting the difficulty in transitioning from preclinical (rodent) to clinical (human) trials. Krystal et al [136] showed that LY354740 pre-treatment in healthy humans reduced the working memory impairment induced by the NMDAR antagonist ketamine [136]. The first published clinical trial with the mGluR2/3 agonist found that LY2140023, the pro-drug of LY404039, was similar in efficacy to olanzapine, but without the metabolic side effects [137]. However, a follow-up study was inconclusive as neither olanzapine nor LY2140023 showed any difference in treatment efficacy compared to the placebo [138]. Liu et al [139] used the data from the Patil [137] and Kinon [138] studies to identify if specific genetic markers appeared to influence the response to LY2140023 [139]. They identified 23 SNPs associated with treatment response, 16 of which were associated with the serotonin 5HT2A gene, highlighting the potential of pharmacogenomics in future therapy. mGluR2/5HT2A receptor complexes have been hypothesised to contribute to this differential response [45].

Animal studies have since suggested that the therapeutic efficacy of the mGluR2/3 agonists is due to the activation of mGluR2 [112]. The highly conserved nature of the glutamate binding site in mGluR2/3 makes it difficult to develop orthosteric agonists specific for mGluR2 only. The development of mGluR2 PAMs therefore allow specificity to mGluR2 whilst also minimizing any desensitization issues that can occur with orthosteric agonists. Preliminary studies show that PAMs specific for mGluR2 do have antipsychotic potential in animal models of SZ [113, 140].

mGluR2/3 antagonists and NAMs, including LY341495 and MGS0039, have been researched for their antidepressant and anxiolytic potential. mGluR2/3 antagonists bind to the pre-synaptic mGlu2/3 receptors and increase pre-synaptic glutamate release. Post-synaptically, glutamatergic activation of the AMPA receptor and its downstream activation of mammalian target of rapamycin (mTOR) signaling, is thought to be the primary therapeutic mechanism of these mGluR2/3 antagonists [141]. Pre-clinical studies in rodent models have shown promising results. LY341495 and MGS0039 have been shown to reduce immobility time in forced swim and tail suspension tests in naive rodents [142–146]. Furthermore, Yoshimuzu et al [147] reported that subchronic MGS0039 administration was able to reduce the number of escape failures in a learned helplessness paradigm of depressive behaviour. Subchronic MGS0039 treatment was also shown to significantly attenuate learning deficits in the passive avoidance test in the olfactory bulbectomy rat model of depression [148]. In addition, MSG0039 has been shown to attenuate depressive-like behaviour in the social isolation reared mouse model of depression [58]. LY341495 has also been trialed in several rodent models of depression with promising results [144]. The administration of LY341495 has been shown to reduce immobility times in the forced swim and tail suspension tests in various genetic and pharmacological rodent models of depression [141]. The repeated positive findings of mGluR2/3 antagonists suggest these could be promising as future antidepressants and anxiolytics.

Group III Novel Therapeutics

Similar to Group II mGluRs, Group III mGluRs are located pre-synaptically, where they regulate glutamate release. However, in contrast to Group II mGluRs, which are located away from the active zone, mGluR4,6,7,8 are located at the active zone [149]. Little is known about Group III mGluRs and accordingly, there has been limited research into their potential as a target for novel SZ and MD treatments. However, several recent studies have explored the potential of mGluR4-preferring agonists and mGluR7 PAMs for the treatment of SZ and MD respectively [150–152]. Wieronska et al [150] reported that the mGluR4 preferring agonist, LSP1-2111, attenuated MK-801 and amphetamine-induced hyperlocomotion, providing support for its potential as a novel antipsychotic for the treatment of positive SZ-like symptoms. They subsequently showed that this drug inhibited MK-801 induced deficits in social interaction (a measure of negative symptoms), and improved MK-801 induced deficits in novel object recognition (a measure of cognitive symptoms). This was further supported by the study of Slawsinska et al [151], which reported potential antipsychotic effects of the specific mGluR4 PAMs, Lu AF21934 and Lu AF32615, which were able to attenuate MK-801-induced behaviours reminiscent of positive, negative and cognitive SZ-like symptoms. Collectively these preliminary data suggest that mGluR4 agonists have potential as novel therapeutics to treat positive, negative and cognitive SZ symptoms.

The mGluR7 PAM, AMN082, has shown promising antidepressant properties. Several studies investigating AMN082 have demonstrated its ability to reduce immobility time in the forced swim and tail suspension tests, highlighting its potential as a novel antidepressant [83, 153–157]. Bradley et al [153] reported that behaviourally efficacious doses of AMN082 modulate AMPA and NMDA receptor subunit phosphorylation in the hippocampus, suggesting this could contribute to the antidepressant effects of AMN082 [153]. While AMN082 shows potential as an antidepressant, it has also been shown that this drug can be metabolised in the liver to form ligands for serotonin, dopamine and norepinephrine transporter proteins, indicating that the action of AMN082 may not be entirely due to mGluR7 interactions [154]. This therefore requires further investigation. While the mGluR7 PAM shows promise as an antidepressant, it should be noted that it has also been shown to potentiate MK-801 induced hyperlocomotion at the same doses required to produce its antidepressant effects [150]. This suggests that this target might be plagued by the same issues as ketamine, which has been shown to produce profound antidepressant effects in treatment resistant patients, but has the potential to cause psychotomimetic effects [101].

Conclusion

Evidence to date suggests that mGluRs are involved in the pathophysiology of SZ and MD. Despite this, the extent of these alterations is largely unknown, and there is very little characterisation of mGluR function, regulation, and protein-protein interactions in these disorders. SZ and MD are heterogeneous disorders and the evidence from post-mortem brain and genetics studies suggests that mGluRs could be altered in specific subsets of patients. In addition to their role in regulating glutamatergic signaling at maturation, these receptors are also important for neurodevelopment. Since SZ and MD are considered to be neurodevelopmental disorders that might arise from alterations in neuronal differentiation, migration and synaptic connections, alterations in the mGluR system during critical periods in brain development could contribute to the underlying pathogenesis of these disorders, and should be further investigated. This is obviously very difficult to explore in humans, and thus valid animal models of SZ and MD provide immense value.

Further supporting the involvement of mGluRs in the pathophysiology of SZ and MD, are the findings of changes to mGluRs in rodent brains following antipsychotic and antidepressant treatment. While there is no consensus regarding the specific effects of these drugs on the mGluR system, evidence from animal studies suggests that these therapeutics can produce brain region-specific alterations in the expression of certain mGluRs, although these effects are minimal compared to the effects on their main monoaminergic targets. Investigating this area is quite complex as the effects of these therapeutics differ not only between drug classes, but also between individual drugs of the same category, drug dose, treatment duration, as well as the animal model employed. Studies in humans have shown that genetic mutations in mGluRs have the potential to predict antipsychotic efficacy, suggesting an interaction between mGluRs and current antipsychotics. These findings in humans reinforce the importance to understand the mGluR systems in SZ and MD and in relation to current therapeutics. As novel mGluR based drugs develop, it will be important to investigate current therapeutics adjunct with these novel treatments, as many patients transitioning to novel therapeutics will likely have a history of prior therapeutic treatment that may influence the efficacy of novel therapeutics.

Considering that SZ and MD can share some overlapping symptomatology, with depressive symptoms being common in patients with SZ, and psychotic symptoms being present in a subset of MD patients [158], and that co-pharmacy of both antipsychotics and antidepressants is common in these disorders, it is interesting that opposing regulation of the glutamatergic system is implicated for the treatment of SZ compared to MD. For example, the NMDAR antagonist ketamine shows antidepressant effects, while the NMDAR agonist d-serine shows antipsychotic effects. Furthermore, mGluR5 and mGluR2 PAMs are currently in preclinical investigations for the treatment of SZ, while in contrast mGluR5 and mGluR2/3 NAMs show promise for the treatment of MD. The problem with this opposition is the potential that to treat one disorder, symptoms may be promoted of the other. For example, rodent studies have shown that mGluR5 NAMs and mGluR2/3 NAMs do have the ability to produce SZ-like behaviours, similar to ketamine, while also demonstrating antidepressant properties. However, specialist drug design using the premise of signaling bias has shown that it is possible to design mGluR5 NAMs that do not show SZ-like behaviours in rodents, while retaining their antidepressant potential.

It is primarily Group I and II mGluRs that are being investigated for the treatment of SZ and MD, although Group III mGluRs do show some potential, particularly mGluR4 PAMs for SZ and mGluR7 PAMs for MD. Group II mGluR agonists have progressed from preclinical rodent studies to clinical trials, with mixed results to date from the clinical trials. These studies highlight the difficulty in translating preclinical rodent findings to the human population. While rodent preclinical studies are invaluable to develop our understanding of novel therapeutics, it is imperative that a model closer to the clinical situation is developed. Human induced pluripotent stem cells may be a beneficial avenue to enable drug screening in a human-based cellular system, in which the pharmacogenomics effects can also be elucidated [21, 159]. It is acknowledged, however, that while this may provide opportunities to understand some of the molecular interactions, it will not provide direct evidence of the behavioural effects of novel therapeutics. For this purpose, a multi-model approach should be implemented.

In conclusion, direct and indirect evidence suggests that mGluRs are involved in the pathophysiology of SZ and MD. Despite this, their specific involvement remains unclear and requires further investigation. With the limitations and side effects of current antipsychotics and antidepressants, the new generation therapeutics targeting the mGluR system show potential for the treatment of SZ and MD. The challenge for mGluR research and drug development hereon in will be to assess, by means of genomics or other biomarkers, the part glutamatergic dysregulation or dysfunction plays in diagnostic sub-populations of these disorders, and assessing the impact this might have on symptom profile and therapeutic response.

Acknowledgments

This work was supported by the Schizophrenia Research Institute, utilising infrastructure funding from the NSW Ministry of Health. Ms Natalie Matosin and Mr Jeremy Lum are supported by Ian Scott Scholarships from Australian Rotary Health.

References

[1] Weickert CS, Fung SJ, Catts VS, Schofield PR, Allen KM et al. (2013) Molecular evidence of N-methyl-D-aspartate receptor hypofunction in schizophrenia. *Mol. Psychiatry* 18:1185-1192.

[2] Tokita K, Yamaji T, Hashimoto K (2012) Roles of glutamate signaling in preclinical and/or mechanistic models of depression. *Pharmacol. Biochem. Behav.* 100:688–704.

[3] Hashimoto K, Malchow B, Falkai P, Schmitt A (2013) Glutamate modulators as potential therapeutic drugs in schizophrenia and affective disorders. *Eur. Arch. Psychiatry Clin. Neurosci.* 263:367–377.

[4] Matosin N, Newell KA (2013) Metabotropic glutamate receptor 5 in the pathology and treatment of schizophrenia. *Neurosci. Biobehav. Rev.* 37:256–268.

[5] Lesage A, Steckler T (2010) Metabotropic glutamate mGlu1 receptor stimulation and blockade: therapeutic opportunities in psychiatric illness. *Eur. J. Pharmacol.* 639:2–16.

[6] Lavreysen H, Pereira SN, Leysen JE, Langlois X, Lesage ASJ (2004) Metabotropic glutamate 1 receptor distribution and occupancy in the rat brain: a quantitative autoradiographic study using [3H]R214127. *Neuropharmacology* 46:609–619.

[7] Zysk JR, Spear N, Fieles W, Stein MM, Sygowski LS, et al. (2013) In vitro binding of a radio-labeled positive allosteric modulator for metabotropic glutamate receptor subtype 5. *Synapse* 67:135–144.

[8] Volk DW, Eggan SM, Lewis DA (2010) Alterations in metabotropic glutamate receptor 1a and regulator of G protein signaling 4 in the prefrontal cortex in schizophrenia. *Am. J. Psychiatry* 167:1489–1498.

[9] Gupta DS, McCullumsmith RE, Beneyto M, Haroutunian V, Davis KL et al. (2005) Metabotropic glutamate receptor protein expression in the prefrontal cortex and striatum in schizophrenia. *Synapse* 57:123–131.

[10] Ayoub MA, Angelicheva D, Vile D, Chandler D, Morar B et al. (2012) Deleterious GRM1 mutations in schizophrenia. *PLoS ONE* 7:e32849.

[11] Frank RAW, McRae AF, Pocklington AJ, van de Lagemaat LN, Navarro P et al. (2011) Clustered coding variants in the glutamate receptor complexes of individuals with schizophrenia and bipolar disorder. *PLoS ONE* 6:e19011.

[12] Brody SA, Conquet F, Geyer MA (2003) Disruption of prepulse inhibition in mice lacking mGluR1. *Eur. J. Neurosci.* 18:3361–3366.

[13] Matosin N, Frank E, Deng C, Huang X-F, Newell KA (2013) Metabotropic glutamate receptor 5 binding and protein expression in schizophrenia and following antipsychotic drug treatment. *Schizophr. Res.* 146:170–176.

[14] Spellmann I, Rujescu D, Musil R, Mayr A, Giegling I et al. (2011) Homer-1 polymorphisms are associated with psychopathology and response to treatment in schizophrenic patients. *J. Psychiatr. Res.* 45:234–241.

[15] Engmann O, Hortobágyi T, Pidsley R, Troakes C, Bernstein H-G et al. (2011) Schizophrenia is associated with dysregulation of a Cdk5 activator that regulates synaptic protein expression and cognition. *Brain* 134:2408–2421.

[16] Szumlinski KK, Lominac KD, Kleschen MJ, Oleson EB, Dehoff MH et al. (2005) Behavioral and neurochemical phenotyping of Homer1 mutant mice: possible relevance to schizophrenia. *Genes Brain Behav* 4:273–288.

[17] Iasevoli F, Tomasetti C, de Bartolomeis A (2012) Scaffolding proteins of the post-synaptic density contribute to synaptic plasticity by regulating receptor localization and distribution: relevance for neuropsychiatric diseases. *Neurochem. Res.* 38:1–22.

[18] Thomas U (2002) Modulation of synaptic signalling complexes by Homer proteins. *J. Neurochem.* 81:407–413.

[19] Mion S, Corti C, Neki A, Shigemoto R, Corsi M et al. (2001) Bidirectional regulation of neurite elaboration by alternatively spliced metabotropic glutamate receptor 5 (mGluR5) isoforms. *Mol. Cell Neurosci.* 17:957–972.

[20] Joly C, Gomeza J, Brabet I, Curry K, Bockaert J et al. (1995) Molecular, functional, and pharmacological characterization of the metabotropic glutamate receptor type 5 splice variants: comparison with mGluR1. *J. Neurosci.* 15:3970–3981.

[21] Newell KA (2013) Metabotropic glutamate receptor 5 in schizophrenia: emerging evidence for the development of antipsychotic drugs. *Future Med. Chem.* 5:1471–1474.

[22] Mao L-M, Liu X-Y, Zhang G-C, Chu X-P, Fibuch EE et al. (2008) Phosphorylation of group I metabotropic glutamate receptors (mGluR1/5) in vitro and in vivo. *Neuropharmacology* 55:403–408.

[23] Orlando LR, Dunah AW, Standaert DG, Young AB (2002) Tyrosine phosphorylation of the metabotropic glutamate receptor mGluR5 in striatal neurons. *Neuropharmacology* 43:161–173.

[24] Kozlovsky N, Scarr E, Dean B, Agam G (2006) Postmortem brain calcineurin protein levels in schizophrenia patients are not different from controls. *Schizophr. Res.* 83:173–177.

[25] Wang H, Westin L, Nong Y, Birnbaum S, Bendor J et al. (2009) Norbin is an endogenous regulator of metabotropic glutamate receptor 5 signaling. *Science* 326:1554–1557.

[26] Miyakawa T, Leiter LM, Gerber DJ, Gainetdinov RR, Sotnikova TD et al. (2003) Conditional calcineurin knockout mice exhibit multiple abnormal behaviors related to schizophrenia. *Proc. Natl. Acad. Sci.* 100:8987–8992.

[27] Timms AE, Dorschner MO, Wechsler J, Choi KY, Kirkwood R, et al. (2013) Support for the N-methyl-D-aspartate receptor hypofunction hypothesis of schizophrenia from exome sequencing in multiplex families. *JAMA Psychiatry* ll:1–9.

[28] Ayalew M, Le-Niculescu H, Levey DF, Jain N, Changala B et al. (2012) Convergent functional genomics of schizophrenia: from comprehensive understanding to genetic risk prediction. *Mol. Psychiatry* 17:887–905.

[29] Gray L, van den Buuse M, Scarr E, Dean B, Hannan AJ (2009) Clozapine reverses schizophrenia-related behaviours in the metabotropic glutamate receptor 5 knockout mouse: association with N-methyl-D-aspartic acid receptor up-regulation. *Int. J. Neuropsychopharmacol.* 12:45–60.

[30] Deschwanden A, Karolewicz B, Feyissa AM, Treyer V, Ametamey SM et al. (2011) Reduced metabotropic glutamate receptor 5 density in major depression determined by [^{11}C] ABP688 PET and postmortem study. *Am. J. Psychiatry* 168:727.

[31] Matosin N, Fernandez-Enright F, Frank E, Deng C, Wong J et al. (2013) Metabotropic glutamate receptor 2/3 and 5 in schizophrenia and allied mood disorders: evidence supporting mGluR-based therapeutics. 12th Biennial Australasian Schizophrenia Conference. Melbourne, Australia.

[32] Feyissa AM, Chandran A, Stockmeier CA, Karolewicz B (2009) Reduced levels of NR2A and NR2B subunits of NMDA receptor and PSD-95 in the prefrontal cortex in major depression. *Prog. Neuropsychopharmacol. Biol. Psychiatry* 33:70–75.

[33] Rietschel M, Mattheisen M, Frank J, Treutlein J, Degenhardt F et al. (2010) Genome-wide association-, replication-, and neuroimaging study implicates HOMER1 in the etiology of major depression. *Biol. Psychiatry* 68:578–585.

[34] Kovačević T, Skelin I, Minuzzi L, Rosa-Neto P, Diksic M (2012) Reduced metabotropic glutamate receptor 5 in the Flinders Sensitive Line of rats, an animal model of depression: An autoradiographic study. *Brain Res. Bull.* 87:406–412.

[35] Wierońska JM, Legutko B, Dudys D, Pilc A (2008) Olfactory bulbectomy and amitriptyline treatment influences mGlu receptors expression in the mouse brain hippocampus. *Pharmacol. Rep.* 60:844–855.

[36] Menke A, Sämann P, Kloiber S, Czamara D, Lucae S et al. (2012) Polymorphisms within the metabotropic glutamate receptor 1 gene are associated with depression phenotypes. *Psychoneuroendocrinology* 37:565–575.

[37] Kew JNC, Kemp JA (2005) Ionotropic and metabotropic glutamate receptor structure and pharmacology. *Psychopharmacology* 179:4–29.

[38] Spooren W, Ballard T, Gasparini F, Amalric M, Mutel V, et al. (2003) Insight into the function of Group I and Group II metabotropic glutamate (mGlu) receptors: behavioural characterization and implications for the treatment of CNS disorders. *Behav. Pharmacol.* 14:257.

[39] Harrison PJ, Lyon L, Sartorius LJ, Burnet P, Lane TA (2008) Review: The group II metabotropic glutamate receptor 3 (mGluR3, mGlu3, GRM3): expression, function and involvement in schizophrenia. *J. Psychopharmacol.* 22:308–322.

[40] Lian J, Huang X-F, Pai N, Deng C (2013) Effects of olanzapine and betahistine co-treatment on serotonin transporter, 5-HT2A and dopamine D2 receptor binding density. *Prog. Neuropsychopharmacol. Biol. Psychiatry* 47:62–68.

[41] Miller JM, Hesselgrave N, Ogden RT, Zanderigo F, Oquendo MA et al. (2013) Brain serotonin 1A receptor binding as a predictor of treatment outcome in major depressive disorder. *Biol. Psychiatry* 74:760–767.

[42] Frank E, Newell KA, Huang XF (2011) Density of metabotropic glutamate receptors 2 and 3 (mGluR2/3) in the dorsolateral prefrontal cortex does not differ with schizophrenia diagnosis but decreases with age. *Schizophr. Res.* 128:56–60.

[43] Corti C, Crepaldi L, Mion S, Roth AL, Xuereb JH et al. (2007) Altered dimerization of metabotropic glutamate receptor 3 in schizophrenia. *Biol. Psychiatry* 62:747–755.

[44] Ghose S, Gleason KA, Potts BW, Lewis-Amezcua K, Tamminga CA (2009) Differential expression of metabotropic glutamate receptors 2 and 3 in schizophrenia: a mechanism for antipsychotic drug action? *Am. J. Psychiatry* 166:812–820.

[45] González-Maeso J, Ang RL, Yuen T, Chan P, Weisstaub NV et al. (2008) Identification of a serotonin/glutamate receptor complex implicated in psychosis. *Nature* 452:93–97.

[46] Chen Q, He G, Chen Q, Wu S, Xu Y, et al. (2005) A case-control study of the relationship between the metabotropic glutamate receptor 3 gene and schizophrenia in the Chinese population. *Schizophr. Res.* 73:21–26.

[47] Egan MF, Straub RE, Goldberg TE, Yakub I, Callicott JH et al. (2004) Variation in GRM3 affects cognition, prefrontal glutamate, and risk for schizophrenia. *Proc. Natl. Acad. Sci.* 101:12604–12609.

[48] Fujii Y, Shibata H, Kikuta R, Makino C, Tani A et al. (2003) Positive associations of polymorphisms in the metabotropic glutamate receptor type 3 gene (GRM3) with schizophrenia. *Psychiatr. Genet.* 13:71–76.

[49] Marti SB, Cichon S, Propping P, Nothen M (2002) Metabotropic glutamate receptor 3 (GRM3) gene variation is not associated with schizophrenia or bipolar affective disorder in the German population. *Am. J. Med. Genet.* 114:46–50.

[50] Norton N, Williams HJ, Dwyer S, Ivanov D, Preece AC et al. (2005) No evidence for association between polymorphisms in GRM3 and schizophrenia. *BMC Psychiatry* 5:23.

[51] Tochigi M, Suga M, Ohashi J, Otowa T, Yamasue H et al. (2006) No association between the metabotropic glutamate receptor type 3 gene (GRM3) and schizophrenia in a Japanese population. *Schizophr. Res.* 88:260–264.

[52] Baune BT, Suslow T, Beśte C, Birosova E, Domschke K et al. (2010) Association between genetic variants of the metabotropic glutamate receptor 3 (GRM3) and cognitive set shifting in healthy individuals. *Genes Brain Behav.* 9:459–466.

[53] Bishop JR, Ellingrod VL, Moline J, Miller D (2005) Association between the polymorphic GRM3 gene and negative symptom improvement during olanzapine treatment. *Schizophr. Res.* 77:253–260.

[54] Tascedda F, Blom JMC, Brunello N, Zolin K, Gennarelli M et al. (2001) Modulation of glutamate receptors in response to the novel antipsychotic olanzapine in rats. *Biol. Psychiatry* 50:117–122.

[55] Zhao J, Bao A-M, Qi X-R, Kamphuis W, Luchetti S et al. (2012) Gene expression of GABA and glutamate pathway markers in the prefrontal cortex of non-suicidal elderly depressed patients. *J. Affect. Disord.* 138:494–502.

[56] Feyissa AM, Woolverton WL, Miguel-Hidalgo JJ, Wang Z, Kyle PB et al. (2010) Elevated level of metabotropic glutamate receptor 2/3 in the prefrontal cortex in major depression. *Prog. Neuropsychopharmacol. Biol. Psychiatry* 34:279–283.

[57] Tsunoka T, Kishi T, Ikeda M, Kitajima T, Yamanouchi Y et al. (2009) Association analysis of Group II metabotropic glutamate receptor genes (GRM2 and GRM3) with mood disorders and fluvoxamine response in a Japanese population. *Prog. Neuropsychopharmacol. Biol. Psychiatry* 33:875–879.

[58] Kawasaki T, Ago Y, Yano K, Araki R, Washida Y et al. (2011) Increased binding of cortical and hippocampal group II metabotropic glutamate receptors in isolation-reared mice. *Neuropharmacology* 60:397–404.

[59] Matrisciano F, Caruso A, Orlando R, Marchiafava M, Bruno V et al. (2008) Defective group-II metaboropic glutamate receptors in the hippocampus of spontaneously depressed rats. *Neuropharmacology* 55:525–531.

[60] Kaffman A, Krystal JJ (2012) New Frontiers in Animal Research of Psychiatric Illness. In: Kobeissy FH (ed) *Psychiatric Disorders*. Totowa, NJ: Humana Press, Vol. 829. pp. 3–30.

[61] Nakajima Y, Iwakabe H, Akazawa C, Nawa H, Shigemoto R et al. (1993) Molecular characterization of a novel retinal metabotropic glutamate receptor mGluR6 with a high agonist selectivity for L-2-amino-4-phosphonobutyrate. *J. Biol. Chem.* 268:11868–11873.

[62] Pilc A, Wierońska JM, Skolnick P (2013) Glutamate-based antidepressants: preclinical psychopharmacology. *Biol. Psychiatry* 73:1125–1132.

[63] Shigemoto R, Kulik A, Roberts JD, Ohishi H, Nusser Z et al. (1996) Target-cell-specific concentration of a metabotropic glutamate receptor in the presynaptic active zone. *Nature* 381:523–525.

[64] Makoff A, Lelchuk R, Oxer M, Harrington K, Emson P (1996) Molecular characterization and localization of human metabotropic glutamate receptor type 4. *Mol. Brain Res.* 37:239–248.

[65] Richardson-Burns SM, Haroutunian V, Davis KL, Watson SJ, Meador-Woodruff JH (2000) Metabotropic glutamate receptor mRNA expression in the schizophrenic thalamus. *Biol. Psychiatry* 47:22–28.

[66] Shibata H, Tani A, Chikuhara T, Kikuta R, Sakai M et al. (2009) Association study of polymorphisms in the group III metabotropic glutamate receptor genes, GRM4 and GRM7, with schizophrenia. *Psychiatry Res.* 167:88–96.

[67] Ohtsuki T, Toru M, Arinami T (2001) Mutation screening of the metabotropic glutamate receptor mGluR4 (GRM4) gene in patients with schizophrenia. *Psychiatr. Genet.* 11:79–83.

[68] Iscru E, Goddyn H, Ahmed T, Callaerts-Vegh Z, D'Hooge R et al. (2013) Improved spatial learning is associated with increased hippocampal but not prefrontal long-term potentiation in mGluR4 knockout mice. *Genes Brain Behav.* 12:615–625.

[69] O'Connor RM, Pusceddu MM, Dinan TG, Cryan JF (2013) Impact of early-life stress, on group III mGlu receptor levels in the rat hippocampus: Effects of ketamine, electroconvulsive shock therapy and fluoxetine treatment. *Neuropharmacology* 66:236–241.

[70] Okamoto N, Hori S, Akazawa C, Hayashi Y, Shigemoto R et al. (1994) Molecular characterization of a new metabotropic glutamate receptor mGluR7 coupled to inhibitory cyclic AMP signal transduction. *J. Biol. Chem.* 269:1231–1236.

[71] Makoff A, Pilling C, Harrington K, Emson P (1996) Human metabotropic glutamate receptor type 7: molecular cloning and mRNA distribution in the CNS. *Brain Res. Mol. Brain Res.* 40:165–170.

[72] Schoepp DD (2001) Unveiling the functions of presynaptic metabotropic glutamate receptors in the central nervous system. *J. Pharmacol. Exp. Ther.* 299:12–20.

[73] Ohtsuki T, Koga M, Ishiguro H, Horiuchi Y, Arai M et al. (2008) A polymorphism of the metabotropic glutamate receptor mGluR7 (GRM7) gene is associated with schizophrenia. *Schizophr. Res.* 101:9–16.

[74] Ganda C, Schwab SG, Amir N, Heriani H, Irmansyah I et al. (2009) A family-based association study of DNA sequence variants in GRM7 with schizophrenia in an Indonesian population. *Int. J. Neuropsychopharmacol.* 12:1283–1289.

[75] Bolonna AA, Kerwin RW, Munro J, Arranz MJ, Makoff AJ (2001) Polymorphisms in the genes for mGluR types 7 and 8: association studies with schizophrenia. *Schizophr. Res.* 47:99–103.

[76] Callaerts-Vegh Z, Beckers T, Ball SM, Baeyens F, Callaerts PF et al. (2006) Concomitant deficits in working memory and fear extinction are functionally dissociated from reduced anxiety in metabotropic glutamate receptor 7-deficient mice. *J. Neurosci.* 26:6573–6582.

[77] Hikichi H, Murai T, Okuda S, Maehara S, Satow A et al. (2010) Effects of a novel metabotropic glutamate receptor 7 negative allosteric modulator, 6-(4-methoxyphenyl)-

5-methyl-3-pyridin-4-ylisoxazonolo[4,5-c]pyridin-4(5H)-one (MMPIP), on the central nervous system in rodents. *Eur. J. Pharmacol.* 639:106–114.

[78] Wierońska JM, Brański P, Siwek A, Dybala M, Nowak G et al. (2010) GABAergic dysfunction in mGlu7 receptor-deficient mice as reflected by decreased levels of glutamic acid decarboxylase 65 and 67 kDa and increased reelin proteins in the hippocampus. *Brain Res.* 1334:12–24.

[79] Konradi C, Yang CK, Zimmerman EI, Lohmann KM, Gresch P et al. (2011) Hippocampal interneurons are abnormal in schizophrenia. *Schizophr. Res.* 131:165–173.

[80] Shyn SI, Shi J, Kraft JB, Potash JB, Knowles JA et al. (2011) Novel loci for major depression identified by genome-wide association study of Sequenced Treatment Alternatives to Relieve Depression and meta-analysis of three studies. *Mol. Psychiatry* 16:202–215.

[81] Cryan JF, Kelly PH, Neijt HC, Sansig G, Flor PJ et al. (2003) Antidepressant and anxiolytic-like effects in mice lacking the group III metabotropic glutamate receptor mGluR7. *Eur. J. Neurosci.* 17:2409–2417.

[82] Mitsukawa K, Mombereau C, Lötscher E, Uzunov DP, van der Putten H et al. (2005) Metabotropic glutamate receptor subtype 7 ablation causes dysregulation of the HPA axis and increases hippocampal BDNF protein levels: implications for stress-related psychiatric disorders. *Neuropsychopharmacology* 31:1112–1122.

[83] Palucha A, Klak K, Branski P, Putten H, Flor PJ et al. (2007) Activation of the mGlu7 receptor elicits antidepressant-like effects in mice. *Psychopharmacology* 194:555–562.

[84] O'Mahony CM, Bravo JA, Dinan TG, Cryan JF (2010) Comparison of hippocampal metabotropic glutamate receptor 7 (mGlu7) mRNA levels in two animal models of depression. *Neurosci. Lett.* 482:137–141.

[85] Corti C, Restituito S, Rimland JM, Brabet I, Corsi M, et al. (1998) Cloning and characterization of alternative mRNA forms for the rat metabotropic glutamate receptors mGluR7 and mGluR8. *Eur. J. Neurosci.* 10:3629–3641.

[86] Saugstad JA, Kinzie JM, Shinohara MM, Segerson TP, Westbrook GL (1997) Cloning and expression of rat metabotropic glutamate receptor 8 reveals a distinct pharmacological profile. *Mol. Pharmacol.* 51:119–125.

[87] Takaki H, Kikuta R, Shibata H, Ninomiya H, Tashiro N et al. (2004) Positive associations of polymorphisms in the metabotropic glutamate receptor type 8 gene (GRM8) with schizophrenia. *Am. J. Med. Genet. B Neuropsychiatr. Genet.* 128B:6–14.

[88] Robbins MJ, Starr KR, Honey A, Soffin EM, Rourke C et al. (2007) Evaluation of the mGlu8 receptor as a putative therapeutic target in schizophrenia. *Brain Res.* 1152:215–227.

[89] Linden A-M, Johnson B., Peters S., Shannon H., Tian M et al. (2002) Increased anxiety-related behavior in mice deficient for metabotropic glutamate 8 (mGlu8) receptor. *Neuropharmacology* 43:251–259.

[90] O'Connor RM, Pusceddu MM, O'Leary OF, Savignac HM, Bravo JA et al. (2013) Hippocampal group III mGlu receptor mRNA levels are not altered in specific mouse models of stress, depression and antidepressant action. *Pharmacol. Biochem. Behav.* 103:561–567.

[91] Terracciano A, Tanaka T, Sutin AR, Sanna S, Deiana B, et al. (2010) Genome-wide association scan of trait depression. *Biol. Psychiatry* 68:811–817.

[92] Fatemi SH, Folsom TD, Rooney RJ, Thuras PD (2013) mRNA and protein expression for novel GABAA receptors θ and ρ2 are altered in schizophrenia and mood disorders; relevance to FMRP-mGluR5 signaling pathway. *Transl. Psychiatry* 3:e271.

[93] Corti C, Xuereb JH, Crepaldi L, Corsi M, Michielin F et al. (2011) Altered levels of glutamatergic receptors and Na+/K+ ATPase-α1 in the prefrontal cortex of subjects with schizophrenia. *Schizophr. Res.* 128:7–14.

[94] Sartorius LJ, Weinberger DR, Hyde TM, Harrison PJ, Kleinman JE et al. (2008) Expression of a GRM3 splice variant is increased in the dorsolateral prefrontal cortex of individuals carrying a schizophrenia risk SNP. *Neuropsychopharmacology* 33:2626–2634.

[95] Crook JM, Akil M, Law BCW, Hyde TM, Kleinman JE (2002) Comparative analysis of group II metabotropic glutamate receptor immunoreactivity in Brodmann's area 46 of the dorsolateral prefrontal cortex from patients with schizophrenia and normal subjects. *Mol. Psychiatry* 7:157–164.

[96] Ohnuma T, Tessler S, Arai H, Faull RLM, McKenna PJ, et al. (2000) Gene expression of metabotropic glutamate receptor 5 and excitatory amino acid transporter 2 in the schizophrenic hippocampus. *Mol. Brain Res.* 85:24–31.

[97] Ohnuma T, Augood SJ, Arai H, McKenna PJ, Emson PC (1998) Expression of the human excitatory amino acid transporter 2 and metabotropic glutamate receptors 3 and 5 in the prefrontal cortex from normal individuals and patients with schizophrenia. *Mol. Brain Res.* 56:207–217.

[98] Kapur S, Mamo D (2003) Half a century of antipsychotics and still a central role for dopamine D2 receptors. *Prog. Neuropsychopharmacol. Biol. Psychiatry* 27:1081–1090.

[99] Iasevoli F, Tomasetti C, Marmo F, Bravi D, Arnt J et al. (2010) Divergent acute and chronic modulation of glutamatergic postsynaptic density genes expression by the antipsychotics haloperidol and sertindole. *Psychopharmacology* 212:329–344.

[100] Kammermeier PJ, Worley PF (2007) Homer 1a uncouples metabotropic glutamate receptor 5 from postsynaptic effectors. *Proc. Natl. Acad. Sci.* 104:6055–6060.

[101] Hashimoto K (2009) Emerging role of glutamate in the pathophysiology of major depressive disorder. *Brain Res. Rev.* 61:105–123.

[102] Śmialowska M, Szewczyk B, Brański P, Wierońska JM, Pa\lucha A et al. (2002) Effect of chronic imipramine or electroconvulsive shock on the expression of mGluR1a and mGluR5a immunoreactivity in rat brain hippocampus. *Neuropharmacology* 42:1016–1023.

[103] Matrisciano F, Storto M, Ngomba RT, Cappuccio I, Caricasole A et al. (2002) Imipramine treatment up-regulates the expression and function of mGlu2/3 metabotropic glutamate receptors in the rat hippocampus. *Neuropharmacology* 42:1008–1015.

[104] Musazzi L, Treccani G, Mallei A, Popoli M (2013) The action of antidepressants on the glutamate system: regulation of glutamate release and glutamate receptors. *Biol. Psychiatry* 73:1180-1188.

[105] Zahorodna A, Bijak M (1999) An antidepressant-induced decrease in the responsiveness of hippocampal neurons to group I metabotropic glutamate receptor activation. *Eur. J. Pharmacol.* 386:173–179.

[106] Hughes ZA, Neal SJ, Smith DL, Sukoff Rizzo SJ, Pulicicchio CM et al. (2013) Negative allosteric modulation of metabotropic glutamate receptor 5 results in broad

spectrum activity relevant to treatment resistant depression. *Neuropharmacology* 66:202–214.

[107] Molinaro G, Traficante A, Riozzi B, Menna LD, Curto M et al. (2009) Activation of mGlu2/3 metabotropic glutamate receptors negatively regulates the stimulation of inositol phospholipid hydrolysis mediated by 5-hydroxytryptamine2A serotonin receptors in the frontal cortex of living mice. *Mol. Pharmacol.* 76:379–387.

[108] Pałucha A, Brański P, Kĺak K, Sowa M (2007) Chronic imipramine treatment reduces inhibitory properties of group II mGlu receptors without affecting their density or affinity. *Pharmacol. Rep.* 59:525–530.

[109] Wierońska JM, Kłak K, Pałucha A, Brański P, Pilc A (2007) Citalopram influences mGlu7, but not mGlu4 receptors' expression in the rat brain hippocampus and cortex. *Brain Res.* 1184:88–95.

[110] Weston-Green K, Huang X-F, Deng C (2013) Second generation antipsychotic-induced type 2 diabetes: a role for the muscarinic M3 receptor. *CNS Drugs* 27:1069–1080.

[111] Geddes AE, Huang XF, Newell KA (2011) Reciprocal signalling between NR2 subunits of the NMDA receptor and Neuregulin1 and their role in schizophrenia. *Prog. Neuropsychopharmacol. Biol. Psychiatry* 4:896–904.

[112] Menniti FS, Lindsley CW, Conn PJ, Pandit J, Zagouras P et al. (2013) Allosteric modulators for the treatment of schizophrenia: targeting glutamatergic networks. *Curr. Top Med. Chem.* 13:26–54.

[113] Vinson P, Conn JP (2012) Metabotropic glutamate receptors as therapeutic targets for schizophrenia. *Neuropharmacology* 62:1461–1472.

[114] Price RB, Nock MK, Charney DS, Mathew SJ (2009) Effects of intravenous ketamine on explicit and implicit measures of suicidality in treatment-resistant depression. *Biol. Psychiatry* 66:522–526.

[115] Moghaddam B, Krystal JH (2012) Capturing the angel in angel dust: twenty years of translational neuroscience studies of NMDA receptor antagonists in animals and humans. *Schizophr. Bull.* 38:942–949.

[116] Biber K, Laurie DJ, Berthele A, Sommer B, Tölle TR et al. (1999) Expression and signaling of group I metabotropic glutamate receptors in astrocytes and microglia. *J. Neurochem.* 72:1671–1680.

[117] Fillman SG, Cloonan N, Catts VS, Miller LC, Wong J et al. (2013) Increased inflammatory markers identified in the dorsolateral prefrontal cortex of individuals with schizophrenia. *Mol. Psychiatry* 18:206–214.

[118] Byrnes KR, Stoica B, Loane DJ, Riccio A, Davis MI et al. (2009) Metabotropic glutamate receptor 5 activation inhibits microglial associated inflammation and neurotoxicity. *Glia* 57:550–560.

[119] Loane DJ, Stoica BA, Pajoohesh-Ganji A, Byrnes KR, Faden AI (2009) Activation of metabotropic glutamate receptor 5 modulates microglial reactivity and neurotoxicity by inhibiting NADPH oxidase. *J. Biol. Chem.* 284:15629–15639.

[120] Ohta H, Kawamoto H, Suzuki G (2010) mGluR1 negative allosteric modulators: an alternative metabotropic approach for the treatment of schizophrenia. In: Skolnick P (ed). *Glutamate-based Therapies for Psychiatric Disorders. Milestones in Drug Therapy*. Springer. pp. 117–131.

[121] De Vry J, Horváth E, Schreiber R (2001) Neuroprotective and behavioral effects of the selective metabotropic glutamate mGlu1 receptor antagonist BAY 36-7620. *Eur. J. Pharmacol.* 428:203–214.

[122] Pietraszek M, Gravius A, Schafer D, Weil T, Trifanova D et al. (2005) mGluR5, but not mGluR1, antagonist modifies MK-801-induced locomotor activity and deficit of prepulse inhibition. *Neuropharmacology* 49:73–85.

[123] Belozertseva IV, Kos T, Popik P, Danysz W, Bespalov AY (2007) Antidepressant-like effects of mGluR1 and mGluR5 antagonists in the rat forced swim and the mouse tail suspension tests. *Eur. Neuropsychopharmacol.* 17:172–179.

[124] Rosenbrock H, Kramer G, Hobson S, Koros E, Grundl M et al. (2010) Functional interaction of metabotropic glutamate receptor 5 and NMDA-receptor by a metabotropic glutamate receptor 5 positive allosteric modulator. *Eur. J. Pharmacol.* 639:40–46.

[125] Gastambide F, Gilmour G, Robbins TW, Tricklebank MD (2013) The $mGlu_5$ positive allosteric modulator LSN2463359 differentially modulates motor, instrumental and cognitive effects of NMDA receptor antagonists in the rat. *Neuropharmacology* 64:240–247.

[126] Gastambide F, Cotel M-C, Gilmour G, O'Neill MJ, Robbins TW et al. (2012) Selective remediation of reversal learning deficits in the neurodevelopmental MAM model of schizophrenia by a novel mGlu5 positive allosteric modulator. *Neuropsychopharmacol.* 37:1057–1066.

[127] Gilmour G, Broad LM, Wafford KA, Britton T, Colvin EM et al. (2013) In vitro characterisation of the novel positive allosteric modulators of the $mGlu_5$ receptor, LSN2463359 and LSN2814617, and their effects on sleep architecture and operant responding in the rat. *Neuropharmacology* 64:224–239.

[128] O'Brien JA, Lemaire W, Wittmann M, Jacobson MA, Ha SN et al. (2004) A novel selective allosteric modulator potentiates the activity of native metabotropic glutamate receptor subtype 5 in rat forebrain. *J. Pharmacol. Exp. Ther.* 309:568–577.

[129] Zhang Y, Rodriguez AL, Conn PJ (2005) Allosteric potentiators of metabotropic glutamate receptor subtype 5 have differential effects on different signaling pathways in cortical astrocytes. *J. Pharmacol. Exp. Ther.* 315:1212–1219.

[130] Noetzel MJ, Gregory KJ, Vinson PN, Manka JT, Stauffer SR et al. (2013) A novel metabotropic glutamate receptor 5 positive allosteric modulator acts at a unique site and confers stimulus bias to mGlu5 signaling. *Mol. Pharmacol.* 83:835–847.

[131] Ayala JE, Chen Y, Banko JL, Sheffler DJ, Williams R et al. (2009) mGluR5 positive allosteric modulators facilitate both hippocampal LTP and LTD and enhance spatial learning. *Neuropsychopharmacology* 34:2057–2071.

[132] Pecknold JC, McClure DJ, Appeltauer L, Wrzesinski L, Allan T (1982) Treatment of anxiety using fenobam (a nonbenzodiazepine) in a double-blind standard (diazepam) placebo-controlled study. *J. Clin. Psychopharmacol.* 2:129–133.

[133] Liu CY, Jiang XX, Zhu YH, Wei DN (2012) Metabotropic glutamate receptor 5 antagonist 2-methyl-6-(phenylethynyl)pyridine produces antidepressant effects in rats: Role of brain-derived neurotrophic factor. *Neuroscience* 223:219–224.

[134] Rodriguez AL, Grier MD, Jones CK, Herman EJ, Kane AS et al. (2010) Discovery of novel allosteric modulators of metabotropic glutamate receptor subtype 5 reveals

chemical and functional diversity and in vivo activity in rat behavioral models of anxiolytic and antipsychotic activity. *Mol. Pharmacol.* 78:1105–1123.

[135] Fell MJ, McKinzie DL, Monn JA, Svensson KA (2012) Group II metabotropic glutamate receptor agonists and positive allosteric modulators as novel treatments for schizophrenia. *Neuropharmacology* 62:1473–1483.

[136] Krystal JH, Abi-Saab W, Perry E, D'Souza DC, Liu N et al. (2005) Preliminary evidence of attenuation of the disruptive effects of the NMDA glutamate receptor antagonist, ketamine, on working memory by pretreatment with the group II metabotropic glutamate receptor agonist, LY354740, in healthy human subjects. *Psychopharmacology* 179:303–309.

[137] Patil ST, Zhang L, Martenyi F, Lowe SL, Jackson KA et al. (2007) Activation of mGlu2/3 receptors as a new approach to treat schizophrenia: A randomized Phase 2 clinical trial. *Nat. Med.* 13:1102–1107.

[138] Kinon B, Zhang L, Millen B, Osuntokun O, Williams J, et al. (2011) A multicenter, inpatient, phase 2, double-blind, placebo-controlled dose-ranging study of LY2140023 monohydrate in patients with DSM-IV schizophrenia. *J. Clin. Psychopharmacol.* 31:349–355.

[139] Liu W, Downing ACM, Munsie LM, Chen P, Reed MR et al. (2012) Pharmacogenetic analysis of the mGlu2/3 agonist LY2140023 monohydrate in the treatment of schizophrenia. *Pharmacogenomics J.* 12:246–254.

[140] Galici R, Echemendia NG, Rodriguez AL, Conn PJ (2005) A selective allosteric potentiator of metabotropic glutamate (mGlu) 2 receptors has effects similar to an orthosteric mGlu2/3 receptor agonist in mouse models predictive of antipsychotic activity. *J. Pharmacol. Exp. Ther.* 315:1181–1187.

[141] Chaki S, Ago Y, Palucha-Paniewiera A, Matrisciano F, Pilc A (2013) mGlu2/3 and mGlu5 receptors: Potential targets for novel antidepressants. *Neuropharmacology* 66:40–52.

[142] Chaki S, Yoshikawa R, Hirota S, Shimazaki T, Maeda M et al. (2004) MGS0039: a potent and selective group II metabotropic glutamate receptor antagonist with antidepressant-like activity. *Neuropharmacology* 46:457–467.

[143] Dwyer JM, Lepack AE, Duman RS (2013) mGluR2/3 blockade produces rapid and long-lasting reversal of anhedonia caused by chronic stress exposure. *J. Mol. Psychiatry* 1:15.

[144] Koike H, Iijima M, Chaki S (2013) Effects of ketamine and LY341495 on the depressive-like behavior of repeated corticosterone-injected rats. *Pharmacol. Biochem. Behav.* 107:20–23.

[145] Campo B, Kalinichev M, Lambeng N, El Yacoubi M, Royer-Urios I et al. (2011) Characterization of an mGluR2/3 negative allosteric modulator in rodent models of depression. *J. Neurogenet.* 25:152–166.

[146] Karasawa J, Shimazaki T, Kawashima N, Chaki S (2005) AMPA receptor stimulation mediates the antidepressant-like effect of a group II metabotropic glutamate receptor antagonist. *Brain Res.* 1042:92–98.

[147] Yoshimizu T, Shimazaki T, Ito A, Chaki S (2006) An mGluR2/3 antagonist, MGS0039, exerts antidepressant and anxiolytic effects in behavioral models in rats. *Psychopharmacology* 186:587–593.

[148] Pałucha-Poniewiera A, Wierońska JM, Brański P, Stachowicz K, Chaki S et al. (2010) On the mechanism of the antidepressant-like action of group II mGlu receptor antagonist, MGS0039. *Psychopharmacology* 212:523–535.

[149] Vizi E, Fekete A, Karoly R, Mike A (2010) Non-synaptic receptors and transporters involved in brain functions and targets of drug treatment. *Br. J. Pharmacol.* 160:785–809.

[150] Wierońska JM, Stachowicz K, Acher F, Lech T, Pilc A (2011) Opposing efficacy of group III mGlu receptor activators, LSP1-2111 and AMN082, in animal models of positive symptoms of schizophrenia. *Psychopharmacology* 220:481–494.

[151] Sławińska A, Wierońska JM, Stachowicz K, Marciniak M, Łasoń-Tyburkiewicz M et al. (2013) The antipsychotic-like effects of positive allosteric modulators of metabotropic glutamate mGlu4 receptors in rodents. *Br. J. Pharmacol.* 169:1824–1839.

[152] Wierońska JM, Acher FC, Sławińska A, Gruca P, Łasoń-Tyburkiewicz M et al. (2013) The antipsychotic-like effects of the mGlu group III orthosteric agonist, LSP1-2111, involves 5-HT1A signalling. *Psychopharmacology* 227:711–725.

[153] Bradley SR, Uslaner JM, Flick RB, Lee A, Groover KM et al. (2012) The mGluR7 allosteric agonist AMN082 produces antidepressant-like effects by modulating glutamatergic signaling. *Pharmacol. Biochem. Behav.* 101:35–40.

[154] Sukoff Rizzo SJ, Leonard SK, Gilbert A, Dollings P, Smith DL et al. (2011) The metabotropic glutamate receptor 7 allosteric modulator AMN082: a monoaminergic agent in disguise? *J. Pharmacol. Exp. Ther.* 338:345–352.

[155] Klakotskaia D, Ramsey AK, Fowler SW, Serfozo P, Simonyi A et al. (2013) Effects of group II and III metabotropic glutamate receptor ligands on conditioned taste aversion learning. *Behav. Brain Res.* 253:9–16.

[156] Pałucha-Poniewiera A, Pilc A (2013) A selective mGlu7 receptor antagonist MMPIP reversed antidepressant-like effects of AMN082 in rats. *Behav. Brain Res.* 238:109–112.

[157] O'Connor RM, Cryan JF (2013) The effects of mGlu7 receptor modulation in behavioural models sensitive to antidepressant action in two mouse strains: *Behav. Pharmacol.* 24:105–113.

[158] Buckley PF, Miller BJ, Lehrer DS, Castle DJ (2009) Psychiatric Comorbidities and Schizophrenia. *Schizophr. Bull.* 35:383–402.

[159] Filippich C, Wolvetang EJ, Mowry BJ (2013) Will brain cells derived from induced pluripotent stem cells or directly converted from somatic cells (iNs) be useful for schizophrenia research? *Schizophr. Bull.* 39:948-954.

In: Metabotropic Glutamate Receptors
Editor: M. Foster Olive

ISBN: 978-1-63117-451-3

Chapter 6

Metabotropic Glutamate Receptor Subtype 1 (mGluR1) as a Therapeutic Target for Schizophrenia

Gentaroh Suzuki[1,*], Hiroshi Kawamoto[2,†] and Hisashi Ohta[2,‡]
[1]Central Pharmaceutical Research Institute, Japan Tobacco Inc., Takatsuki, Japan
[2]Tsukuba Research Institute, Banyu Pharmaceutical Co. Ltd., Tsukuba, Japan

Abstract

Metabotropic glutamate receptors (mGluRs) are G protein-coupled receptors activated by L-glutamate in the central nervous system, suggesting their roles in physiological and pathophysiological function in the brain. mGluRs consist of eight receptor subtypes (mGluR1-8), and they are subdivided into three groups based on sequence identity and pharmacological properties, including G-protein coupling and signal transduction.

We identified novel mGluR1-selective negative allosteric modulators (NAMs) by random screening of a chemical collection with functional assays. Medicinal chemistry efforts resulted in development of optimized compounds with high affinity and selectivity to mGluR1 and suitable drug metabolism/pharmacokinetic (DMPK) profiles, including metabolic stability and brain penetrability. In vivo assays based on mGluR1 agonist-induced behavior and ex vivo receptor occupancy assay revealed that systemically administered mGluR1 NAMs inhibited native mGluR1 expressed in the brain of rodents, depending on mGluR1 occupancy by the compounds. Involvement of mGluR1 in the pathophysiology of schizophrenia was investigated by both behavioral and biochemical approaches with the mGluR1 NAMs. In the behavioral studies, the mGluR1 NAMs inhibited methamphetamine (METH)- and ketamine (KET)-induced hyperlocomotion in mice. Furthermore, the mGluR1 NAMs improved prepulse inhibition in rats impaired by

* Corresponding author: Gentaroh Suzuki. Central Pharmaceutical Research Institute, Japan Tobacco Inc., 1-1, Murasaki-cho, Takatsuki, Osaka 569-1125, Japan. E-mail: gentaro.suzuki@jt.com.
† Hiroshi Kawamoto. Present address: Taisho Pharmaceutical Co., Ltd., Saitama, Japan.
‡ Hisashi Ohta. Present address: Torii Pharmaceutical Co., Ltd., Tokyo, Japan.

METH or KET as well as social interaction impaired by MK-801. In contrast, the mGluR1 NAMs caused neither catalepsy nor abnormal gait in rodents.

Selective blockade of mGluR1 also suggested unique and shared actions on Fos expression in various brain regions in comparison with atypical antipsychotics. These behavioral and biochemical studies with the mGluR1 NAMs suggested potential antipsychotic effects by blockade of mGluR1 with less adverse effects, including extra-pyramidal side effects. The hypothesis that mGluR1 is involved in the pathophysiology of schizophrenia was strengthened by antipsychotic activities of multiple, structurally diverse mGluR1 NAMs. Furthermore, a positron emission tomography (PET) tracer has been successfully developed, and it may be suitable to quantify mGluR1 occupancy by mGluR1 NAMs and guide dose selection for clinical use. In this chapter, we provide an overview of the experimental evidence that mGluR1 could play a role in the pathophysiology of schizophrenia and discuss the potential utility for its treatment by mGluR1 NAMs on the basis of the latest findings.

Keywords: Metabotropic glutamate receptor, mGluR1, G protein-coupled receptor, negative allosteric modulator, schizophrenia, FTIDC, CFMTI, MK-1312, MK-5435

Abbreviations

CFMTI	2-cyclopropyl-5-[1-(2-fluoro-3-pyridinyl)-5-methyl-1*H*-1,2,3-triazol-4-yl]-2,3-dihydro-1*H*-isoindol-1-one
FTIDC	4-[1-(2-fluoropyridin-3-yl)-5-methyl-1*H*-1,2,3-triazol-4-yl]-*N*-isopropyl-*N*-methyl-3,6-dihydropyridine-1(2*H*)-carboxamide
KET	ketamine
MPEP	2-methyl-6-(phenylethynyl)pyridine
METH	methamphetamine
PCP	phencyclidine.

Introduction

Schizophrenia is a major mental illness that affects approximately 1% of the world's population. Symptoms of schizophrenia are classified into three categories: positive symptoms (including hallucinations, delusions, and thought disturbance), negative symptoms (including emotional blunting and social withdrawal), and cognitive impairments (including impaired working memory and attention) [1].

Typical antipsychotics such as haloperidol and chlorpromazine are used to treat positive symptoms; however, these drugs elicit extrapyramidal side effects (EPSs) at their clinically effective doses [2]. Atypical antipsychotics, including clozapine, olanzapine, and risperidone, are likely to be more efficacious than haloperidol [3, 4] and cause fewer extrapyramidal side effects [5]. These atypical antipsychotics had been thought to ameliorate negative symptoms [6]; however, recent studies have reported that these classes of drugs are less effective on negative symptoms and on cognitive impairment [7, 8]. Furthermore, atypical antipsychotics such as clozapine induced more weight gain and were significantly more sedating than was haloperidol [4]. Consequently, there is an urgent need to develop novel compounds that demonstrate increased efficacy against negative symptoms and cognitive dysfunction with less adverse effects [9].

Dysregulation of dopaminergic neurotransmission with hyperfunction of the mesolimbic dopamine system is known to be the primary etiology of schizophrenia. This hypothesis is supported by the fact that all current medications for schizophrenia are primarily based on blocking dopamine D2 receptor [10]. On the other hand, increasing evidence supports the idea that dysfunction of glutamatergic neurotransmission with hypofunction of N-methyl-D-aspartate receptor (NMDAR) is involved in pathophysiology of schizophrenia. Pharmacological evidence for the hypothesis based on NMDAR hypofunction in schizophrenia centers on findings that blockade of NMDAR by noncompetitive antagonists, such as ketamine (KET) and phencyclidine (PCP), induces schizophrenia-like symptoms including negative symptoms and cognitive deficits in normal healthy subjects [11-14].

L-Glutamate is the major excitatory neurotransmitter in the mammalian central nervous system (CNS) and acts on ionotropic glutamate receptors (iGluRs), including NMDAR, and metabotropic glutamate receptors (mGluRs). mGluRs belong to a family of G protein-coupled receptors, and they are involved in the modulation of neuronal excitability and neuro-transmitter release. Eight receptor subtypes (mGluR1-mGluR8) have been identified, and are classified into three groups based on sequence homology, pharmacological profile, and signal transduction pathways. mGluR1 and mGluR5 are classified as group I mGluRs, and are coupled to phospholipase C via Gq proteins and subsequent intracellular calcium mobilization. Group II mGluRs (mGluR2 and mGluR3) and group III mGluRs (mGluR4, mGluR6, mGluR7, and mGluR8) are negatively coupled to adenylate cyclase via Gi/o proteins, resulting in decreased levels of intracellular cAMP [15, 16].

Targeting the NMDAR or other iGluRs directly for new treatments of schizophrenia may not be considered to be a viable option due to their widespread roles in the fast synaptic transmission throughout the CNS and the potential toxicity of overactivation of NMDAR [17]. Meanwhile, each mGluR subtype is differentially distributed at presynaptic and/or postsynaptic sites in the CNS and is thought to contribute to fine-tuning of synaptic function [18, 19]. Accordingly, another option for modulating dysfunction of glutamatergic neuro-transmission is to target mGluRs. Noncompetitive NMDAR antagonists such as KET and PCP increased L-glutamate release in the prefrontal cortex (PFC) of rodents. The PCP-induced increase in L-glutamate release in the mPFC correlated with hyperlocomotion and cognitive impairment elicited by PCP in rats [20-22]. These results suggested that hyperglutamategic state underlying hypofunction of NMDAR would be involved in symptoms of schizophrenia. Group II mGluRs (mGluR2 and mGluR3) are presynaptically and postsynatpcally distributed in the brain [23-25] and group II mGluR-selective agonists, including LY354740 and LY404039, inhibited L-glutamate release via a presynaptic mechanism [26-30].

Furthermore, these group II mGluR-selective agonists showed antipsychotic efficacy in various rodent models [22, 31-33]. In line with this finding, the phase II clinical trial of LY2140023 (the oral prodrug of group II mGluR agonist, LY404039) showed significant antipsychotic efficacy for both positive and negative symptoms with no major side effects [34]. Meanwhile, the efficacy of LY2140023 remains inconclusive since both LY2140023 and the active control, olanzapine, did not separate from placebo in the treatment of patients with schizophrenia in another phase II clinical trial [35].

An alternative strategy to modulate a hyperglutamategic state based on dysfunction of NMDAR is to inhibit postsynaptic mGluRs.

Group I mGluRs (mGluR1 and mGluR5) could be candidates for this approach, since these receptor subtypes are predominantly localized at postsynaptic sites [36, 37]. However, an mGluR5-selective NAM, 2-methyl-6-(phenyl-ethynyl)pyridine (MPEP), augmented PCP-induced deficits in prepulse inhibition (PPI) and hyperlocomotion [38], as illustrated by the fact that mGluR5 potentiates NMDAR function in forebrain regions [39]. On the other hand, it remained unclear whether mGluR1 was involved in pathophysiology of schizophrenia.

Thus, we investigated the potential of mGluR1 as a therapeutic target for schizophrenia using mGluR1-selective NAMs. In this chapter, we provide an overview of the experimental evidence that mGluR1 could play a role in the pathophysiology of schizophrenia and discuss the potential utility for its treatment by mGluR1 NAMs on the basis of the latest findings.

Development of mGluR1-Selective NAMs

mGluR1-selective ligands are useful to elucidate physiological and/or pathophysiological roles of mGluR1. Earlier approaches to identify mGluR1-selective ligands were random screenings of chemical collections using binding assays with an orthosteric radioligand such as [^{3}H]L-quisqualate and/or chemical derivatizations from L-glutamate and its analogues. Nevertheless, these challenges were unsuccessful in identifying selective mGluR1-selective ligands, presumably due to the highly conserved amino acid sequences of the L-glutamate-binding site in mGluR1 and mGluR5 [40].

In contrast, functional assays based on detection of intracellular Ca^{2+} increases mediated via mGluR1 activation resulted in identification of subtype-selective compounds [41-43]. The success of this approach may stem from the fact that functional assays can identify compounds interacting with sites different from the L-glutamate binding site, such as the transmembrane domain, which is less conserved than the L-glutamate binding site.

In order to identify mGluR1-selective antagonists with novel chemical scaffolds, we performed high throughput screening (HTS) based on inhibition of L-glutamate-induced Ca^{2+} mobilization in CHO cells expressing human mGluR1a [44]. The high throughput screening campaign identified several hit compounds with non-amino acid structures. One of them was compound 1, which had a unique tetrahydropyridine structure (Figure 1). Compound 1 had a potent antagonistic activity against human mGluR1a (IC_{50} = 9.8 nM) and moderate selectivity toward human mGluR5 (IC_{50} = 820 nM).

Meanwhile, compound 1 showed very high lipophilicity ($logD_{7.4}$ >4.0), considering that the optimal range of $logD_{7.4}$ values in orally active CNS drugs is generally thought to be between 1 and 3 [45]. In an attempt to reduce the lipophilicity of the compound, compound 2 was identified as a less lipophilic mGluR1 antagonist with 2.9 of $logD_{7.4}$. However, the IC_{50} values of compound 2 toward human mGluR1a and human mGluR5 were 7.0 nM and 170 nM, respectively, suggesting less selectivity than compound 1 [46]. In the course of further optimization, 4-[1-(2-fluoropyridin-3-yl)-5-methyl-1*H*-1,2,3-triazol-4-yl]-*N*-isopropyl-*N*-methyl-3,6-dihydropyridine-1(2*H*)-carboxamide (FTIDC) was obtained as a potent and selective mGluR1 antagonist with appropriate lipophilicity (Figure 1).

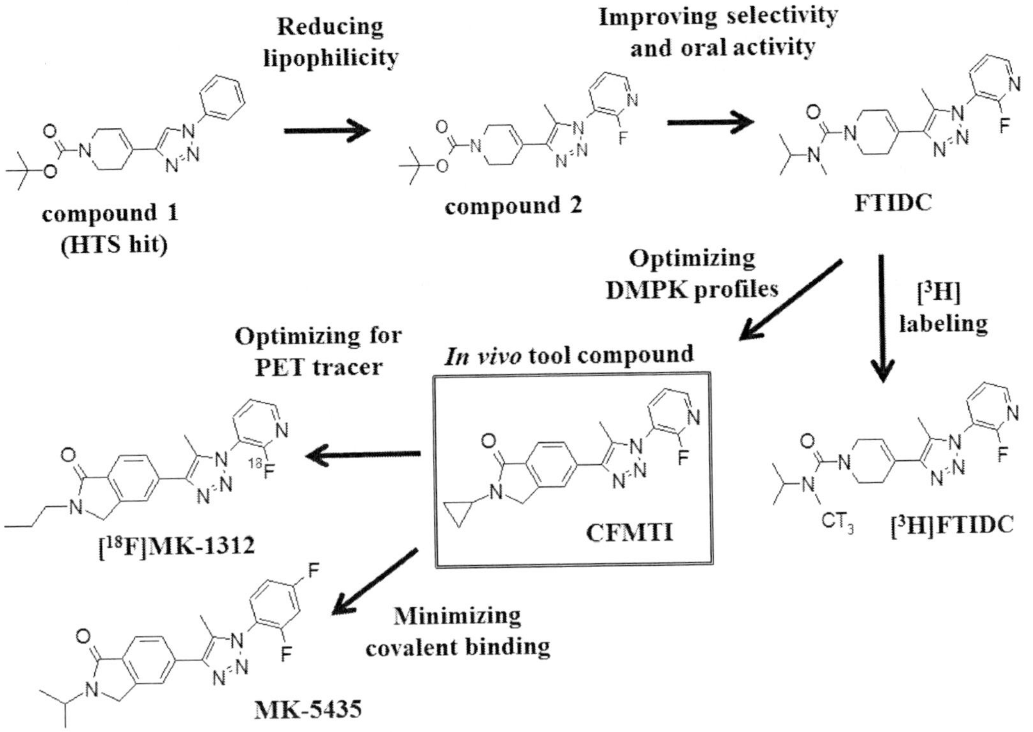

Figure 1. Chemical structures of mGluR1-selective NAMs and process of optimization from the HTS hit compound. Information was collected from previous reports [44, 46-48, 50, 98].

Table 1. Activities of FTIDC and CFMTI against mGluR subtypes

Subtypes	Species	Assay	FTIDC IC_{50} (nM)	CFMTI IC_{50} (nM)
mGluR1a	human	Ca^{2+}	5.8	2.6
mGluR1a	human	[^{3}H]Quis	>10,000	>10,000
mGluR1a	rat	Ca^{2+}	5.8	2.3
mGluR5	human	Ca^{2+}	6,200	5,400
mGluR2	human	[^{35}S]GTPγS	>10,000	>10,000
mGluR4	human	[^{35}S]GTPγS	>10,000	>10,000
mGluR6	human	[^{35}S]GTPγS	>10,000	>10,000
mGluR7	human	[^{35}S]GTPγS	>10,000	>10,000
mGluR8	human	[^{35}S]GTPγS	>10,000	>10,000

Ca^{2+}, inhibition on L-glutamate-induced Ca^{2+} mobilization in CHO cells expressing mGluR1a or mGluR5; [^{3}H]Quis, inhibition of [^{3}H]quisqualic acid binding to membranes prepared from CHO cells expressing mGluR1a; [^{35}S]GTPγS, inhibition of agonist-induced [^{35}S]GTPγS binding to membranes prepared from CHO cells expressing mGluR2, mGluR4, mGluR6, mGluR7, or mGluR8. Data were collected from previous reports [44, 48].

FTIDC inhibited L-glutamate-induced intracellular Ca^{2+} mobilization in CHO cells expressing human mGluR1a or rat mGluR1a with similar potency, suggesting no species difference on inhibitory activity toward mGluR1. FTIDC is approximately 1000 times more selective towards mGluR1a than towards mGluR5. FTIDC also showed neither agonist nor antagonist activity towards group II and group III mGluRs, including mGluR2, mGluR4, mGluR6, mGluR7, and mGluR8 (Table 1). FTIDC reduced the maximum response of agonist concentration-response curves in Ca^{2+} mobilization assay using CHO cells expressing human mGluR1a, whereas [^{3}H]L-quisqualate binding to human mGluR1a was not inhibited by FTIDC. These results suggested that FTIDC is an mGluR1-selective NAM [44].

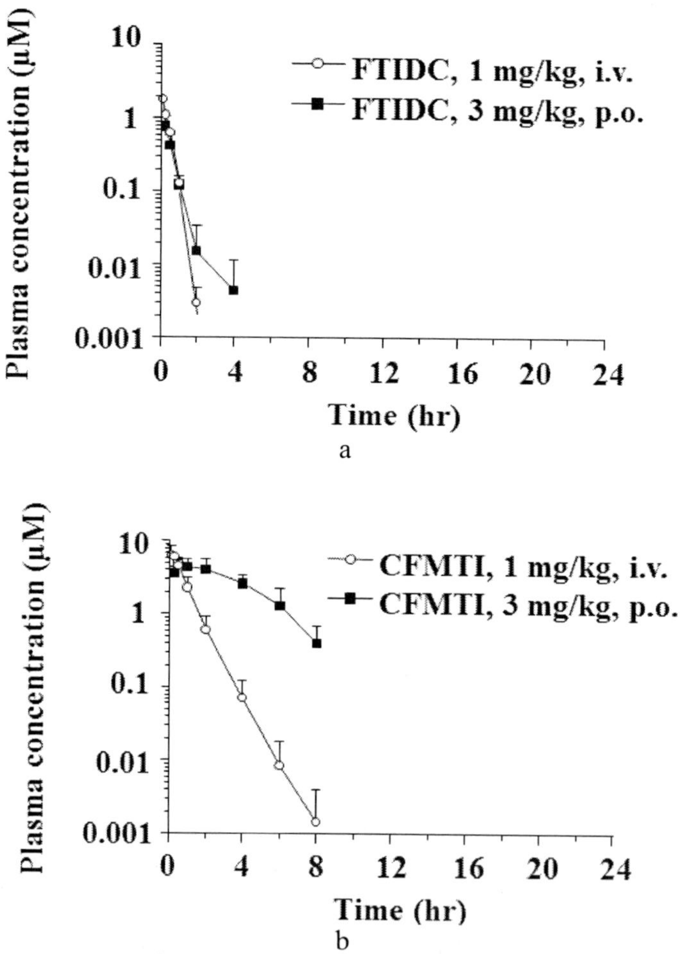

Figure 2. Plasma concentration-time profiles of FTIDC and CFMTI in rats. FTIDC (A) and CFMTI (B) were intravenously or orally administered to rats, and the plasma concentrations were determined at each time point [48]. The oral bioavailability (F), plasma half-life (T1/2), and plasma clearance (CLp) of FTIDC were 18%, 0.2 h, and 57 mL/min/kg, respectively. The oral bioavailability (F), plasma half-life (T1/2), and plasma clearance (CLp) of CFMTI were 100%, 0.6 h, and 8.1 mL/min/kg, respectively [46].

However, FTIDC indicated short half-life and low oral bioavailability (Figure 2A). Thus, FTIDC was further optimized for improvement of drug metabolism/pharmacokinetic (DMPK) profiles [47]. In the course of the optimization process, the isoindolinone derivative 2-cyclopropyl-5-[1-(2-fluoro-3-pyridinyl)-5-methyl-1*H*-1,2,3-triazol-4-yl]-2,3-dihydro-1*H*-isoindol-1-one (CFMTI) was identified as a compound with preferable DMPK profiles (Figure 1 and Figure 2B). CFMTI noncompetitively inhibited L-glutamate-induced intracellular Ca^{2+} mobilization in CHO cells expressing human and rat mGluR1a, with IC_{50} values of 2.6 and 2.3 nM, respectively (Figure 3). In addition, CFMTI, like FTIDC, did not displace [^{3}H]L-quisqualate binding to human mGluR1a. The selectivity of CFMTI to mGluR1 over mGluR5 was 2000-fold, and CFMTI at 10 μM showed no significant activity toward other mGluR subtypes (TABLE 1) or toward 166 target molecules, such as enzymes, neurotransmitter receptors, transporters, and ion channels [48]. Taken together, CFMTI is a more potent and selective mGluR1 NAM with better DMPK profile than FTIDC.

In Vivo Measurement of mGluR1 Activity and Receptor Occupancy

Intracerebroventricular administration of a mGluR1/mGluR5 dual agonist, (S)-3,5-dihydroxyphenylglycine ((S)-3,5-DHPG), in mice causes face-washing behavior, which is defined as wiping of the face from the ear to the mouth with the forelimbs.

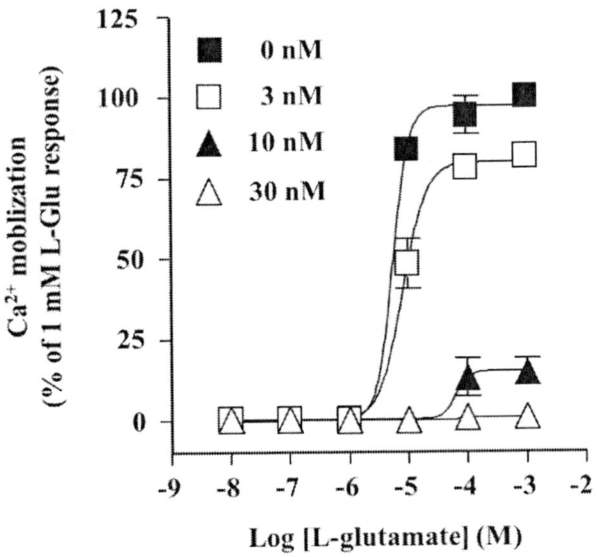

Reproduced from [48]. Copyright© 2009 with permission from American Society for Pharmacology and Experimental Therapeutics.

Figure 3. Effect of CFMTI on concentration-response curves for L-glutamate-induced increases in intracellular Ca^{2+} mobilization via human mGluR1a. Concentration response curves for L-glutamate in the absence or presence of CFMTI at 3 to 30 nM in CHO cells expressing human mGluR1a. Results are expressed as a percentage of the response to 1 mM L-glutamate, and they are mean ± S.D. of the representative experiment in three independent experiments performed in triplicate. L-Glu, L-glutamate.

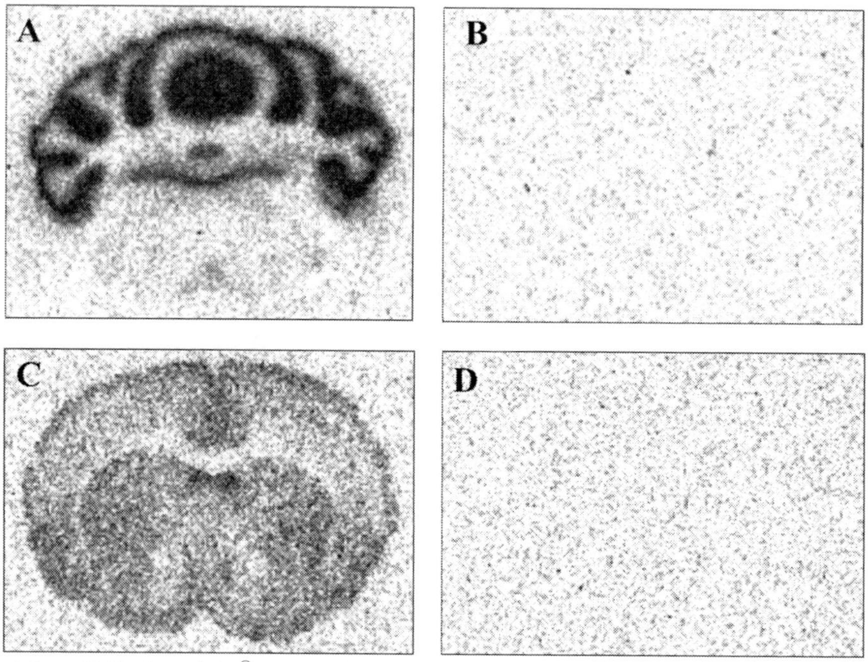

Figure 4. [^{3}H]FTIDC binding to brain sections from wild-type and mGluR1 knockout mice. Coronal brain sections at the cerebellar (A and B) and the striatal (C and D) levels were prepared from wild-type (A and C) and mGluR1 knockout (B and D) mice. High specificity of [^{3}H]FTIDC binding to mGluR1 was demonstrated by comparison of [^{3}H]FTIDC binding to brain sections prepared from wild-type and mGluR1 knockout mice.

This behavior, like facial grooming, was inhibited by structurally diverse mGluR1 NAMs, including FTIDC and (3-ethyl-2-methyl-quinolin-6-yl)-(4-methoxy-cyclohexyl)-methanone methanesulfonate (EMQMCM), but not by an mGluR5 NAM, MPEP [49, 50].

These results indicate that (S)-3,5-DHPG-induced face-washing behavior is due to activation of mGluR1 in the brain. This *in vivo* assay is useful to evaluate whether peripherally administered compounds block native mGluR1 activation in the brain of mice. Thus, we utilized the (S)-3,5-DHPG-induced face-washing behavior in mice for *in vivo* measurement of mGluR1 activity. If efficacy of the compound in an animal model is based on blockade of mGluR1, significant receptor occupancy by the compound must be observed at the doses showing efficacy. On the other hand, if the compound does not show efficacy in the animal model at the doses at which it significantly occupies mGluR1 in the brain, one can infer that mGluR1 is not likely to be involved in neuronal functions related to the animal model. Accordingly, determination of mGluR1 occupancy by compounds is important to appropriately interpret results of *in vivo* experiment with compounds.

The tritiated mGluR1-selective NAM [^{3}H]FTIDC was synthesized as a radioligand to determine mGluR1 occupancy by compounds (Figure 1). [^{3}H]FTIDC is a radioligand with high affinity for membranes prepared from CHO cells expressing human mGluR1a (K_D = 2.1 nM) and mouse cerebellum (K_D = 3.7 nM). [^{3}H]FTIDC bound to mGluR1 was displaced by structurally unrelated mGluR1-selective NAMs, indicating that the allosteric binding pocket of FTIDC is shared with different mGluR1 NAMs. The binding specificity of [^{3}H]FTIDC toward mGluR1 was demonstrated by a lack of significant binding to brain sections prepared from mGluR1 knockout mice (Figure 4), indicating that [^{3}H]FTIDC was an excellent radioligand for receptor occupancy assay.

In *ex vivo* mGluR1 occupancy assay, brain sections prepared from mice administered compounds were used to determine receptor occupancy level. The brain sections were incubated with [^{3}H]FTIDC, and radioactivity based on specific [^{3}H]FTIDC binding was measured. Comparing with brain sections from mice treated with vehicle, reduced [^{3}H]FTIDC binding to brain sections from mice administered the compound indicates mGluR1 occupancy by peripherally administered compounds. Intraperitoneally administered FTIDC into mice occupied mGluR1 in the cerebellum and the striatum in a dose-dependent manner. The receptor occupancy level by FTIDC was clearly correlated with the plasma concentration of FTIDC [50]. Furthermore, correlation analysis indicated a linear relationship between *in vivo* activity of structurally unrelated mGluR1-selective NAMs [43, 51] and receptor occupancy levels by the compounds in the brain (Figure 5) [50]. Taken together, the *ex vivo* receptor occupancy assay using [^{3}H]FTIDC could be a useful tool for determining direct relationships between receptor occupancy levels and various *in vivo* pharmacological effects by mGluR1 NAMs.

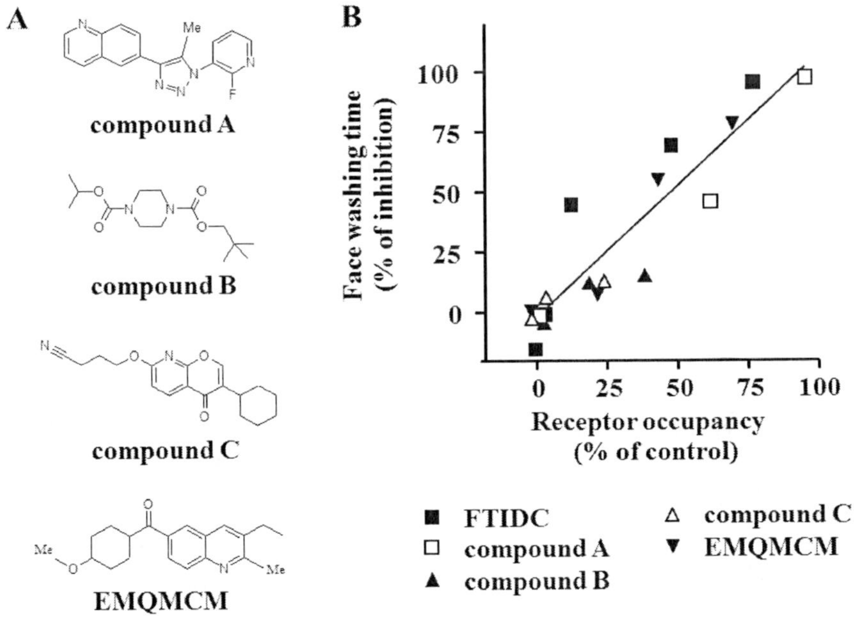

Figure 5. Relationship between the degree of mGluR1 occupancy and *in vivo* activities by mGluR1 NAMs with diverse chemical structures. (A) Chemical structures of mGluR1 NAMs, compounds A, B, and C, and EMQMCM. (B) Correlation of receptor occupancy with inhibition of (S)-3,5-DHPG-induced face washing behavior by various mGluR1 NAMs. The percent inhibition of (S)-3,5-DHPG-induced face washing behavior was plotted against the receptor occupancy level in the striatum at the corresponding dose of FTIDC (closed squares), compound A (open squares), compound B (closed triangles), compound C (open triangles), and EMQMCM (inverted closed triangles).

mGluR1 As a Therapeutic Target for Schizophrenia

Hyperfunction of the mesolimbic dopamine system is known to be the primary etiology of schizophrenia [10]. METH increases extracellular dopamine levels through effects on the dopamine transporter (DAT) and results in neurochemical changes and psychiatric effects [52]. The characteristics of METH-associated psychosis are at least partially shared with those of schizophrenia [53]. Thus, METH-induced behavioral changes in rodents, including METH-induced hyperlocomotion in mice, have been used to screen compounds for antipsychotic activity. FTIDC inhibited METH-induced hyperlocomotion in mice [54]. This result suggested that mGluR1 could modulate behavioral changes based on a hyperdopaminergic state mimicking some aspects of schizophrenia. Furthermore, FTIDC ameliorated METH-induced deficits in PPI in rats [54]. PPI is considered a translational measure of "sensorimotor gating" with the cross-species nature from rodents to humans and its deficits were seen in schizophrenia [55].

These findings indicated that mGluR1 could be involved in pathophysiology of schizophrenia. Therefore, the potential roles of mGluR1 in schizophrenia were further validated using the optimized mGluR1-selective NAM with better DMPK profiles, CFMTI. Oral administration of CFMTI dose-dependently inhibited (S)-3,5-DHPG-induced face washing behavior in mice. Statistically significant inhibition of the behavior was observed at doses of more than 1 mg/kg (p.o.) of CFMTI and the maximum inhibition was obtained at 30 mg/kg (p.o.), suggesting that orally administered CFMTI at 1-30 mg/kg significantly blocked activation of mGluR1 in the brain in a dose-dependent manner (Table 2). Consistent with the results, CFMTI in a similar dose range caused mGluR1 occupancy by the compound in the brain, and almost full receptor occupancy (94%) in the striatum was observed at 30 mg/kg (p.o.) [48].

Table 2. Summary of *in vivo* tests for CFMTI

Test	Species	Results
Agonist-induced mGluR1 activity (DHPG)		
Face-washing behavior	mouse	Effective at 1-30 mg/kg
Hyperdopaminergic model (METH)		
Hyperlocomotion	mouse	Effective at 1-10 mg/kg
PPI deficit	rat	Effective at 10 mg/kg
HypoNMDAR model (KET/MK-801)		
Hyperlocomotion	mouse	Effective at 30 mg/kg
PPI deficit	rat	Effective at 3-10 mg/kg
Inhibition of social interaction	rat	Effective at 10 mg/kg
Adverse effects		
Spontaneous locomotor activity	mouse	Not caused at 0.3-30 mg/kg
Spontaneous PPI	rat	Not caused at 3-10 mg/kg
Catalepsy	rat	Not caused at 30-100 mg/kg
Rotarod	mouse	Not caused at 3-100 mg/kg
Object location memory	mouse	Not caused at 3-10 mg/kg

DHPG, (S)-3,5-DHPG; METH, methamphetamine; KET, ketamine. Data were collected from a previous report [48].

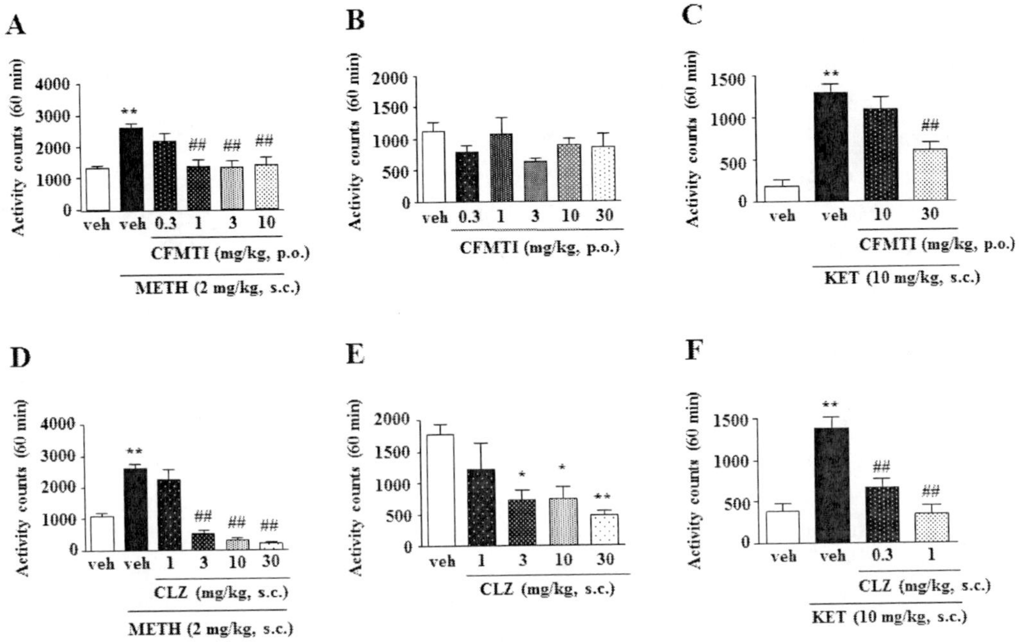

Figure 6. Effects of CFMTI and clozapine on locomotor activity. Methamphetamine (METH)-induced hyperlocomotion (A, D) and spontaneous locomotor activity (B, E) were evaluated in mice administered CFMTI (A, B) and clozapine (D, E) under nonhabituated conditions. Ketamine (KET)-induced hyperlocomotion (C, F) was evaluated in mice administered CFMTI (C) and clozapine (F) under habituated conditions. Data are presented as means ± S.E.M. veh, vehicle; CLZ, clozapine. $^{*}p < 0.05$ and $^{**}p < 0.01$ versus vehicle-treated group; $^{\#\#}p < 0.01$ versus METH or KET-treated group (one-way analysis of variance followed by Dunnett's test).

Taken together, oral dosing of at least 1 mg/kg of CFMTI could elicit pharmacological effects based on mGluR1 blockade, and 30 mg/kg of CFMTI seems to be sufficient to obtain maximal pharmacological effects through inhibition of mGluR1.

In line with the relationship between the dose range of CFMTI and mGluR1 occupancy, the compound significantly inhibited METH-induced hyperlocomotion in the dose range of 1-10 mg/kg (p.o.), whereas CFMTI had no significant effect on spontaneous locomotor activity even at 30 mg/kg (p.o.) (Figure 6, A and B). Meanwhile, clozapine inhibited METH-induced hyperlocomotion at doses in the range of 3-30 mg/kg (s.c.), in which the drug inhibited spontaneous locomotor activity (Figure 6, D and E). Like METH-induced hyperlocomotion, deficits in PPI elicited by METH were alleviated by CFMTI at 10 mg/kg, whereas the compound did not affect spontaneous PPI in rats (Table 2).

These results indicated that the effects of CFMTI on hyperlocomotion and deficits in PPI elicited by METH were due to antipsychotic-like effects based on mGluR1 blockade in the brain, but not due to motor impairment or action on basal sensorimotor gating.

Alternative pharmacological animal models of schizophrenia are based on hypofunction of NMDAR. Noncompetitive NMDAR antagonists, such as KET, PCP, and MK-801, induced schizophrenia-like symptoms including negative symptoms and cognitive deficits in normal healthy subjects [11-14].

These noncompetitive NMDAR antagonists caused not only hyperlocomotion and PPI deficits, but also social withdrawal and cognitive impairment [9], suggesting that NMDAR hypofunction-based animal models may be particularly relevant to persistent, poor outcome forms of schizophrenia [56].

Therefore, we evaluated effects of CFMTI on these animal models with KET and MK-801 in order to elucidate the roles of mGluR1 in the pathology of schizophrenia based on NMDAR hypofunction. Systemic administration of CFMTI at 30 mg/kg, like clozapine, inhibited KET-induced hyperlocomotion in mice (Figure 6, C and F). CFMTI at 3-10 mg/kg also reversed KET-induced deficits in PPI in rats (Table 2). These results indicated that mGluR1 could play roles in the hypoNMDAR state as well as in hyperdopaminergic state of schizophrenia. Furthermore, CFMTI significantly improved social withdrawal induced by MK-801 at 10 mg/kg, indicating that mGluR1 blockade could ameliorate negative symptoms of schizophrenia such as social withdrawal (Table 2).

DBA/2J mice exhibited lower levels of spontaneous PPI than other mouse strains [57, 58]. Lower spontaneous PPI in the DBA/2J mice was improved by administration of antipsychotics such as clozapine, risperidone, and haloperidol, suggesting that DBA/2J mouse may be a spontaneous animal model to evaluate antipsychotic activities of compounds [57, 58]. This basal level of PPI in DBA/2J mice was normalized by CFMTI [59], suggesting that blockade of mGluR1 had antipsychotic effects in a non-pharmacological, spontaneous rodent model of schizophrenia.

We have not yet evaluated effects of CFMTI on cognitive deficits caused by noncompetitive NMDAR antagonists. On the other hand, a very recent study reported that blockade of mGluR1 by another mGluR1-selective NAM, JNJ16259685, improved social memory impairment elicited by MK-801 [60]. Further studies will be required to demonstrate whether mGluR1 blockade by mGluR1-selective NAMs improve all aspects of schizophrenia-like symptoms, including cognitive deficits, in rodent models.

Profiling of mGluR1 Blockade by Fos Expression Pattern

c-Fos is an immediate early gene. Acute expression of its gene product, Fos, is directly and indirectly induced by various physiological stimuli [61]. Induction of Fos in neurons is a useful marker to map changes in neuronal activity, since neuronal activation induces Fos expression [62]. Clinically effective antipsychotic drugs induce Fos expression in the shell of the nucleus accumbens (NAcc-shell) in rodents, suggesting that the NAcc-shell could be an important region for their action. The potency of EPSs caused by antipsychotics is correlated with Fos induction in the dorsolateral striatum (dlSTR). Fos expression in medial prefrontal cortex (mPFC) could be associated with potency for the treatment of negative symptoms in schizophrenia [63, 64]. Thus, evaluation of Fos expression patterns could be useful for identification of neuroanatomical substrates, classification of novel compounds, and prediction of their therapeutic utility [65]. CFMTI induced Fos expression in the NAcc-shell, but not in the dlSTR [48], similar to atypical antipsychotic drugs such as clozapine and olanzapine [63, 64]. These results were consistent with the observation that CFMTI showed a significant antipsychotic-like activity without EPSs in rodent models (Table 2)

As mentioned above, Fos expression in the mPFC could be related to potency for ameliorating the negative symptoms of schizophrenia, and CFMTI induced Fos expression in the mPFC. The activity of CFMTI on Fos expression in the mPFC might explain why CFMTI ameliorated deficits in social interaction in rats that are believed to mimic negative symptoms such as social withdrawal in schizophrenic patients.

In the previous study, clozapine predominantly increased Fos expression in glutamatergic pyramidal cells, and in a small population of presumptive GABAergic interneurons in the mPFC [66]. CFMTI also significantly increased Fos expression in both vesicular glutamate transporter 1 (VGLUT1; a marker for glutamatergic neurons) -positive and -negative neurons, while the majority of neurons expressing Fos were VGLUT1-positive (Figure 7A). These results suggest that CFMTI preferentially activates glutamatergic neurons in the mPFC, as does clozapine [67].

The hyperglutamatergic state in the PFC through hypofunction of NMDAR may be due to inhibition of GABAergic interneurons in the PFC caused by blockade of NMDAR-dependent excitatory input and subsequent disinhibition of glutamatergic neurons in the PFC [14, 68]. The neurochemical approach based on Fos expression in the brain revealed that CFMTI activated glutamatergic and non-glutamatergic (presumably GABAergic) neurons in the PFC of rats (Figure 7A). The activity of CFMTI on the PFC might appropriately restore the imbalance between excitatory and inhibitory inputs in the PFC to an appropriate level, resulting in antipsychotic-like activity of CFMTI. However, validation of this hypothesis by further studies, including electrophysiological analyses and determinations of neuro-transmitters, is necessary.

Comparative analysis of Fos expression revealed that Fos expression patterns induced by CFMTI and clozapine were shared with many brain regions, including the NAcc-shell, the mPFC, the central amygdaloid nucleus (CeA), the locus coeruleus (LC), and the bed nucleus of the stria terminalis (BNST). Meanwhile, there were characteristic differences in Fos induction in the lateral prefrontal cortex (lPFC) and the lateral hypothalamic/perifornical area (LH/PFA) between CFMTI and clozapine. CFMTI, but not clozapine, induced Fos expression in the lPFC. Positron emission tomography (PET) studies indicated decreased blood flow in the PFC of schizophrenic patients, including the dorsolateral (dlPFC) and the orbital (oPFC) prefrontal cortices [69]. The dlPFC and the oPFC in primates were suggested to correspond anatomically to the mPFC and lPFC in rodents, respectively [70]. Therefore, the activity of CFMTI on Fos induction in the lPFC might contribute to unique antipsychotic-like effects of mGluR1 blockade, whereas functional equivalence between primate oPFC and rodent lPFC remains unproven.

Clozapine and olanzapine induced Fos expression in the LC [71, 72], and increased the firing rate of noradrenergic neurons in the LC [72, 73]. Furthermore, the PFC was innervated by noradrenergic neurons of the LC [74]. Indeed, clozapine and olanzapine increased norepinephrine release in the mPFC [75, 76]. Collectively, these data indicate that the LC could be an alternative target for atypical antipsychotics. CFMTI at 10-30 mg/kg increases Fos expression in noradrenergic neurons of the LC (Figure 7, B and C) [67], suggesting that mGluR1 blockade by CFMTI could activate noradrenergic neurons in the LC. This presumable activation of noradrenergic neurons by CFMTI might result in increased norepinephrine release in the PFC.

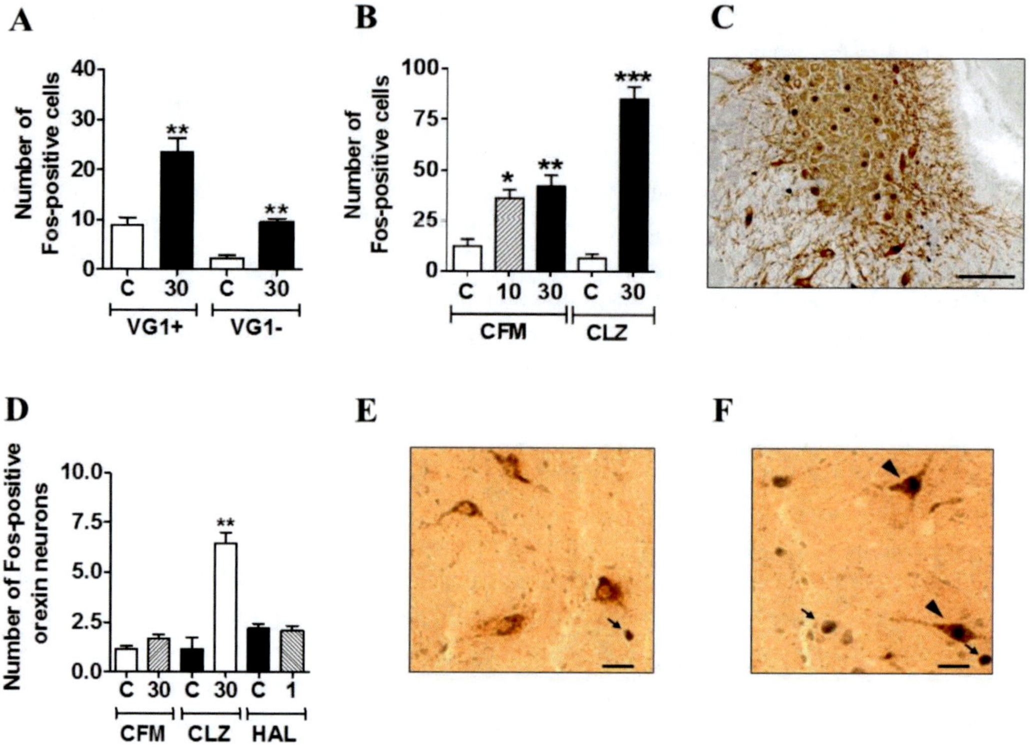

Figure 7. Fos expression in various brain regions of rats administered CFMTI, clozapine, or haloperidol. (A) The numbers of Fos-positive cells among VGLUT1 (a marker for glutamatergic neurons)-positive (VG1+) and -negative (VG1-) cells were counted in the medial prefrontal cortex (mPFC) of rats administered vehicle or CFMTI at 30 mg/kg (p.o.). Statistical analysis was performed using Student's *t* test. $^{**}p < 0.01$ versus vehicle control group. (B) The number of Fos-positive cells was counted in the locus coeruleus (LC) of rats injected with vehicle (C), CFMTI (CFM), or clozapine (CLZ). Statistical analyses were conducted with one-way analysis of variance followed by post hoc multiple comparison test (Dunnett's test) for CFMTI, and Student's *t* test for clozapine. $^{*}p < 0.05$, $^{**}p < 0.01$, $^{***}p < 0.001$ versus vehicle control group. (C) Representative photomicrograph showing dual labeling of Fos protein in nuclei and dopamine β-hydroxylase (DBH) protein (a marker for noradrenergic neurons) in cytoplasm in the LC of rats that received 30 mg/kg CFMTI (p.o.). Scale bars = 100 μm. (D) The number of Fos-positive cells in orexin neurons was counted in the lateral hypothalamic/perifornical area (LH/PFA) of rats that received vehicle, CFMTI (CFM, 30 mg/kg, p.o.), clozapine (CLZ, 30 mg/kg, i.p.), or haloperidol (HAL, 1 mg/kg, s.c.). Statistical analysis was performed using Student's *t* test. $^{**}p < 0.01$ versus vehicle control group. (E) Representative photomicrograph showing dual labeling of Fos protein in nuclei and orexin peptide in cytoplasm in the LH/PFA of rats that received 30 mg/kg CFMTI. (F) Representative photomicrograph showing dual labeling of Fos protein in nuclei and orexin peptide in cytoplasm in the LH/PFA of rats that received 30 mg/kg clozapine. Arrowheads and arrows indicate Fos/orexin-double positive cells and Fos-positive/orexin-negative cells, respectively. Scale bars = 25 μm. Data are expressed as the mean ± SD of the number of Fos-positive cells (n = 3-5). C, 1, 10, and 30 represent vehicle control, 1 mg/kg, 10 mg/kg, and 30 mg/kg, respectively.

Activity of CFMTI on the LC may contribute to normalizing imbalances in the PFC function, whereas it will be necessary to evaluate the effects of mGluR1 blockade on norepinephrine release in the PFC as part of validation of the hypothesis.

Atypical antipsychotics may have efficacy in reducing negative symptoms as well as positive symptoms with less EPS liability; however, one of the current concerns with atypical antipsychotics is their tendency to cause significant weight gain [4, 77, 78]. The weight gain could result in an increased risk of metabolic disorders such as cardiovascular diseases and type II diabetes. However, whether antipsychotic drug-induced weight gain in rodent models can predict the clinical situation remains controversial [79]. On the other hand, the weight gain liability of antipsychotic drugs has been correlated with the degree of Fos induction in orexin neurons of the LH/PFA, known as the classical "feeding area" [80].

Thus, Fos expression in orexin neurons of the LH/PFA could be a maker for predicting the weight gain liability of antipsychotic drugs. Like haloperidol with minimal weight gain liability, CFMTI did not induce Fos expression in orexin neurons of the LH/PFA, whereas its expression was induced by clozapine, which shows significant weight gain liability (Figure 7, D-F). These results suggest that CFMTI might have less weight gain liability in comparison with current atypical antipsychotic drugs that typically cause weight gain [67].

Potential Adverse Effects of mGluR1 Blockade

Typical antipsychotics such as haloperidol cause EPSs such as parkinsonism-like motor abnormalities [4]. Rodents catalepsy model is the standard preclinical test used to predict EPS liability [81, 82]. In the rat catalepsy test, CFMTI did not cause catalepsy even at 100 mg/kg (p.o.), suggesting less EPS liability by mGluR1 blockade.

A rotarod test of motor coordination/sedation is commonly used to predict clinical sedation caused by novel drugs. Many atypical antipsychotic drugs, including clozapine, were associated with much fewer EPSs than haloperidol, whereas clozapine was significantly more sedating than haloperidol [4]. Consistent with the clinical observations, clozapine caused motor incoordination in mouse rotarod test at doses at which antipsychotic effects were demonstrated [31, 48]. In contrast, CFMTI at doses of 3 to 100 mg/kg had negligible effects on motor coordination in the rotarod test, indicating that blockade of mGluR1 did not cause motor incoordination (Figure 8A).

As in the previous reports, clozapine elicited motor incoordination at doses known to exert antipsychotic effects (Figure 8B). mGluR1 knockout mice showed cerebellar symptoms such as ataxic gait and impaired synaptic plasticity in the cerebellum [83, 84]. Anatomical analysis of mGluR1 knockout mice revealed that about one third of Purkinje cells were abnormally innervated by multiple climbing fibers even in adulthood [85]. Therefore, cerebellar ataxia in mGluR1 knockout mice had been thought to be due to lack of mGluR1 during the postnatal developmental stage. However, conditional knockout of mGluR1 expressed in Purkinje cells of adult mice also impaired motor coordination [86]. Autoantibodies in Hodgkin's disease patients with severe cerebellar ataxia reacted specifically with mGluR1 and blocked agonist-response of recombinant mGluR1 expressed in CHO cells [87]. The injection of autoantibodies into the cerebellar subarachnoid space of mice caused severe, reversible ataxia [87].

These results suggested that mGluR1 was essential for cerebellar function, not only during postnatal development, but also in adulthood.

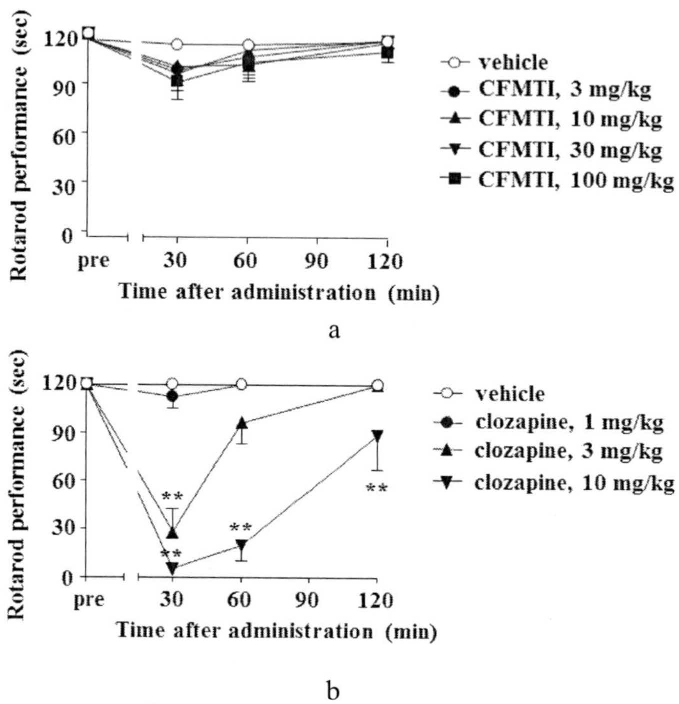

Reproduced from [48]. Copyright© 2009 with permission from American Society for Pharmacology and Experimental Therapeutics.

Figure 8. Effects on rotarod performance in mice. (A) Effects of CFMTI. (B) Effects of clozapine. Data are presented as means ± S.E.M. ** $p < 0.01$ versus vehicle-treated group at each time point (one-way repeated measures analysis of variance followed by Dunnett's test).

However, CFMTI did not inhibit spontaneous locomotor activity even at 30 mg/kg (Figure 6B), at which mGluR1 occupancy level by the compound was 89% in the cerebellum. Other mGluR1 NAMs, LY456236 and FTIDC, also did not inhibit spontaneous locomotor activity [48, 88]. Collectively, pharmacological blockade of mGluR1-selective NAMs is unlikely to cause severe motor impairment, unlike in mGluR1 knockout mice. In addition, another behavioral study indicated that mGluR1 knockout mice exhibited spontaneous PPI deficits [89], whereas CFMTI did not affect spontaneous PPI in rats, as mentioned above (Table 2). *In vivo* experiments using mGluR1 NAMs JNJ16259685 and EMQMCM indicated that blockade of mGluR1 might cause memory impairment [90-92], as did earlier studies of mGluR1 knockout mice [84, 93]. In these studies, memory impairments related to mGluR1 were evaluated using aversive learning procedures such as water maze, passive avoidance, and contextual fear conditioning. However, taking into consideration that mGluR1 blockade could be involved in analgesic and anxiolytic effects [54, 88, 94, 95], effects of mGluR1 blockade on aversive learning do not necessarily indicate memory impairment [96]. In line with the discussion, blockade of mGluR1 by EMQMCM in rats did not impair, but rather improved working memory in the delayed non-matching to position (DNMTP) task, a non-aversive memory paradigm [96]. Therefore, effects of mGluR1 blockade with CFMTI on memory function were evaluated using another non-aversive memory paradigm, object location memory task [97].

In the object location memory task, CFMTI did not impair spatial memory task at a dose showing antipsychotic activities in animal models (Table 2). However, further validation will be necessary to estimate the liability to memory impairment.

Further Development of mGluR1 NAMs and Biomarkers for Proof-of-Concept

CFMTI showed antipsychotic effects in various rodent models, including negative symptoms, whereas CFMTI did not cause catalepsy related to EPSs, indicating that antipsychotic effects via mGluR1 blockade were similar to those of atypical antipsychotics such as clozapine.

CFMTI did not inhibit rotarod performance related to motor coordination and sedation, while clozapine showed impairment in rotarod test. These promising results prompted us to conduct a "proof-of-concept" study in humans. In the course of detailed characterization of CFMTI, the compound was proven to generate a glutathione (GSH) adduct at fluoropyridine part in rat hepatocytes incubation [47]. This result suggested that CFMTI had the possibility to bind irreversibly to biomolecules such as proteins and DNA, giving rise to drug-induced toxicities. In the process of further optimization of compounds, MK-5435 was identified as an mGluR1-selective NAM with less liability of covalent binding (Figure 1) [47, 98]. The IC_{50} values of MK-5435 toward human mGluR1 and rat mGluR1 were 4.3 nM and 3.6 nM, respectively, whereas its IC_{50} value toward human mGluR5 was 7000 nM. Furthermore, MK-5435 showed antipsychotic activities in rat PPI model and mouse hyperlocomotion model [47].

Ex vivo receptor occupancy assay with [^{3}H]FTIDC was useful to clarify mGluR1 occupancy level sufficiently to achieve their pharmacological effects, whereas this approach cannot be applied to humans due to the invasive nature of the procedures. PET is a non-invasive medical imaging tool widely used in both clinical and preclinical research [99]. The non-invasive property of PET allows us to determine drug concentrations and receptor occupancy in protected and isolated tissue compartments such as the brain in humans and non-human primates *in vivo* [100].

An mGluR1 PET tracer, [^{18}F]MK-1312 was developed as a biomarker to study the interaction of MK-5435 with mGluR1 expressed in the brain of non-human primates and ultimately in humans (Figure 1). *In vitro* autoradiographic studies with [^{18}F]MK-1312 in rhesus monkey and human brain slices revealed that the cerebellum showed the highest binding of the PET tracer, followed by a moderate binding in the thalamus and the hippocampus, with the lowest binding in the cortical areas and the caudate/putamen [98]. The distribution of [^{18}F]MK-1312 was consistent with the reported localization of mGluR1 in human brain [37, 101]. The PET studies in rhesus monkey with [^{18}F]MK-1312 showed high brain uptake and a regional distribution consistent with results of *in vitro* autoradiography (Figure 9A). Systemic administration of MK-5435 reduced [^{18}F]MK-1312 uptake in all regions of gray matter in a plasma concentration-dependent manner (Figure 9B). Receptor occupancy studies using [^{18}F]MK-1312 indicated that plasma concentrations of MK-5435 to occupy 50% and 75% of mGluR1 in the cerebellum were 76 nM and 285 nM, respectively [98].

Considering that more than about 75% occupancy of mGluR1 in mice showed almost full inhibition of (S)-3,5-DHPG-induced face-washing behavior, evaluation of antipsychotic effects by MK-5435 in non-human primates would require doses achieving 285 nM of MK-5435 as minimum plasma concentrations.

PET study with $[^{18}F]$MK-1312 could be suitable for clinical use to quantify mGluR1 occupancy by MK-5435, and data relating plasma concentration to mGluR1 occupancy level could guide dose selection of MK-5435 [98]. The hypothesis that mGluR1 is involved in pathophysiology of schizophrenia could be strengthened if multiple, structurally diverse mGluR1 NAMs indicate antipsychotic activities in animal models. If only mGluR1 NAMs belonging to single chemical scaffold are effective in animal models related to schizophrenia, we cannot completely exclude the possibility that the class of compounds exerts antipsychotic effects mediated by unknown targets rather than mGluR1.

As mentioned above, the potential of mGluR1 as a therapeutic target for schizophrenia was evaluated mainly using tetrahydropyridine derivatives such as FTIDC and isoindolinone analogs such as CFMTI and MK-5435. In addition to those classes of compounds, mGluR1-selective NAMs with quinoline structure also showed antipsychotic effects in rodent models [102]. These three classes of compounds are structurally diverse, while they have an aryl-triazole group in common. Therefore, it is still necessary to demonstrate that mGluR1-NAMs with completely distinct chemical structures have antipsychotic effects in animal models for schizophrenia, in order to strengthen our hypothesis.

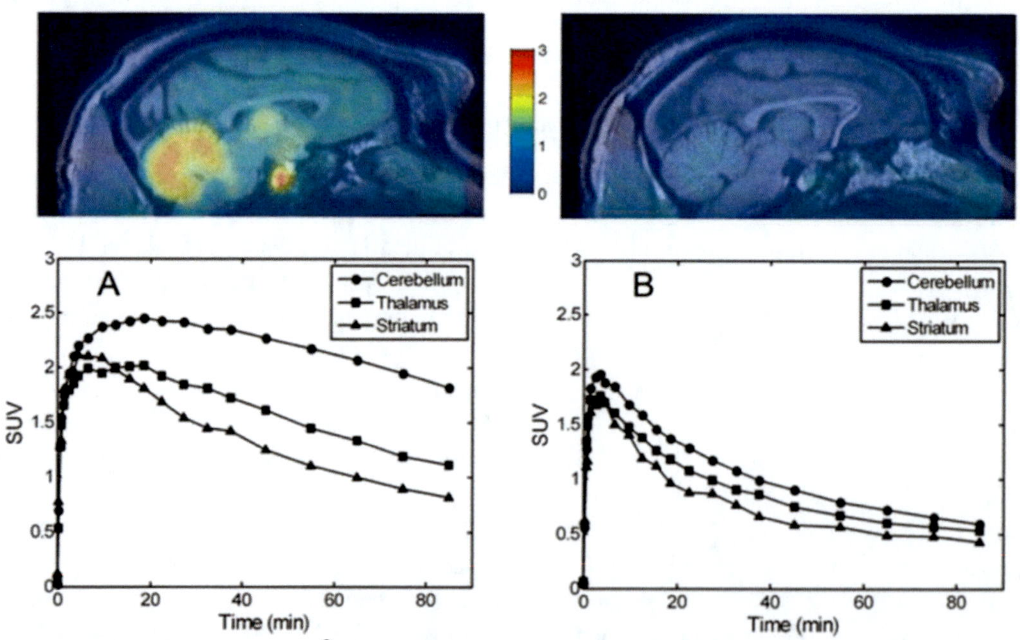

Figure 9. PET/MR aligned and summed PET images (sagittal view) and tissue time-activity curves (TACs) in rhesus monkey brain for $[^{18}F]$MK-1312 under baseline conditions (left image and panel A) and in the presence of MK-5435 (right image and panel B): closed circles, cerebellum; closed squares, thalamus; closed triangles, striatum.

An HTS campaign for mGluR1 NAMs was conducted to obtain structurally novel compounds and identified a novel biaryl amide compound without a triazole group. Optimization of the HTS hit compound resulted in compound 27 [103]. The compound inhibited recombinant human mGluR1 at IC_{50} of 5.1 nM with high selectivity toward human mGluR5 (IC_{50} = 7000 nM) and suitable DMPK profiles. Furthermore, oral administration of compound 27 showed antipsychotic activities on rat PPI model and mouse hyperlocomotion model.

Those studies, using multiple mGluR1-selective NAMs with diverse structures, strongly suggest that mGluR1 is involved in the pathophysiology of schizophrenia. Furthermore, [^{18}F]FITM (^{18}F-labeled compound 27) and [^{11}C]ITMM (a close analog of compound 27) were developed and characterized as mGluR1 PET tracers by other research groups [104-107]. Very recently, the first human PET studies of mGluR1 were conducted using [^{11}C]ITMM [108]. PET imaging with [^{11}C]ITMM may be useful for dose selection of mGluR1 NAMs in clinical trials.

Conclusion

The HTS campaign targeting allosteric modulators identified mGluR1-selective hit compounds with novel structures. One of the optimized compounds, FTIDC, selectively inhibited mGluR1 in a noncompetitive manner, suggesting that FTIDC was an mGluR1 NAM. By further medicinal chemistry efforts, CFMTI was synthesized as an mGluR1-selective NAM having more preferable DMPK profile. *In vivo* assays based on mGluR1 agonist-induced behaviors and *ex vivo* receptor occupancy assay revealed that systemically administered FTIDC and CFMTI inhibited native mGluR1 expressed in the brain of rodents, depending on mGluR1 occupancy by the compounds. CFMTI inhibited METH-induced hyperlocomotion and deficits in PPI that were rodent models for schizophrenia underlying a hyperdopaminergic state. CFMTI was also effective on KET-induced hyperlocomotion and PPI deficits in rodents mimicking a hypoNMDAR state of schizophrenia. Furthermore, mGluR1 blockade by CFMTI improved MK-801-induced inhibition of social interaction in rats, suggesting efficacy of CFMTI in negative symptoms of schizophrenia. Meanwhile, CFMTI did not induce catalepsy in rats, indicating less EPS liability, unlike haloperidol. In addition, mGluR1 blockade had no impact on rotarod performance in mice, in contrast to clozapine. Phenotype analysis of mGluR1 knockout mice had suggested that full inhibition of mGluR1 may cause motor incoordination, memory impairment, and deficits in sensorimotor gating. However, pharmacological blockade of mGuR1 by CFMTI did not cause such adverse effects.

mGluR1 blockade by CFMTI induced Fos expression in the brain regions including the NAcc-shell, the mPFC, and the LC, which were involved in the pathophysiology of schizophrenia. The Fos expression patterns induced by CFMTI were similar to those of atypical antipsychotics, clozapine and olanzapine, illustrating that mGluR1 blockade by CFMTI was effective in rodent models of schizophrenia, including negative symptoms, and did not cause EPSs. In contrast to clozapine, CFMTI did not induce Fos expression in orexin neurons in the LH/PFA, which is positively correlated with weight gain liability.

In addition, PET imaging using [^{18}F]MK-1312 could help provide guidelines for selecting appropriate doses of a more advanced compound, MK-5435. Furthermore, the hypothesis that mGluR1 is involved in the pathophysiology of schizophrenia was strengthened by the results that multiple, structurally diverse mGluR1 NAMs showed antipsychotic efficacy in animal models.

In conclusion, our pharmacological approach using mGluR1-selective NAMs indicated that mGluR1 could play roles in the pathophysiology of schizophrenia and could be a novel therapeutic target for treatment of schizophrenia.

Acknowledgments

The authors wish to thank Dr. Eric Hostetler at Merck Research Laboratories for PET studies. We also thank Akio Satow, Toshifumi Kimura, Dr. Shunsuke Maehara, and Hirohiko Hikichi for pharmacological studies, Mikiko Hata for DMPK studies, and Satoru Ito for medicinal chemistry.

Our studies were conducted at Tsukuba Research Institute, BANYU Pharmaceutical Co. Ltd. (Currently, MSD K.K.) in collaboration with Merck Research Laboratories.

References

[1] Andreasen, N. C. (2000) Schizophrenia: the fundamental questions. *Brain Res. Brain Res. Rev.* 31:106–112.

[2] Marder, S. R. (1999) Limitations of dopamine-D2 antagonists and the search for novel antipsychotic strategies. *Neuropsychopharmacology* 21:S117–S121.

[3] Davis, J. M., Chen, N., Glick, I. D. (2003) A meta-analysis of the efficacy of second-generation antipsychotics. *Arch. Gen. Psychiatry* 60:553–564.

[4] Leucht, S., Corves, C., Arbter, D., Engel, R. R., Li, C., Davis, J. M. (2009) Second-generation versus first-generation antipsychotic drugs for schizophrenia: a meta-analysis. *Lancet* 373:31–41.

[5] Geddes, J., Freemantle, N., Harrison, P., Bebbington, P. (2000) Atypical antipsychotics in the treatment of schizophrenia: systematic overview and meta-regression analysis. *BMJ* 321:1371–1376.

[6] King, D. J. (1998) Drug treatment of the negative symptoms of schizophrenia. *Eur. Neuropsychopharmacol.* 8:33–42.

[7] Kirkpatrick, B., Fenton, W. S., Carpenter, W. T. Jr, Marder, S. R. (2006) The NIMH-MATRICS consensus statement on negative symptoms. *Schizophr. Bull.* 32:214–219.

[8] Keefe, R. S., Bilder, R. M., Davis, S. M., Harvey, P. D., Palmer, B. W., Gold, J. M., Meltzer, H. Y., Green, M. F., Capuano, G., Stroup, T. S., McEvoy, J. P., Swartz, M. S., Rosenheck, R. A., Perkins, D. O., Davis, C. E., Hsiao, J. K., Lieberman, J. A.; CATIE Investigators; Neurocognitive Working Group (2007) Neurocognitive effects of antipsychotic medications in patients with chronic schizophrenia in the CATIE Trial. *Arch. Gen. Psychiatry* 64:633–647.

[9] Jones, C. A., Watson, D. J., Fone, K. C. (2011) Animal models of schizophrenia. *Br. J. Pharmacol.* 164:1162–1194.

[10] Carlsson, A. (2001) *Neurotransmitters – Dopamine and Beyond. Current Issues in the Psychopharmacology of Schizophrenia* (Breier, A., Tran, P. V., Herrea, J. M., Tollefson, G. D., Bymaster, F. P., eds), pp. 3–11, Lippincott Williams and Wilkins Healthcare, Philadelphia, PA.

[11] Ghoneim, M. M., Hinrichs, J. V., Mewaldt, S. P., Petersen, R. C. (1985) Ketamine: behavioral effects of subanesthetic doses. *J. Clin. Psychopharmacol.* 5:70–77.

[12] Krystal, J. H., Karper, L. P., Seibyl, J. P., Freeman, G. K., Delaney, R., Bremner, J. D., Heninger, G. R., Bowers, M. B. Jr, Charney, D. S. (1994) Subanesthetic effects of the noncompetitive NMDA antagonist, ketamine, in humans. Psychotomimetic, perceptual, cognitive, and neuroendocrine responses. *Arch. Gen. Psychiatry* 51:199–214.

[13] Morris, B. J., Cochran, S. M., Pratt, J. A. (2005) PCP: from pharmacology to modelling schizophrenia. *Curr. Opin. Pharmacol.* 5:101–106.

[14] Paz, R. D., Tardito, S., Atzori, M., Tseng, K. Y. (2008) Glutamatergic dysfunction in schizophrenia: from basic neuroscience to clinical psychopharmacology. *Eur. Neuro psychopharmacol.* 18:773–786.

[15] Conn, P. J., Pin, J. P. (1997) Pharmacology and functions of metabotropic glutamate receptors. *Annu. Rev. Pharmacol. Toxicol.* 37:205–237.

[16] Schoepp, D. D., Jane, D. E., Monn, J. A. (1999) Pharmacological agents acting at subtypes of metabotropic glutamate receptors. *Neuropharmacology* 38:1431–1476.

[17] Vinson, P. N., Conn, P. J. (2012) Metabotropic glutamate receptors as therapeutic targets for schizophrenia. *Neuropharmacology* 62:1461–1472.

[18] Niswender, C. M., Conn, P. J. (2010) Metabotropic glutamate receptors: physiology, pharmacology, and disease. *Annu. Rev. Pharmacol. Toxicol.* 50:295–322.

[19] Nicoletti, F., Bockaert, J., Collingridge, G. L., Conn, P. J., Ferraguti, F., Schoepp, D. D., Wroblewski, J. T., Pin, J. P. (2011) Metabotropic glutamate receptors: from the workbench to the bedside. *Neuropharmacology* 60:1017–1041.

[20] Moghaddam, B., Adams, B., Verma, A., Daly, D. (1997) Activation of glutamatergic neurotransmission by ketamine: a novel step in the pathway from NMDA receptor blockade to dopaminergic and cognitive disruptions associated with the prefrontal cortex. *J. Neurosci.* 17:2921–2927.

[21] Adams, B., Moghaddam, B. (1998) Corticolimbic dopamine neurotransmission is temporally dissociated from the cognitive and locomotor effects of phencyclidine. *J. Neurosci.* 18:5545–5554.

[22] Moghaddam, B., Adams, B. W. (1998) Reversal of phencyclidine effects by a group II metabotropic glutamate receptor agonist in rats. *Science* 281:1349–1352.

[23] Neki, A., Ohishi, H., Kaneko, T., Shigemoto, R., Nakanishi, S., Mizuno, N. (1996) Pre- and postsynaptic localization of a metabotropic glutamate receptor, mGluR2, in the rat brain: an immunohistochemical study with a monoclonal antibody. *Neurosci. Lett.* 202: 197–200.

[24] Petralia, R. S., Wang, Y. X., Niedzielski, A. S., Wenthold, R. J. (1996) The metabotropic glutamate receptors, mGluR2 and mGluR3, show unique postsynaptic, presynaptic and glial localizations. *Neuroscience* 71:949–976.

[25] Tamaru, Y., Nomura, S., Mizuno, N., Shigemoto, R. (2001) Distribution of metabotropic glutamate receptor mGluR3 in the mouse CNS: differential location relative to pre- and postsynaptic sites. *Neuroscience* 106:481–503.

[26] Battaglia, G., Monn, J. A., Schoepp, D. D. (1997) In vivo inhibition of veratridine-evoked release of striatal excitatory amino acids by the group II metabotropic glutamate receptor agonist LY354740 in rats. *Neurosci. Lett.* 229:161–164.

[27] Marek, G. J., Wright, R. A., Schoepp, D. D., Monn, J. A., Aghajanian, G. K. (2000) Physiological antagonism between 5-hydroxytryptamine(2A) and group II metabotropic glutamate receptors in prefrontal cortex. *J. Pharmacol. Exp. Ther.* 292:76–87.

[28] Kew, J. N., Ducarre, J. M., Pflimlin, M. C., Mutel, V., Kemp, J. A. (2001) Activity-dependent presynaptic autoinhibition by group II metabotropic glutamate receptors at the perforant path inputs to the dentate gyrus and CA1. *Neuropharmacology* 40:20–27.

[29] Alexander, G. M., Godwin, D. W. (2005) Presynaptic inhibition of corticothalamic feedback by metabotropic glutamate receptors. *J. Neurophysiol.* 94:163–175.

[30] Rorick-Kehn, L. M., Johnson, B. G., Burkey, J. L., Wright, R. A., Calligaro, D. O., Marek, G. J., Nisenbaum, E. S., Catlow, J. T., Kingston, A. E., Giera, D. D., Herin, M. F., Monn, J. A., McKinzie, D. L., Schoepp, D. D. (2007) Pharmacological and pharmacokinetic properties of a structurally novel, potent, and selective metabotropic glutamate 2/3 receptor agonist: in vitro characterization of agonist (-)-(1R,4S,5S,6S)-4-amino-2-sulfonylbicyclo[3.1.0]-hexane-4,6-dicarboxylic acid (LY404039). *J. Pharmacol. Exp. Ther.* 321:308–317.

[31] Cartmell, J., Monn, J. A., Schoepp, D. D. (1999) The metabotropic glutamate 2/3 receptor agonists LY354740 and LY379268 selectively attenuate phencyclidine versus d-amphetamine motor behaviors in rats. *J. Pharmacol. Exp. Ther.* 291:161–170.

[32] Cartmell, J., Monn, J. A., Schoepp, D. D. (2000) The mGlu(2/3) receptor agonist LY379268 selectively blocks amphetamine ambulations and rearing. *Eur. J. Pharmacol.* 400:221–224.

[33] Rorick-Kehn, L. M., Johnson, B. G., Knitowski, K. M., Salhoff, C. R., Witkin, J. M., Perry, K. W., Griffey, K. I., Tizzano, J. P., Monn, J. A., McKinzie, D. L., Schoepp, D. D. (2007) In vivo pharmacological characterization of the structurally novel, potent, selective mGlu2/3 receptor agonist LY404039 in animal models of psychiatric disorders. *Psychopharmacology* (Berl.) 193:121–136.

[34] Patil, S. T., Zhang, L., Martenyi, F., Lowe, S. L., Jackson, K. A., Andreev, B. V., Avedisova, A. S., Bardenstein, L. M., Gurovich, I. Y., Morozova, M. A., Mosolov, S. N., Neznanov, N. G., Reznik, A. M., Smulevich, A. B., Tochilov, V. A., Johnson, B. G., Monn, J. A., Schoepp, D. D. (2007) Activation of mGlu2/3 receptors as a new approach to treat schizophrenia: a randomized Phase 2 clinical trial. *Nat. Med.* 13:1102–1107.

[35] Kinon, B. J., Zhang, L., Millen, B. A., Osuntokun, O. O., Williams, J. E., Kollack-Walker, S., Jackson, K., Kryzhanovskaya, L., Jarkova, N.; HBBI Study Group (2011) A multicenter, inpatient, phase 2, double-blind, placebo-controlled dose-ranging study of LY2140023 monohydrate in patients with DSM-IV schizophrenia. *J. Clin. Psychopharmacol.* 31:349–355.

[36] Lujan, R., Nusser, Z., Roberts, J. D., Shigemoto, R., Somogyi, P. (1996) Perisynaptic location of metabotropic glutamate receptors mGluR1 and mGluR5 on dendrites and dendritic spines in the rat hippocampus. *Eur. J. Neurosci.* 8:1488–1500.

[37] Ferraguti, F., Crepaldi, L., Nicoletti, F. (2008) Metabotropic glutamate 1 receptor: Current concepts and perspectives. *Pharmacol. Rev.* 60:536–581.

[38] Henry, S. A., Lehmann-Masten, V., Gasparini, F., Geyer, M. A., Markou, A. (2002) The mGluR5 antagonist MPEP, but not the mGluR2/3 agonist LY314582, augments PCP effects on prepulse inhibition and locomotor activity. *Neuropharmacology* 43: 1199–1209.

[39] Conn, P. J., Lindsley, C. W., Jones, C. K. (2009) Activation of metabotropic glutamate receptors as a novel approach for the treatment of schizophrenia. *Trends Pharmacol. Sci.* 30:25–31.

[40] Kunishima, N., Shimada, Y., Tsuji, Y., Sato, T., Yamamoto, M., Kumasaka, T., Nakanishi, S., Jingami, H., Morikawa, K. (2000) Structural basis of glutamate recognition by a dimeric metabotropic glutamate receptor. *Nature* 407:971–977.

[41] Varney, M. A., Suto, C. M. (2000) Discovery of subtype-selective metabotropic glutamate receptor ligands using functional HTS assays. *Drug Discov. Today: Suppl.* 1: 20–26.

[42] Lavreysen, H., Wouters, R., Bischoff, F., Nóbrega Pereira, S., Langlois, X., Blokland, S., Somers, M., Dillen, L., Lesage, A. S. (2004) JNJ16259685, a highly potent, selective and systemically active mGlu1 receptor antagonist. *Neuropharmacology* 47: 961–972.

[43] Suzuki, G., Kawamoto, H., Ohta, H. (2010) Development of a beta-lactamase reporter gene assay for metabotropic glutamate receptor 1 by using coexpression of glutamate transporter. *J. Biomol. Screen.* 15:148–158.

[44] Suzuki, G., Kimura, T., Satow, A., Kaneko, N., Fukuda, J., Hikichi, H., Sakai, N., Maehara, S., Kawagoe-Takaki, H., Hata, M., Azuma, T., Ito, S., Kawamoto, H., Ohta, H. (2007) Pharmacological characterization of a new, orally active and potent allosteric metabotropic glutamate receptor 1 antagonist, 4-[1-(2-fluoropyridin-3-yl)-5-methyl-1H-1,2,3-triazol-4-yl]-N-isopropyl-N-methyl-3,6-dihydropyridine-1(2H)-carboxamide (FTIDC). *J. Pharmacol. Exp. Ther.* 321:1144–1153.

[45] Comer, J. E. A. (2003) *Drug Bioavailability: Estimation of Solubility, Permeability, Absorption and Bioavailability* (van de Waterbeemd, H., Lennernäs, H., Artursson, P., eds), pp. 21–45, Wiley-VCH, Weinheim, Germany.

[46] Ito, S., Satoh, A., Nagatomi, Y., Hirata, Y., Suzuki, G., Kimura, T., Satow, A., Maehara, S., Hikichi, H., Hata, M., Kawamoto, H., Ohta, H. (2008) Discovery and biological profile of 4-(1-aryltriazol-4-yl)-tetrahydropyridines as an orally active new class of metabotropic glutamate receptor 1 antagonist. *Bioorg. Med. Chem.* 16:9817–9829.

[47] Ito, S., Hirata, Y., Nagatomi, Y., Satoh, A., Suzuki, G., Kimura, T., Satow, A., Maehara, S., Hikichi, H., Hata, M., Ohta, H., Kawamoto, H. (2009) Discovery and biological profile of isoindolinone derivatives as novel metabotropic glutamate receptor 1 antagonists: a potential treatment for psychotic disorders. *Bioorg. Med. Chem. Lett.* 19: 5310–5313.

[48] Satow, A., Suzuki, G., Maehara, S., Hikichi, H., Murai, T., Murai, T., Kawagoe-Takaki, H., Hata, M., Ito, S., Ozaki, S., Kawamoto, H., Ohta, H. (2009) Unique antipsychotic activities of the selective metabotropic glutamate receptor 1 allosteric antagonist 2-cyclopropyl-5-[1-(2-fluoro-3-pyridinyl)-5-methyl-1H-1,2,3-triazol-4-yl]-2,3-dihydro-1H-isoindol-1-one. *J. Pharmacol. Exp. Ther.* 330:179–190.

[49] Hikichi, H., Iwahori, Y., Murai, T., Maehara, S., Satow, A., Ohta, H. (2008) Face-washing behavior induced by the group I metabotropic glutamate receptor agonist (S)-3,5-DHPG in mice is mediated by mGlu1 receptor. *Eur. J. Pharmacol.* 586:212–216.

[50] Suzuki, G., Kawagoe-Takaki, H., Inoue, T., Kimura, T., Hikichi, H., Murai, T., Satow, A., Hata, M., Maehara, S., Ito, S., Kawamoto, H., Ozaki, S., Ohta, H. (2009) Correlation of receptor occupancy of metabotropic glutamate receptor subtype 1 (mGluR1) in mouse brain with in vivo activity of allosteric mGluR1 antagonists. *J. Pharmacol. Sci.* 110:315–325.

[51] Fukuda, J., Suzuki, G., Kimura, T., Nagatomi, Y., Ito, S., Kawamoto, H., Ozaki, S., Ohta, H. (2009) Identification of a novel transmembrane domain involved in the negative modulation of mGluR1 using a newly discovered allosteric mGluR1 antagonist, 3-cyclohexyl-5-fluoro-6-methyl-7-(2-morpholin-4-ylethoxy)-4H-chromen-4-one. *Neuropharmacology* 57:438–445.

[52] Rothman, R. B., Baumann, M. H. (2003) Monoamine transporters and psychostimulant drugs. *Eur. J. Pharmacol.* 479:23–40.

[53] Grant, K. M., LeVan, T. D., Wells, S. M., Li, M., Stoltenberg, S. F., Gendelman, H. E., Carlo, G., Bevins, R. A. (2012) Methamphetamine-associated psychosis. *J. Neuroimmune Pharmacol.* 7:113–139.

[54] Satow, A., Maehara, S., Ise, S., Hikichi, H., Fukushima, M., Suzuki, G., Kimura, T., Tanak, T., Ito, S., Kawamoto, H., Ohta, H. (2008) Pharmacological effects of the metabotropic glutamate receptor 1 antagonist compared with those of the metabotropic glutamate receptor 5 antagonist and metabotropic glutamate receptor 2/3 agonist in rodents: detailed investigations with a selective allosteric metabotropic glutamate receptor 1 antagonist, FTIDC [4-[1-(2-fluoropyridine-3-yl)-5-methyl-1H-1,2,3-triazol-4-yl]-N-isopropyl-N-methyl-3,6-dihydropyridine-1(2H)-carboxamide]. *J. Pharmacol. Exp. Ther.* 326:577–586.

[55] Geyer, M. A. (2008) Developing translational animal models for symptoms of schizophrenia or bipolar mania. *Neurotox. Res.* 14:71–78.

[56] Moghaddam, B., Javitt, D. (2012) From revolution to evolution: the glutamate hypothesis of schizophrenia and its implication for treatment. *Neuropsychopharmacology* 37:4–15.

[57] McCaughran, J. Jr, Mahjubi, E., Decena, E., Hitzemann, R. (1997) Genetics, haloperidol-induced catalepsy and haloperidol-induced changes in acoustic startle and prepulse inhibition. *Psychopharmacology* (Berl.) 134:131–139.

[58] Olivier, B., Leahy, C., Mullen, T., Paylor, R., Groppi, V. E., Sarnyai, Z., Brunner, D. (2001) The DBA/2J strain and prepulse inhibition of startle: a model system to test antipsychotics? *Psychopharmacology* (Berl.) 156:284–290.

[59] Hikichi, H., Nishino, M., Fukushima, M., Satow, A., Maehara, S., Kawamoto, H., Ohta, H. (2010) Pharmacological effects of metabotropic glutamate receptor ligands on prepulse inhibition in DBA/2J mice. *Eur. J. Pharmacol.* 639:99–105.

[60] Hikichi, H., Kaku, A., Karasawa, J., Chaki, S. (2013) Stimulation of metabotropic glutamate (mGlu) 2 receptor and blockade of mGlu1 receptor improve social memory impairment elicited by MK-801 in rats. *J. Pharmacol. Sci.* 122:10–16.

[61] Herdegen, T., Leah, J. D. (1998) Inducible and constitutive transcription factors in the mammalian nervous system: control of gene expression by Jun, Fos and Krox, and CREB/ATF proteins. *Brain Res. Brain Res. Rev.* 28:370–490.

[62] Ananth, J., Burgoyne, K. S., Gadasalli, R., Aquino, S. (2001) How do the atypical antipsychotics work? *J. Psychiatry Neurosci.* 26:385–394.

[63] Robertson, G. S., Matsumura, H., Fibiger, H. C. (1994) Induction patterns of Fos-like immunoreactivity in the forebrain as predictors of atypical antipsychotic activity. *J. Pharmacol. Exp. Ther.* 271:1058–1066.

[64] Robertson, G. S., Fibiger, H. C. (1996) Effects of olanzapine on regional c-Fos expression in rat forebrain. *Neuropsychopharmacology* 14:105–110.

[65] Sumner, B. E., Cruise, L. A., Slattery, D. A., Hill, D. R., Shahid, M., Henry, B. (2004) Testing the validity of c-fos expression profiling to aid the therapeutic classification of psychoactive drugs. *Psychopharmacology* (Berl.) 171:306–321.

[66] Deutch, A. Y., Duman, R. S. (1996) The effects of antipsychotic drugs on Fos protein expression in the prefrontal cortex: cellular localization and pharmacological characterization. *Neuroscience* 70:377–389.

[67] Suzuki, G., Satow, A., Ohta, H. (2010) Effect of CFMTI, an allosteric metabotropic glutamate receptor 1 antagonist with antipsychotic activity, on Fos expression in regions of the brain related to schizophrenia. *Neuroscience* 168:787–796.

[68] Pratt, J. A., Winchester, C., Egerton, A., Cochran, S. M., Morris, B. J. (2008) Modelling prefrontal cortex deficits in schizophrenia: implications for treatment. *Br. J. Pharmacol.* 153 (Suppl. 1):S465–S470.

[69] Andreasen, N. C., O'Leary, D. S., Flaum, M., Nopoulos, P., Watkins, G. L., Boles Ponto, L. L., Hichwa, R. D. (1997) Hypofrontality in schizophrenia: distributed dysfunctional circuits in neuroleptic-naïve patients. *Lancet* 349:1730–1734.

[70] Lacroix, L., Spinelli, S., White, W., Feldon, J. (2000) The effects of ibotenic acid lesions of the medial and lateral prefrontal cortex on latent inhibition, prepulse inhibition and amphetamine-induced hyperlocomotion. *Neuroscience* 97:459–468.

[71] Ohashi, K., Hamamura, T., Lee, Y., Fujiwara, Y., Suzuki, H., Kuroda, S. (2000) Clozapine- and olanzapine-induced Fos expression in the rat medial prefrontal cortex is mediated by beta-adrenoceptors. *Neuropsychopharmacology* 23:162–169.

[72] Dawe, G. S., Huff, K. D., Vandergriff, J. L., Sharp, T., O'Neill, M. J., Rasmussen, K. (2001) Olanzapine activates the rat locus coeruleus: in vivo electrophysiology and c-Fos immunoreactivity. *Biol. Psychiatry* 50:510–520.

[73] Ramirez, O. A., Wang, R. Y. (1986) Locus coeruleus norepinephrine-containing neurons: effects produced by acute and subchronic treatment with antipsychotic drugs and amphetamine. *Brain Res.* 362:165–170.

[74] Arnsten, A. F. (2004) Adrenergic targets for the treatment of cognitive deficits in schizophrenia. *Psychopharmacology* (Berl.) 174:25–31.

[75] Li, X. M., Perry, K. W., Wong, D. T., Bymaster, F. P. (1998) Olanzapine increases in vivo dopamine and norepinephrine release in rat prefrontal cortex, nucleus accumbens and striatum. *Psychopharmacology* (Berl.) 136:153–161.

[76] Westerink, B. H., de Boer, P., de Vries, J. B., Kruse, C. G., Long, S. K. (1998) Antipsychotic drugs induce similar effects on the release of dopamine and noradrenaline in the medial prefrontal cortex of the rat brain. *Eur. J. Pharmacol.* 361: 27–33.

[77] Allison, D. B., Casey, D. E. (2001) Antipsychotic-induced weight gain: a review of the literature. *J. Clin. Psychiatry* 62 (Suppl. 7):22–31.

[78] Nasrallah, H. (2003) A review of the effect of atypical antipsychotics on weight. *Psychoneuroendocrinology* 28 (Suppl. 1):83–96.

[79] Goudie, A. J., Cooper, G. D., Halford, J. C. (2005) Antipsychotic-induced weight gain. *Diabetes Obes. Metab*. 7:478–487.

[80] Fadel, J., Bubser, M., Deutch, A. Y. (2002) Differential activation of orexin neurons by antipsychotic drugs associated with weight gain. *J. Neurosci*. 22:6742–6746.

[81] Wadenberg, M. L. (1996) Serotonergic mechanisms in neuroleptic-induced catalepsy in the rat. *Neurosci. Biobehav. Rev*. 20:325–339.

[82] Wadenberg, M. L., Soliman, A., VanderSpek, S. C., Kapur, S. (2001) Dopamine D(2) receptor occupancy is a common mechanism underlying animal models of antipsychotics and their clinical effects. *Neuropsychopharmacology* 25:633–641.

[83] Aiba, A., Kano, M., Chen, C., Stanton, M. E., Fox, G. D., Herrup, K., Zwingman, T. A., Tonegawa, S. (1994) Deficient cerebellar long-term depression and impaired motor learning in mGluR1 mutant mice. *Cell* 79:377–388.

[84] Conquet, F., Bashir, Z. I., Davies, C. H., Daniel, H., Ferraguti, F., Bordi, F., Franz-Bacon, K., Reggiani, A., Matarese, V., Condé, F., Collingridge, G. L., Crépel, F. (1994) Motor deficit and impairment of synaptic plasticity in mice lacking mGluR1. *Nature* 372: 237–243.

[85] Kano, M., Hashimoto, K., Kurihara, H., Watanabe, M., Inoue, Y., Aiba, A., Tonegawa, S. (1997) Persistent multiple climbing fiber innervation of cerebellar Purkinje cells in mice lacking mGluR1. *Neuron* 18:71–79.

[86] Nakao, H., Nakao, K., Kano, M., Aiba, A. (2007) Metabotropic glutamate receptor subtype-1 is essential for motor coordination in the adult cerebellum. *Neurosci. Res*. 57: 538–543.

[87] Sillevis Smitt, P., Kinoshita, A., De Leeuw, B., Moll, W., Coesmans, M., Jaarsma, D., Henzen-Logmans, S., Vecht, C., De Zeeuw, C., Sekiyama, N., Nakanishi, S., Shigemoto, R. (2000) Paraneoplastic cerebellar ataxia due to autoantibodies against a glutamate receptor. *N. Engl. J. Med*. 342:21–27.

[88] Varty, G. B., Grilli, M., Forlani, A., Fredduzzi, S., Grzelak, M. E., Guthrie, D. H., Hodgson, R. A., Lu, S. X., Nicolussi, E., Pond, A. J., Parker, E. M., Hunter, J. C., Higgins, G. A., Reggiani, A., Bertorelli, R. (2005) The antinociceptive and anxiolytic-like effects of the metabotropic glutamate receptor 5 (mGluR5) antagonists, MPEP and MTEP, and the mGluR1 antagonist, LY456236, in rodents: a comparison of efficacy and side-effect profiles. *Psychopharmacology* (Berl.) 179:207–217.

[89] Brody, S. A., Conquet, F., Geyer, M. A. (2003) Disruption of prepulse inhibition in mice lacking mGluR1. *Eur. J. Neurosci*. 18:3361–3366.

[90] Steckler, T., Oliveira, A. F., Van Dyck, C., Van Craenendonck, H., Mateus, A. M., Langlois, X., Lesage, A. S., Prickaerts, J. (2005) Metabotropic glutamate receptor 1 blockade impairs acquisition and retention in a spatial water maze task. *Behav. Brain Res*. 164:52–60.

[91] Gravius, A., Pietraszek, M., Schäfer, D., Schmidt, W. J., Danysz, W. (2005) Effects of mGlu1 and mGlu5 receptor antagonists on negatively reinforced learning. *Behav. Pharmacol*. 16:113–121.

[92] Gravius, A., Barberi, C., Schäfer, D., Schmidt, W. J., Danysz, W. (2006) The role of group I metabotropic glutamate receptors in acquisition and expression of contextual

and auditory fear conditioning in rats - a comparison. *Neuropharmacology* 51:1146–1155.

[93] Aiba, A., Chen, C., Herrup, K., Rosenmund, C., Stevens, C. F., Tonegawa, S. (1994) Reduced hippocampal long-term potentiation and context-specific deficit in associative learning in mGluR1 mutant mice. *Cell* 79:365–375.

[94] Steckler, T., Lavreysen, H., Oliveira, A. M., Aerts, N., Van Craenendonck, H., Prickaerts, J., Megens, A., Lesage, A. S. (2005) Effects of mGlu1 receptor blockade on anxiety-related behaviour in the rat lick suppression test. *Psychopharmacology* (Berl.) 179:198–206.

[95] Sevostianova, N., Danysz, W. (2006) Analgesic effects of mGlu1 and mGlu5 receptor antagonists in the rat formalin test. *Neuropharmacology* 51:623–630.

[96] Sukhotina, I. A., Dravolina, O. A., Novitskaya, Y., Zvartau, E. E., Danysz, W., Bespalov, A. Y. (2008) Effects of mGlu1 receptor blockade on working memory, time estimation, and impulsivity in rats. *Psychopharmacology* (Berl.) 196:211–220.

[97] Murai, T., Okuda, S., Tanaka, T., Ohta, H. (2007) Characteristics of object location memory in mice: Behavioral and pharmacological studies. *Physiol. Behav.* 90:116-124.

[98] Hostetler, E. D., Eng, W., Joshi, A. D., Sanabria-Bohórquez, S., Kawamoto, H., Ito, S., O'Malley, S., Krause, S., Ryan, C., Patel, S., Williams, M., Riffel, K., Suzuki, G., Ozaki, S., Ohta, H., Cook, J., Burns, H. D., Hargreaves, R. (2011) Synthesis, characterization, and monkey PET studies of [^{18}F]MK-1312, a PET tracer for quantification of mGluR1 receptor occupancy by MK-5435. *Synapse* 65:125–135.

[99] Peng, B. H., Levin, C. S. (2010) Recent developments in PET instrumentation. *Curr. Pharm. Biotechnol.* 11:555–571.

[100] Matthews, P. M., Rabiner, E. A., Passchier, J., Gunn, R. N. (2012) Positron emission tomography molecular imaging for drug development. *Br. J. Clin. Pharmacol.* 73:175–186.

[101] Ferraguti, F., Shigemoto, R. (2006) Metabotropic glutamate receptors. *Cell Tissue Res.* 326:483–504.

[102] Kawamoto, H., Ito, S., Satoh, A., Nagatomi, Y., Hirata, Y., Suzuki, G., Kimura, T., Satow, A., Maehara, S., Hikichi, H., Hata, M., Ohta, H. (2009) Abstracts of Papers, *The 129th Annual meeting of the Pharmaceutical Society of Japan*. 26Q-am214.

[103] Satoh, A., Nagatomi, Y., Hirata, Y., Ito, S., Suzuki, G., Kimura, T., Maehara, S., Hikichi, H., Satow, A., Hata, M., Ohta, H., Kawamoto, H. (2009) Discovery and in vitro profiles of 4-fluoro-N-[4-[6-(isopropylamino)pyrimidin-4-yl]-1,3-thiazol-2-yl]-N-methylbenzamide as novel class of an orally active metabotropic glutamate receptor 1 (mGluR1) antagonist. *Bioorg. Med. Chem. Lett.* 19:5464–5468.

[104] Yamasaki, T., Fujinaga, M., Yoshida, Y., Kumata, K., Yui, J., Kawamura, K., Hatori, A., Fukumura, T., Zhang, M. R. (2011) Radiosynthesis and preliminary evaluation of 4-[^{18}F]fluoro-N-[4-[6-(isopropylamino)pyrimidin-4-yl]-1,3-thiazol-2-yl]-N-methylbenzamide as a new positron emission tomography ligand for metabotropic glutamate receptor subtype 1. *Bioorg. Med. Chem. Lett.* 21:2998–3001.

[105] Yamasaki, T., Fujinaga, M., Kawamura, K., Yui, J., Hatori, A., Ohya, T., Xie, L., Wakizaka, H., Yoshida, Y., Fukumura, T., Zhang, M. R. (2012) In vivo measurement of the affinity and density of metabotropic glutamate receptor subtype 1 in rat brain using ^{18}F-FITM in small-animal PET. *J. Nucl. Med.* 53:1601–1607.

[106] Yamasaki, T., Fujinaga, M., Maeda, J., Kawamura, K., Yui, J., Hatori, A., Yoshida, Y., Nagai, Y., Tokunaga, M., Higuchi, M., Suhara, T., Fukumura, T., Zhang, M. R. (2012) Imaging for metabotropic glutamate receptor subtype 1 in rat and monkey brains using PET with [^{18}F]FITM. *Eur. J. Nucl. Med. Mol. Imaging* 39:632–641.

[107] Toyohara, J., Sakata, M., Fujinaga, M., Yamasaki, T., Oda, K., Ishii, K., Zhang, M. R., Moriguchi Jeckel, C. M., Ishiwata, K. (2012) Preclinical and the first clinical studies on [^{11}C]ITMM for mapping metabotropic glutamate receptor subtype 1 by positron emission tomography. *Nucl. Med. Biol.* 40:214–220.

[108] Toyohara, J., Sakata, M., Oda, K., Ishii, K., Ito, K., Hiura, M., Fujinaga, M., Yamasaki, T., Zhang, M. R., Ishiwata, K. (2013) Initial Human PET studies of metabotropic glutamate receptor type 1 ligand ^{11}C-ITMM. *J. Nucl. Med.* 54:1302–1307.

In: Metabotropic Glutamate Receptors
Editor: M. Foster Olive

ISBN: 978-1-63117-451-3

Chapter 7

Group I and Group II Metabotropic Glutamate Receptors: Role in Pathophysiology and Treatment of Major Depressive Disorder

***Elizabeth N. Holly*[1]*, Amber L. LaCrosse*[2] *and Todd M. Hillhouse*[3,*]**
[1]Tufts University, Medford, MA, US
[2]Arizona State University, Phoenix, Arizona, US
[3]Department of Psychology, Virginia Commonwealth University, Richmond, VA, US

Abstract

Major depressive disorder (MDD) is a chronic, recurring illness with a lifetime prevalence of 16.5% in the United States [1]. For the past 50 years, drugs that increase the synaptic availability of monoamines (norepinephrine, serotonin, and dopamine) have been used to treat depression [2]. However, these traditional antidepressants take at least 2 weeks for meaningful improvement in core depressive symptoms, suggesting that downstream neural adaptations as a result of monoaminergic elevation may be more responsible for therapeutic effects. Recently, the role of the glutamatergic system in the pathophysiology and treatment of depression has received increased attention.

Disruptions within the glutamatergic system have been observed in individuals suffering from depression, particularly abnormalities in glutamate/glutamine cycling and glutamate receptors. Furthermore, elevated levels of mGluR2/3 and reduced levels of mGluR5 protein have been demonstrated in the prefrontal cortex of patients with MDD, and drugs targeting Group I and Group II mGluRs have shown promise in preclinical studies. This chapter will review Group I and Group II metabotropic glutamate receptors, their role in the pathophysiology of depression, and new potential avenues for treatment.

* Corresponding author: Todd M. Hillhouse, Virginia Commonwealth University, Department of Psychology, 806 West Franklin Street, Richmond, VA 23284. E-mail: hillhousetm@vcu.edu.

Keywords: Metabotropic glutamate receptors, major depressive disorder, mGluR1, mGluR5, mGluR2/3, glutamate, antidepressant

Introduction

Major depressive disorder (MDD) is the most commonly occurring mood disorder in both the United States and European countries with a prevalence rate of 7.1% and 6.9%, respectively [3, 4]. Due to the prevalence, MDD has created a substantial economic burden for these countries as well as others worldwide. For example, the total economic burden costs was USD$83.1 billion for the United States in 2000 [5] and €105.6 billion for European countries in 2004 [6]. Despite the availability of treatment, the World Health Organization has estimated that MDD will be the second leading cause of disability worldwide by 2020 [7]. For more than 50 years, drugs that increase levels of monoamines (norepinephrine, serotonin, and dopamine) at the synapse, such as selective serotonin reuptake inhibitors (SSRIs), serotonin and norepinephrine reuptake inhibitors (SNRIs), and tricyclic antidepressants have been used to treat depression [2]. The primary mechanism of these antidepressant drugs is increasing the presence of monoamines in the synapse, allowing for greater postsynaptic cell activation. Most the clinically available antidepressant drugs work through the same mechanisms of action. However, a major limitation with the current antidepressant drugs is the delayed onset (4-12 weeks) before symptom remission [8]. This suggests that downstream neuronal adaptations, likely a result of increased synaptic monoamines, may be responsible for therapeutic effects [9]. As such, efforts to develop more effective antidepressants with fewer side effects and faster onset of clinical benefits have begun to target alternative neural systems.

Glutamate Dysfunction in MDD

The glutamatergic system has emerged as a potential target to improve clinical efficacy and onset of remission for the treatment of MDD. Both clinical and postmortem research has provided evidence of glutamatergic dysfunction in patients diagnosed with MDD (see Table 1). For example, elevated levels of glutamate, glutamine, and glycine have been found patients with MDD using indirect measures such as plasma, serum, or cerebrospinal fluid [10-16]. Interestingly, antidepressant treatment has been found to reduce the elevated serum glutamate levels in MDD patients [17]. Hashimoto and colleagues found elevated levels of glutamate in the frontal cortex of postmortem tissue from patients with MDD [18].

Additionally, the glia high affinity glutamate transporter (EAAT2), the main source the glutamate reuptake, has been shown to be down-regulated in the postmortem brain tissue in MDD patients [19, 20]. In contrast, several studies have found reduced levels of excitatory amino acids (i.e. glutamate, glutamine, and glycine) using the same direct measures mentioned above [10, 21, 22].

These indirect measures of glutamate in CSF, plasma, and serum are difficult to interpret, as these disruptions may be due to medication, metabolic effects, or the inability of these methods to distinguish between central and peripheral glutamate.

Table 1. Evidence for glutamate system dysfunction in patients with MDD

Interest	Region of Analysis	Change	Reference
Glutamate/ Glutamine	Serum	Increased	[11]
		No difference	[17]
	Plasma	Increased	[10, 13-15]
	CSF	Increased	[12, 16]
		Decreased	[22]
	Occipital cortex	Increased	[23]
	Anterior cingulate cortex	Decreased	[24, 25]
	Amygdala	Decreased	[26]
	Hippocampus	Decreased	[27]
	Dorsolateral prefrontal cortex	Decreased	[28]
	Dorsomedial and ventromedial prefrontal cortex	Decreased	[29]
mGluR5	Hippocampus, frontal, temporal, and parietal cortices	Decreased	[31]
mGluR2/3	Prefrontal cortex	Increased	[32]

As such, proton magnetic resonance spectroscopy (^{1}H-MRS) has been used to more directly measure neurotransmitter concentration, specifically glutamate and glutamine, in adults with severe depression. ^{1}H-MRS studies have shown increased glutamate levels in the occipital cortex [23], but reduced Glx (glutamate and glutamine) in the anterior cingulate cortex [24, 25], amygdala [26], hippocampus [27], left dorsolateral prefrontal cortex [28], dorsomedial and ventromedial prefrontal cortex [29] of adult MDD patients. Moreover, a 19% reduction in glutamate and glutamine has been observed in adolescent patients with severe MDD [30].

In combination with significant alterations in glutamate, levels of glutamatergic receptor expression are altered in patients with MDD. In addition to differences in N-methyl-D-aspartate (NMDA) receptor levels [33-35], several studies have shown a reduction in the expression of type 5 metabotropic glutamate receptors (mGluR5) in patients diagnosed with MDD. A positron emission tomography (PET) scan study found a decrease in mGuR5 distribution volume ratio (DVR) in multiple brain regions including the frontal, temporal, and parietal cortices [31].

Specifically, the study found an 8.8% decrease in DVR for the prefrontal cortex (Brodmann's area 10) in MDD patients compared to healthy controls. In a complementary study using postmortem tissue, this same research group found a reduction in mGluR5 expression in the prefrontal cortex. Together these results suggest the reductions in DVR are likely due to a reduction in mGluR5 expression.

Furthermore, the PET scan study revealed that mGluR5 binding in the hippocampus negatively correlated with the depressive symptoms [31]. Disruptions have also been reported within Group II mGluRs in MDD patients. In a postmortem analysis of MDD patient brain tissue, compared to healthy controls patients with MDD had a 67% increase in type 2/3 mGluRs (mGluR2/3) expression in the prefrontal cortex [32].

Taken together, these results may suggest a bi-directional glutamatergic dysfunction in MDD, with disruptions both in glutamate/glutamine cycling and glutamatergic receptor expression.

Glutamate and mGluR Physiology

Glutamate is the major excitatory neurotransmitter in the central nervous system (CNS), and is present at over 60% of synapses. The fact that glutamate acts as a rapid excitatory transmitter, combined with its prevalence, suggests that it may be involved in abnormal circuitries, and may facilitate or even cause disease symptoms.

Like many neurotransmitters, glutamate acts on two types of receptors, ionotropic and metabotropic. Ionotropic glutamate receptors include α-amino-3-hydroxy-5-methyl-4-isoxazolepropionic acid (AMPA), kainate, and NMDA receptors. These receptors are ion channels that are permeable to cations, and function by allowing Na^+ and Ca^{2+} to enter the cell, thus causing depolarization and other intracellular changes. Eight subtypes of metabotropic glutamate receptors (mGluR1-8) are divided into Group I (mGluR1 and mGluR5), Group II (mGluR2 and mGluR3), and Group III (mGluR4, mGluR6, mGluR7, and mGluR8) based upon their homology and function.

These receptors are G-protein-coupled receptors (GPCRs), and function by either influencing intracellular second messenger formation, and/or releasing G protein subunits which influence ion channel activity [36-38]. Since assessing all glutamate receptors and their implications with major depressive disorder would be outside the scope of this chapter, we will focus only on Group I and Group II metabotropic receptors.

mGluRs are densely expressed in areas of the brain necessary for cognition, such as the PFC, striatum and hippocampus [39]. Group I mGluRs (mGluR1 and mGluR5) are usually found postsynaptically [40-42], and generally function to enhance glutamatergic excitation. Their intracellular signaling mechanisms stimulate phospholipase C (PLC) which forms diacylglycerol (DAG) and inositol triphosphate (IP_3).

DAG activates protein kinase C (PKC), while IP_3 exerts further downstream effects by binding to receptors on the endoplasmic reticulum causing intracellular Ca^{2+} release. PKC exerts further downstream effects by phosphorylating calcium/calmodulinCREB and other signaling molecules which can eventually lead to altered gene transcription [40-42].

mGluR5 receptors are physically linked to NMDA receptors through their intracellular anchors such as post synaptic density (PSD) proteins and synaptic associated protein (SAP) 102 [43].

Through its activation of PKC, mGluR5 activation also increases the phosphorylation of NMDA receptor subunits, thereby indirectly increasing the probability of NMDA receptor channel opening.

This physical and biochemical connection allows the stimulation of Group I mGluRs to activate NMDA receptors without inducing desensitization or excitotoxicity [44, 45]. Increased NMDA receptor functioning may benefit cognition via the upregulation of AMPA receptors which facilitate learning and memory through strengthening LTP [46].

Group II mGluRs include mGluR2 and mGluR3. Unlike Group I mGluRs, Group II mGluRs are predominantly located presynaptically [40-42].

Additionally, the intracellular signaling mechanisms of Group II mGluRs involve releasing inhibitory G protein subunits (G_i/G_o), which inhibit adenylate cyclase (AC) and reduce the formation of cyclic adenosine monophosphate (cAMP), which ultimately results in reduced neurotransmitter release [40-42].

Glutamate acts as the endogenous ligand to all mGluRs, however numerous antagonist and agonist ligands have been synthesized to bind to the orthogonal binding site of mGluRs [47]. Also, numerous positive (PAMs) and negative (NAMs) allosteric modulators have been synthesized to bind within the GPCR transmembrane domains, which modulate receptor functioning by either potentiating or attenuating orthogonal binding [47].

Animal Models of Depression

While robust animal models of depression are lacking, there are some paradigms that have proven sensitive to clinically active antidepressant drugs and may provide insight into the pathophysiology of depression. Describing in detail the various animal models of depression is beyond the scope of this chapter (see [48] for detailed review).

Some models, such as olfactory bulbectomy, do not necessarily result in a depressive or anhedonic-like phenotype, but rather are characterized by their high predictive validity in that they produce distinct symptoms that are exclusively reversed by clinically effective antidepressant drugs.

Other models, such as chronic mild stress, prenatal restraint stress, and chronic corticosterone administration, induce deficits in behavioral tasks that have been linked to anhedonia.

The most common of these behavioral assays to screen for antidepressant activity include the forced swim test (FST) and tail suspension test (TST).

In both assays, immobility has been construed as avolition or "behavioral despair", and many clinically relevant antidepressants have been shown to reduce immobility time.

Disruptions in the mGluR System as a Consequence of Animal Models of Depression

In line with clinical data indicating disruptions in mGluR levels, preclinical studies have also found disruptions within both Group I and Group II mGluRs in animal models of depression (see Table 2). Chronic mild stress has been shown to upregulate mGluR5 in the CA1 region of the hippocampus while mGluR5 is downregulated in the CA3 region [49]. Similarly, hippocampal mGluR5 levels are also reduced after prenatal restraint stress [50] and chronic corticosterone treatment [51], while hippocampal mGluR1a is upregulated after olfactory bulbectomy [52]. Flinders Sensitive Line (FSL) rats are a putative genetic animal model of depression, with these spontaneously depressed rats showing many of the distinct features of MDD.

Much like the receptor changes found in behavioral models of depression, FSL rats also display decreased mGluR5 density in a variety of brain regions, including the hippocampus, ventral tegmental area, and nucleus accumbens [53].

Table 2. mGluR disruptions in animal models of depression

Receptor	Model	Region(s)	Change	Reference
Group I mGluR1	OB	CA1	Increased	[52]
mGluR5	Chronic mild stress	CA1 CA3	Increased Decreased	[49]
	Prenatal restraint stress	Hippocampus	Decreased	[50]
	Chronic corticosterone	Hippocampus	Decreased	[51]
	FSL	Hippocampus, VTA, NAC	Decreased	[53]
Group II				
mGluR2/3	OB	Hippocampus	Decreased	[52]
	Prenatal restraint stress	Hippocampus	Decreased	[50]
	FSL	Hippocampus	Decreased	[54]

mGluR=metabotropic glutamate receptor; OB=olfactory bulbectomy; FSL=Flinder's Sensitive Line; VTA=ventral tegmental area; NAC=nucleus accumbens.

In addition to altering Group I mGluR expression, both olfactory bulbectomy and prenatal restraint stress reduce mGluR2/3 levels in the rodent hippocampus, while FSL rats similarly express reduced mGluR2/3 levels in the hippocampus [50, 52, 54].

Overall, there has been very little work detailing mGluR disruptions in the various animal models of MDD, yet the studies that have been reported show distinct alterations within Group I and Group II receptor expression, in line with postmortem tissue from MDD patients.

Disruptions in the mGluR System as a Consequence of Chronic Antidepressant Administration

In addition to the above-mentioned alterations in receptor expression within animal models, there are a handful of studies that have demonstrated significant alterations in mGluR expression as a result of chronic antidepressant treatment in both animal models of depression and normal, otherwise unmanipulated animals (see Table 3). Chronic imipramine treatment upregulates both mGluR1 and mGluR5 in the rat hippocampus and reverses olfactory bulbectomy-induced upregulation of mGluR1 [52, 55-57].

However, while Group I mGluRs may be upregulated within the hippocampus after chronic antidepressant treatment, Zahorodna and Bijak [58] demonstrated that 14 days of twice daily imipramine (10 mg/kg, ip), hippocampal CA1 neurons attenuated the enhancing effect of dihydroxyphenylglycine (DHPG), a Group I mGluR agonist, on population spikes, depolarization, and decreased afterhypolarization.

This indicates that while there may be alterations within the Group I mGluR distribution as a result of chronic antidepressant treatment, the potential decreased function of the receptors should not be ignored.

Thus far, there have been only three reports of chronic antidepressant treatment on Group II mGluRs.

Table 3. Effects of chronic antidepressant treatment in animals

Receptor	Region(s)	Drug	Change	Reference
Group I	CA1	Imipramine	Decreased function	[58]
mGluR1	Hippocampus	Imipramine	Increased	[55, 56]
	Hippocampus	Amitriptyline	Recovered OB-induced increase	[52]
mGluR5	Hippocampus	Imipramine	Increased	[57]
Group II				
mGluR2/3	Prefrontal cortex	Fluoxetine	None	[32]
	Hippocampus, nucleus accumbens, cortex, striatum	Imipramine	Increased	[55]
	Hippocampus	Amitriptyline	Recovered OB-induced decrease	[52]

mGluR=metabotropic glutamate receptor; OB=olfactory bulbectomy.

In one study, rhesus monkeys were treated chronically with fluoxetine to determine if the previously mentioned mGluR2/3 increases found in human postmortem tissue were a result of antidepressant medication, but no significant changes were found between fluoxetine and placebo treated monkeys. However, others demonstrated widespread mGluR2/3 upregulation following chronic imipramine treatment, with increased expression in the hippocampus, nucleus accumbens, cortex, and corpus striatum [55]. Similarly, the decreased expression of mGluR2/3 in mice after olfactory bulbectomy has been reversed with chronic amitriptyline treatment [52]. Interestingly, much like what has been observed with Group I mGluRs, chronic imipramine also inhibits the function of mGluR2/3 in cortical slices [59]. In light of these recent reports, it seems that the efficacy of SSRIs and tricyclic antidepressants may at least partially depend on alterations within the metabotropic glutamate receptor system. While there may be up or down regulation of receptor expression, it is important to note that chronic imipramine treatment has been demonstrated to inhibit both Group I and Group II mGluR receptor function. Thus, directly targeting or inhibiting these receptors may provide a novel avenue for treatment.

Antidepressant Actions of Drugs that Target mGluRs

Due to the growing preclinical and clinical literature pointing to a key role of glutamatergic systems in the pathophysiology and treatment of MDD, and the disruptions within mGluRs in both MDD patients and preclinical models of depression, mGluRs have proven an attractive target for therapeutic intervention. Thus far, drugs targeting mGluR1 have not received significant attention, in part due to their minimal disruption in the disorder (see Table 4). However, the mGluR1 antagonist 3-ethyl-2-methyl-quinolin-6-yl)-(4-methoxy-cyclohexyl)-methanone methane sulfonate (EMQMCM) has been shown to reduce immobility and increase escape behaviors in the FST [60-62], while another mGluR1 antagonist, 1-aminoindan-1,5-dicarboxylic acid (AIDA), decreases immobility time in the TST [60, 63]. mGluR5, on the other hand, is proving to be a more promising target. Tatarcsyńska and colleagues [64] first showed that the mGluR5 antagonist 2-methyl-6-(phenylethynyl)pyridine (MPEP) decreased immobility time in the TST, while having no effect in the FST. Others have shown that MPEP decreases immobility in both FST and TST, as well as reverses olfactory bulbectomy-induced passive avoidance deficits [60, 65-67].

Another study reported decreased immobility in FST after acute MPEP, and decreased escape failures in a learned helplessness task after chronic MPEP administration [68].

Another selective mGluR5 antagonist, 3-((2-methyl-1,3-thiazol-4-yl)ethynyl)pyridine (MTEP), has been shown to decrease immobility in both TST and FST, and chronic MTEP administration reverses olfactory bulbectomy-induced deficits [69].

Others have also found similar antidepressant effects of MTEP, including reduced immobility in FST and TST after acute treatment [60-62, 65]. A later study demonstrated that these effects are dependent on mGluR5 interactions with NMDA receptors [70]. Recently, a novel mGluR5 negative allosteric modulator, 4-(difluoromethoxy)-3-(pyridin-2-ylethynyl) phenyl)(5H-pyrrolo[3,4-b]pyridin-6(7H)-yl)methanone (GRN-529), has been synthesized, and it has also shown antidepressant activity in both TST and FST [71].

Similarly, there are limited data on the clinical efficacy of mGluR5 ligands for treating MDD. A 5-week double blind study evaluated the anxiolytic and antidepressant effects of fenobam, an mGluR5 NAM [72], in patients with severe anxiety. Fenobam produced both anxiolytic and antidepressant effects compared to vehicle by reducing anxiety and depression symptoms measured by the Hamilton Rating Scale for Anxiety and Hamilton Rating Scale for Depression, respectively [73]. The AstraZeneca mGluR5 NAM AZD2066 underwent Phase II clinical trials, but AZD2066 was found to be no more effective for the treatment of MDD than controls (i.e., placebo and duloxetine).

Table 4. Antidepressant effects of Group I mGluR compounds in preclinical models

Compound	Model/Test	Result	Reference
mGluR1 antagonists			
EMQMCM	FST	Decreased immobility	[60-62]
AIDA	TST	Decreased immobility	[60, 63]
mGluR5 antagonists			
MPEP	FST	Decreased immobility	[65, 68]
	TST	Decreased immobility	[60, 64]
	OB	Reversed passive avoidance deficits	[66, 67]
	Learned helplessness	Reduced escape failures	[68]
MTEP	FST	Decreased immobility	[60-62, 65]
	TST	Decreased immobility	[60, 69]
	OB	Reversed passive avoidance deficits	[69]
mGluR5 NAMs			
GRN-529	TST	Decreased immobility	[71]
	FST	Decreased immobility	[71]

mGluR=metabotropic glutamate receptor; NAM=negative allosteric modulator; FST=forced swim test; TST=tail suspension test; OB=olfactory bulbectomy.

Importantly, no serious side effects were reported (see http://clinicaltrials.gov/ct2/show/study/ NCT01145755). Roche also has developed an mGluR5 NAM (RO4917523) that has completed Phase II clinical trials for treatment-resistant patients with MDD (see http://clinical trials.gov/ct2/show/ study/NCT00809562 – no data available). This same compound is currently undergoing Phase II clinical trials as an adjunctive therapy in patients with MDD and inadequate response to ongoing antidepressant treatment (see http://www.clinicaltrials.gov/ct2/show/ NCT01437657).

Drugs targeting group II mGluRs are also receiving growing attention for possible MDD treatment, particularly for treatment resistant depression (TRD, see Table 5). Chronic corticosterone treatment has been heralded as a model for TRD, as traditional antidepressants such as fluoxetine or desipramine cannot reverse anhedonic-like deficits [74].

The mGluR2/3 antagonists MGS0039 and LY341495 have been shown to decrease the chronic corticosterone-induced increase in immobility in the FST [74, 75]. These two compounds have also decreased immobility in both TST and FST in otherwise untreated rodents [76-78]. Interestingly, AMPA and not serotonin receptor activation is required for these effects [76], and LY341495 induces intracellular pathways similar to those mediating the antidepressant efficacy of ketamine [79]. A negative allosteric modulator of mGluR2/3 (RO4491533) has recently been developed, and both RO4491533 and LY341495 have been shown to decrease immobility time in both TST and FST in C57BL6/J and Helpless [a genetically selective strain of mice demonstrating a depression-like phenotype) mice [80].

mGluR2/3 NAMs and antagonists are also proceeding into clinical trials. The Roche mGluR2/3 NAM (RO4995819; a.k.a. RG1578) has undergone several safety and tolerability Phase I clinical trials (for full list see http://www.roche-trials.com/resultsByProductGet.action?productName=RO4995819), but no published results are currently available. Phase II clinical trials are underway to assess the effects of RO4995819 as an adjunctive treatment in patients with MDD and inadequate response to ongoing antidepressant treatment (see http://www.clinicaltrials.gov/ct2/show/NCT01457677). BrainCells Inc has developed an mGluR2/3 antagonist (BCI-632; a.k.a. MGS0039) and several prodrugs (BCI-838, BCI-1038, BCI-1206, BCI-1283) that have completed Phase I clinical trials (see http://clinicaltrials.gov/show/ NCT 01546051 or for full list see http://www.braincellsinc.com/pipeline/bci-632).

These drugs have been shown to have poor gastrointestinal permeability resulting in low oral bioavailability.

Conclusion

In conclusion, there is mounting evidence for glutamatergic dysfunction within MDD. This includes circulating glutamate levels, NMDA and AMPA dysregulation (not reviewed here), and disruption of mGluR expression and function. Ultimately, both Group I and Group II mGluRs are proving to be very promising targets for therapeutic intervention in MDD.

Group I mGluR manipulation has demonstrated marginal success with the AstraZeneca mGluR5 NAM (AZD2066), which continued to Phase II clinical trials, but ultimately was found to be no more effective for the treatment of MDD than controls. Roche mGluR5 NAM (RO4917523) has shown great promise with the completion of Phase II clinical trials for treatment-resistant patients with MDD.

Table 5. Antidepressant effects of Group II mGluR2/3 compounds in preclinical models

Compound	Model/Test	Result	Reference
Antagonists			
MGS0039	FST after chronic cort.	Decreased the increased immobility	[74]
	FST	Decreased immobility	[76, 78]
	TST	Decreased immobility	[78, 81]
	Learned helplessness	Decreased escape failures	[82]
LY341496	FST after chronic cort.	Decreased the increased immobility	[74, 75]
	FST	Decreased immobility	[77, 78, 80]
	TST	Decreased immobility	[78-80, 83]
NAMs			
RO4491533	FST	Decreased immobility	[80]
	TST	Decreased immobility	[80]

mGluR=metabotropic glutamate receptor; NAM=negative allosteric modulator; FST=forced swim test; TST=tail suspension test; cort.=corticosterone.

At the time of this chapter, it is different to interpret the clinical efficacy of Group II mGluRs in patients with MDD because there is no detailed published data on the clinical trials that have been completed. Even with the lack of published results, the mGluR2/3 NAM (RO4995819) is currently undergoing Phase II clinical trials for the treatment of MDD suggesting positive results obtained during Phase I clinical trials.

Patients diagnosed with MDD have been shown to have both elevated and reduced levels of glutamate. As Group I and Group II mGluR ligands progress in treating the symptoms of MDD, a note of caution should be reported as well. Treatment of MDD with mGluR ligands may need to be individualized for each patient. MDD patients with reduced glutamate or an underactive glutamatergic system may not respond to mGluR5 NAM; however, these mGluR5 NAMs may be a promising treatment for a patient with elevated levels of glutamate. Additionally, previous and current antidepressant treatment may influence the efficacy of mGluR ligands. For example, chronic imipramine treatment produces an upregulation of Group I mGluRs, although their function may be reduced. As a final note, many variables, such as past antidepressant treatment history, responsiveness to placebo, or bioavailability issues may interfere with any positive results. Future studies are necessary to clarify how the glutamate systems pathophysiology relates to MDD.

References

[1] Kessler, R. C., Berglund, P., Demler, O., Jin, R., Koretz, D., Merikangas, K. R., Rush, A. J., Walters, E. E., Wang, P. S. (2003) The epidemiology of major depressive disorder: results from the National Comorbidity Survey Replication (NCS-R). *JAMA* 289: 3095-3105.

[2] Belmaker, R. H. (2008) The future of depression psychopharmacology. *CNS Spectr*. 13: 682-687.

[3] Kessler, R. C., Petukhova, M., Sampson, N. A., Zaslavsky, A. M., Wittchen, H. U. (2012) Twelve-month and lifetime prevalence and lifetime morbid risk of anxiety and mood disorders in the United States. *Int. J. Methods Psychiatr. Res*. 21:169-184.

[4] Wittchen, H. U., Jacobi, F., Rehm, J., Gustavsson, A., Svensson, M., Jonsson, B., Olesen, J., Allgulander, C., Alonso, J., Faravelli, C., Fratiglioni, L., Jennum, P., Lieb, R., Maercker, A., van Os, J., Preisig, M., Salvador-Carulla, L., Simon, R., Steinhausen, H. C. (2011) The size and burden of mental disorders and other disorders of the brain in Europe 2010. *Eur. Neuropsychopharmacol*. 21:655-679.

[5] Greenberg, P. E., Kessler, R. C., Birnbaum, H. G., Leong, S. A., Lowe, S. W., Berglund, P. A., Corey-Lisle, P. K. (2003) The economic burden of depression in the United States: how did it change between 1990 and 2000? *J. Clin. Psychiatry* 64:1465-1475.

[6] Andlin-Sobocki, P., Rossler, W. (2005) Cost of psychotic disorders in Europe. *Eur. J. Neurol*. 12 Suppl. 1:74-77.

[7] Murray, C. J., Lopez, A. D. (1996) Evidence-based health policy--lessons from the Global Burden of Disease Study. *Science* 274:740-743.

[8] Uher, R., Mors, O., Rietschel, M., Rajewska-Rager, A., Petrovic, A., Zobel, A., Henigsberg, N., Mendlewicz, J., Aitchison, K. J., Farmer, A., McGuffin, P. (2011) Early and delayed onset of response to antidepressants in individual trajectories of change during treatment of major depression: a secondary analysis of data from the Genome-Based Therapeutic Drugs for Depression (GENDEP) study. *J. Clin. Psychiatry* 72: 1478-1484.

[9] Hashimoto, K. (2011) The role of glutamate on the action of antidepressants. *Prog. Neuropsychopharmacol. Biol. Psychiatry* 35:1558-1568.

[10] Altamura, C. A., Mauri, M. C., Ferrara, A., Moro, A. R., D'Andrea, G., Zamberlan, F. (1993) Plasma and platelet excitatory amino acids in psychiatric disorders. *Am. J. Psychiatry* 150:1731-1733.

[11] Kim, J. S., Schmid-Burgk, W., Claus, D., Kornhuber, H. H. (1982) Increased serum glutamate in depressed patients. *Arch. Psychiatr. Nervenkr*. 232:299-304.

[12] Levine, J., Panchalingam, K., Rapoport, A., Gershon, S., McClure, R. J., Pettegrew, J. W. (2000) Increased cerebrospinal fluid glutamine levels in depressed patients. *Biol. Psychiatry* 47:586-593.

[13] Mauri, M. C., Ferrara, A., Boscati, L., Bravin, S., Zamberlan, F., Alecci, M., Invernizzi, G. (1998) Plasma and platelet amino acid concentrations in patients affected by major depression and under fluvoxamine treatment. *Neuropsychobiology* 37:124-129.

[14] Mitani, H., Shirayama, Y., Yamada, T., Maeda, K., Ashby, C. R., Jr., Kawahara, R. (2006) Correlation between plasma levels of glutamate, alanine and serine with severity of depression. *Prog. Neuropsychopharmacol. Biol. Psychiatry* 30:1155-1158.

[15] Kucukibrahimoglu, E., Saygin, M. Z., Caliskan, M., Kaplan, O. K., Unsal, C., Goren, M. Z. (2009) The change in plasma GABA, glutamine and glutamate levels in fluoxetine- or S-citalopram-treated female patients with major depression. *Eur. J. Clin. Pharmacol*. 65:571-577.

[16] Garakani, A., Martinez, J. M., Yehuda, R., Gorman, J. M. (2013) Cerebrospinal fluid levels of glutamate and corticotropin releasing hormone in major depression before and after treatment. *J. Affect Disord*. 146:262-265.

[17] Maes, M., Verkerk, R., Vandoolaeghe, E., Lin, A., Scharpe, S. (1998) Serum levels of excitatory amino acids, serine, glycine, histidine, threonine, taurine, alanine and arginine in treatment-resistant depression: modulation by treatment with antidepressants and prediction of clinical responsivity. *Acta Psychiatr. Scand.* 97:302-308.

[18] Hashimoto, K., Sawa, A., Iyo, M. (2007) Increased levels of glutamate in brains from patients with mood disorders. *Biol. Psychiatry* 62:1310-1316.

[19] Choudary, P. V., Molnar, M., Evans, S. J., Tomita, H., Li, J. Z., Vawter, M. P., Myers, R. M., Bunney, W. E., Jr., Akil, H., Watson, S. J., Jones, E. G. (2005) Altered cortical glutamatergic and GABAergic signal transmission with glial involvement in depression. *Proc. Natl. Acad. Sci. US* 102:15653-15658.

[20] Bernard, R., Kerman, I. A., Thompson, R. C., Jones, E. G., Bunney, W. E., Barchas, J. D., Schatzberg, A. F., Myers, R. M., Akil, H., Watson, S. J. (2011) Altered expression of glutamate signaling, growth factor, and glia genes in the locus coeruleus of patients with major depression. *Mol. Psychiatry* 16:634-646.

[21] Altamura, C., Maes, M., Dai, J., Meltzer, H. Y. (1995) Plasma concentrations of excitatory amino acids, serine, glycine, taurine and histidine in major depression. *Eur. Neuropsychopharmacol.* 5 Suppl.:71-75.

[22] Frye, M. A., Tsai, G. E., Huggins, T., Coyle, J. T., Post, R. M. (2007) Low cerebrospinal fluid glutamate and glycine in refractory affective disorder. *Biol. Psychiatry* 61:162-166.

[23] Sanacora, G., Gueorguieva, R., Epperson, C. N., Wu, Y. T., Appel, M., Rothman, D. L., Krystal, J. H., Mason, G. F. (2004) Subtype-specific alterations of gamma-aminobutyric acid and glutamate in patients with major depression. *Arch. Gen. Psychiatry* 61:705-713.

[24] Auer, D. P., Putz, B., Kraft, E., Lipinski, B., Schill, J., Holsboer, F. (2000) Reduced glutamate in the anterior cingulate cortex in depression: an in vivo proton magnetic resonance spectroscopy study. *Biol. Psychiatry* 47:305-313.

[25] Pfleiderer, B., Michael, N., Erfurth, A., Ohrmann, P., Hohmann, U., Wolgast, M., Fiebich, M., Arolt, V., Heindel, W. (2003) Effective electroconvulsive therapy reverses glutamate/glutamine deficit in the left anterior cingulum of unipolar depressed patients. *Psychiatry Res.* 122:185-192.

[26] Michael, N., Erfurth, A., Ohrmann, P., Arolt, V., Heindel, W., Pfleiderer, B. (2003) Neurotrophic effects of electroconvulsive therapy: a proton magnetic resonance study of the left amygdalar region in patients with treatment-resistant depression. *Neuro-psychopharmacology* 28:720-725.

[27] Block, W., Traber, F., von Widdern, O., Metten, M., Schild, H., Maier, W., Zobel, A., Jessen, F. (2009) Proton MR spectroscopy of the hippocampus at 3T in patients with unipolar major depressive disorder: correlates and predictors of treatment response. *Int. J. Neuropsychopharmacol.* 12:415-422.

[28] Michael, N., Erfurth, A., Ohrmann, P., Arolt, V., Heindel, W., Pfleiderer, B. (2003) Metabolic changes within the left dorsolateral prefrontal cortex occurring with electroconvulsive therapy in patients with treatment resistant unipolar depression. *Psychol. Med.* 33:1277-1284.

[29] Hasler, G., van der Veen, J. W., Tumonis, T., Meyers, N., Shen, J., Drevets, W. C. (2007) Reduced prefrontal glutamate/glutamine and gamma-aminobutyric acid levels in

major depression determined using proton magnetic resonance spectroscopy. *Arch. Gen. Psychiatry* 64:193-200.

[30] Mirza, Y., Tang, J., Russell, A., Banerjee, S. P., Bhandari, R., Ivey, J., Rose, M., Moore, G. J., Rosenberg, D. R. (2004) Reduced anterior cingulate cortex glutamatergic concentrations in childhood major depression. *J. Am. Acad. Child Adolesc. Psychiatry* 43:341-348.

[31] Deschwanden, A., Karolewicz, B., Feyissa, A. M., Treyer, V., Ametamey, S. M., Johayem, A., Burger, C., Auberson, Y. P., Sovago, J., Stockmeier, C. A., Buck, A., Hasler, G. (2011) Reduced metabotropic glutamate receptor 5 density in major depression determined by [^{11}C]ABP688 PET and postmortem study. *Am. J. Psychiatry* 168:727-734.

[32] Feyissa, A. M., Woolverton, W. L., Miguel-Hidalgo, J. J., Wang, Z., Kyle, P. B., Hasler, G., Stockmeier, C. A., Iyo, A. H., Karolewicz, B. (2010) Elevated level of metabotropic glutamate receptor 2/3 in the prefrontal cortex in major depression. *Prog. Neuropsychopharmacol. Biol. Psychiatry* 34:279-283.

[33] Beneyto, M., Kristiansen, L. V., Oni-Orisan, A., McCullumsmith, R. E., Meador-Woodruff, J. H. (2007) Abnormal glutamate receptor expression in the medial temporal lobe in schizophrenia and mood disorders. *Neuropsychopharmacology* 32:1888-1902.

[34] Feyissa, A. M., Chandran, A., Stockmeier, C. A., Karolewicz, B. (2009) Reduced levels of NR2A and NR2B subunits of NMDA receptor and PSD-95 in the prefrontal cortex in major depression. *Prog. Neuropsychopharmacol. Biol. Psychiatry* 33:70-75.

[35] Nudmamud-Thanoi, S., Reynolds, G. P. (2004) The NR1 subunit of the glutamate/NMDA receptor in the superior temporal cortex in schizophrenia and affective disorders. *Neurosci. Lett.* 372:173-177.

[36] Pin, J. P., Duvoisin, R. (1995) The metabotropic glutamate receptors: structure and functions. *Neuropharmacology* 34:1-26.

[37] Schoepp, D. D., Conn, P. J. (2002) Metabotropic glutamate receptors. *Pharmacol. Biochem. Behav.* 74:255-256.

[38] Schoepp, D. D., Conn, P. J. (1993) Metabotropic glutamate receptors in brain function and pathology. *Trends Pharmacol. Sci.* 14:13-20.

[39] Stefani, M. R., Moghaddam, B. (2010) Activation of type 5 metabotropic glutamate receptors attenuates deficits in cognitive flexibility induced by NMDA receptor blockade. *Eur. J. Pharmacol.* 639:26-32.

[40] Cunningham, M. D., Ferkany, J. W., Enna, S. J. (1994) Excitatory amino acid receptors: a gallery of new targets for pharmacological intervention. *Life Sci.* 54:135-148.

[41] Hollmann, M., Heinemann, S. (1994) Cloned glutamate receptors. *Annu. Rev. Neurosci.* 17: 31-108.

[42] Ozawa, S., Kamiya, H., Tsuzuki, K. (1998) Glutamate receptors in the mammalian central nervous system. *Prog. Neurobiol.* 54:581-618.

[43] Lau, C. G., Zukin, R. S. (2007) NMDA receptor trafficking in synaptic plasticity and neuropsychiatric disorders. *Nat. Rev. Neurosci.* 8:413-426.

[44] Albensi, B. C. (2007) The NMDA receptor/ion channel complex: a drug target for modulating synaptic plasticity and excitotoxicity. *Curr. Pharm. Des.* 13:3185-3194.

[45] Hardingham, G. E., Bading, H. (2003) The Yin and Yang of NMDA receptor signalling. *Trends Neurosci.* 26:81-89.

[46] Lu, W., Man, H., Ju, W., Trimble, W. S., MacDonald, J. F., Wang, Y. T. (2001) Activation of synaptic NMDA receptors induces membrane insertion of new AMPA receptors and LTP in cultured hippocampal neurons. *Neuron* 29:243-254.

[47] Niswender, C. M., Conn, P. J. (2010) Metabotropic glutamate receptors: physiology, pharmacology, and disease. *Annu. Rev. Pharmacol. Toxicol.* 50:295-322.

[48] Cryan, J. F., Slattery, D. A. (2007) Animal models of mood disorders: Recent developments. *Curr. Opin. Psychiatry* 20:1-7.

[49] Wieronska, J. M., Branski, P., Szewczyk, B., Palucha, A., Papp, M., Gruca, P., Moryl, E., Pilc, A. (2001) Changes in the expression of metabotropic glutamate receptor 5 (mGluR5) in the rat hippocampus in an animal model of depression. *Pol. J. Pharmacol.* 53: 659-662.

[50] Morley-Fletcher, S., Mairesse, J., Soumier, A., Banasr, M., Fagioli, F., Gabriel, C., Mocaer, E., Daszuta, A., McEwen, B., Nicoletti, F., Maccari, S. (2011) Chronic agomelatine treatment corrects behavioral, cellular, and biochemical abnormalities induced by prenatal stress in rats. *Psychopharmacology* (Berl.) 217:301-313.

[51] Iyo, A. H., Feyissa, A. M., Chandran, A., Austin, M. C., Regunathan, S., Karolewicz, B. (2010) Chronic corticosterone administration down-regulates metabotropic glutamate receptor 5 protein expression in the rat hippocampus. *Neuroscience* 169: 1567-1574.

[52] Wieronska, J. M., Legutko, B., Dudys, D., Pilc, A. (2008) Olfactory bulbectomy and amitriptyline treatment influences mGlu receptors expression in the mouse brain hippocampus. *Pharmacol. Rep.* 60:844-855.

[53] Kovacevic, T., Skelin, I., Minuzzi, L., Rosa-Neto, P., Diksic, M. (2012) Reduced metabotropic glutamate receptor 5 in the Flinders Sensitive Line of rats, an animal model of depression: an autoradiographic study. *Brain Res. Bull.* 87:406-412.

[54] Matrisciano, F., Caruso, A., Orlando, R., Marchiafava, M., Bruno, V., Battaglia, G., Gruber, S. H., Melchiorri, D., Tatarelli, R., Girardi, P., Mathe, A. A., Nicoletti, F. (2008) Defective group-II metaboropic glutamate receptors in the hippocampus of spontaneously depressed rats. *Neuropharmacology* 55:525-531.

[55] Matrisciano, F., Storto, M., Ngomba, R. T., Cappuccio, I., Caricasole, A., Scaccianoce, S., Riozzi, B., Melchiorri, D., Nicoletti, F. (2002) Imipramine treatment up-regulates the expression and function of mGlu2/3 metabotropic glutamate receptors in the rat hippocampus. *Neuropharmacology* 42:1008-1015.

[56] Bajkowska, M., Branski, P., Smialowska, M., Pilc, A. (1999) Effect of chronic anti-depressant or electroconvulsive shock treatment on mGluR1a immunoreactivity expression in the rat hippocampus. *Pol. J. Pharmacol.* 51:539-541.

[57] Smialowska, M., Szewczyk, B., Branski, P., Wieronska, J. M., Palucha, A., Bajkowska, M., Pilc, A. (2002) Effect of chronic imipramine or electroconvulsive shock on the expression of mGluR1a and mGluR5a immunoreactivity in rat brain hippocampus. *Neuropharmacology* 42:1016-1023.

[58] Zahorodna, A., Bijak, M. (1999) An antidepressant-induced decrease in the responsiveness of hippocampal neurons to group I metabotropic glutamate receptor activation. *Eur. J. Pharmacol.* 386:173-179.

[59] Palucha, A., Branski, P., Klak, K., Sowa, M. (2007) Chronic imipramine treatment reduces inhibitory properties of group II mGlu receptors without affecting their density or affinity. *Pharmacol. Rep.* 59:525-530.

[60] Belozertseva, I. V., Kos, T., Popik, P., Danysz, W., Bespalov, A. Y. (2007) Antidepressant-like effects of mGluR1 and mGluR5 antagonists in the rat forced swim and the mouse tail suspension tests. *Eur. Neuropsychopharmacol.* 17:172-179.

[61] Molina-Hernandez, M., Tellez-Alcantara, N. P., Perez-Garcia, J., Olivera-Lopez, J. I., Jaramillo-Jaimes, M. T. (2008) Desipramine or glutamate antagonists synergized the antidepressant-like actions of intra-nucleus accumbens infusions of minocycline in male Wistar rats. *Prog. Neuropsychopharmacol. Biol. Psychiatry* 32:1660-1666.

[62] Molina-Hernandez, M., Tellez-Alcantara, N. P., Perez-Garcia, J., Olivera-Lopez, J. I., Jaramillo-Jaimes, M. T. (2008) Antidepressant-like actions of minocycline combined with several glutamate antagonists. *Prog. Neuropsychopharmacol. Biol. Psychiatry* 32: 380-386.

[63] Smolders, I., Clinckers, R., Meurs, A., De Bundel, D., Portelli, J., Ebinger, G., Michotte, Y. (2008) Direct enhancement of hippocampal dopamine or serotonin levels as a pharmacodynamic measure of combined antidepressant-anticonvulsant action. *Neuropharmacology* 54:1017-1028.

[64] Tatarczynska, E., Klodzinska, A., Chojnacka-Wojcik, E., Palucha, A., Gasparini, F., Kuhn, R., Pilc, A. (2001) Potential anxiolytic- and antidepressant-like effects of MPEP, a potent, selective and systemically active mGlu5 receptor antagonist. *Br. J. Pharmacol.* 132:1423-1430.

[65] Li, X., Need, A. B., Baez, M., Witkin, J. M. (2006) Metabotropic glutamate 5 receptor antagonism is associated with antidepressant-like effects in mice. *J. Pharmacol. Exp. Ther.* 319:254-259.

[66] Pilc, A., Klodzinska, A., Branski, P., Nowak, G., Palucha, A., Szewczyk, B., Tatarczynska, E., Chojnacka-Wojcik, E., Wieronska, J. M. (2002) Multiple MPEP administrations evoke anxiolytic- and antidepressant-like effects in rats. *Neuropharmacology* 43:181-187.

[67] Wieronska, J. M., Szewczyk, B., Branski, P., Palucha, A., Pilc, A. (2002) Antidepressant-like effect of MPEP, a potent, selective and systemically active mGlu5 receptor antagonist in the olfactory bulbectomized rats. *Amino Acids* 23:213-216.

[68] Liu, C. Y., Jiang, X. X., Zhu, Y. H., Wei, D. N. (2012) Metabotropic glutamate receptor 5 antagonist 2-methyl-6-(phenylethynyl)pyridine produces antidepressant effects in rats: role of brain-derived neurotrophic factor. *Neuroscience* 223:219-224.

[69] Palucha, A., Branski, P., Szewczyk, B., Wieronska, J. M., Klak, K., Pilc, A. (2005) Potential antidepressant-like effect of MTEP, a potent and highly selective mGluR5 antagonist. *Pharmacol. Biochem. Behav.* 81:901-906.

[70] Palucha-Poniewiera, A., Pilc, A. (2012) Involvement of mGlu5 and NMDA receptors in the antidepressant-like effect of acamprosate in the tail suspension test. *Prog. Neuropsychopharmacol. Biol. Psychiatry* 39:102-106.

[71] Hughes, Z. A., Neal, S. J., Smith, D. L., Sukoff Rizzo, S. J., Pulicicchio, C. M., Lotarski, S., Lu, S., Dwyer, J. M., Brennan, J., Olsen, M., Bender, C. N., Kouranova, E., Andree, T. H., Harrison, J. E., Whiteside, G. T., Springer, D., O'Neil, S. V., Leonard, S. K., Schechter, L. E., Dunlop, J., Rosenzweig-Lipson, S., Ring, R. H. (2013) Negative allosteric modulation of metabotropic glutamate receptor 5 results in broad spectrum activity relevant to treatment resistant depression. *Neuropharmacology* 66: 202-214.

[72] Porter, R. H., Jaeschke, G., Spooren, W., Ballard, T. M., Buttelmann, B., Kolczewski, S., Peters, J. U., Prinssen, E., Wichmann, J., Vieira, E., Muhlemann, A., Gatti, S., Mutel, V., Malherbe, P. (2005) Fenobam: a clinically validated nonbenzodiazepine anxiolytic is a potent, selective, and noncompetitive mGlu5 receptor antagonist with inverse agonist activity. *J. Pharmacol. Exp. Ther.* 315:711-721.

[73] Pecknold, J. C., McClure, D. J., Appeltauer, L., Wrzesinski, L., Allan, T. (1982) Treatment of anxiety using fenobam (a nonbenzodiazepine) in a double-blind standard (diazepam) placebo-controlled study. *J. Clin. Psychopharmacol.* 2:129-133.

[74] Ago, Y., Yano, K., Araki, R., Hiramatsu, N., Kita, Y., Kawasaki, T., Onoe, H., Chaki, S., Nakazato, A., Hashimoto, H., Baba, A., Takuma, K., Matsuda, T. (2013) Metabotropic glutamate 2/3 receptor antagonists improve behavioral and prefrontal dopaminergic alterations in the chronic corticosterone-induced depression model in mice. *Neuropharmacology* 65:29-38.

[75] Koike, H., Iijima, M., Chaki, S. (2013) Effects of ketamine and LY341495 on the depressive-like behavior of repeated corticosterone-injected rats. *Pharmacol. Biochem. Behav.* 107:20-23.

[76] Palucha-Poniewiera, A., Wieronska, J. M., Branski, P., Stachowicz, K., Chaki, S., Pilc, A. (2010) On the mechanism of the antidepressant-like action of group II mGlu receptor antagonist, MGS0039. *Psychopharmacology* (Berl.) 212:523-535.

[77] Bespalov, A. Y., van Gaalen, M. M., Sukhotina, I. A., Wicke, K., Mezler, M., Schoemaker, H., Gross, G. (2008) Behavioral characterization of the mGlu group II/III receptor antagonist, LY-341495, in animal models of anxiety and depression. *Eur. J. Pharmacol.* 592:96-102.

[78] Chaki, S., Yoshikawa, R., Hirota, S., Shimazaki, T., Maeda, M., Kawashima, N., Yoshimizu, T., Yasuhara, A., Sakagami, K., Okuyama, S., Nakanishi, S., Nakazato, A. (2004) MGS0039: a potent and selective group II metabotropic glutamate receptor antagonist with antidepressant-like activity. *Neuropharmacology* 46:457-467.

[79] Koike, H., Fukumoto, K., Iijima, M., Chaki, S. (2013) Role of BDNF/TrkB signaling in antidepressant-like effects of a group II metabotropic glutamate receptor antagonist in animal models of depression. *Behav. Brain Res.* 238:48-52.

[80] Campo, B., Kalinichev, M., Lambeng, N., El Yacoubi, M., Royer-Urios, I., Schneider, M., Legrand, C., Parron, D., Girard, F., Bessif, A., Poli, S., Vaugeois, J. M., Le Poul, E., Celanire, S. (2011) Characterization of an mGluR2/3 negative allosteric modulator in rodent models of depression. *J. Neurogenet.* 25:152-166.

[81] Karasawa, J., Shimazaki, T., Kawashima, N., Chaki, S. (2005) AMPA receptor stimulation mediates the antidepressant-like effect of a group II metabotropic glutamate receptor antagonist. *Brain Res.* 1042:92-98.

[82] Yoshimiz, T., Shimazaki, T., Ito, A., Chaki, S. (2006) An mGluR2/3 antagonist, MGS 0039, exerts antidepressant and anxiolytic effects in behavioral models in rats. *Psychopharmacology* (Berl.) 186:587-593.

[83] Koike, H., Iijima, M., Chaki, S. (2011) Involvement of the mammalian target of rapamycin signaling in the antidepressant-like effect of group II metabotropic glutamate receptor antagonists. *Neuropharmacology* 61:1419-1423.

In: Metabotropic Glutamate Receptors
Editor: M. Foster Olive

ISBN: 978-1-63117-451-3

Chapter 8

Involvement of mGluR5 and mGluR2/3 in Drug and Alcohol Dependence

Alessandra Matzeu[*], *Eva Rebecca Zamora-Martinez and Rémi Martin-Fardon*
The Scripps Research Institute, La Jolla, CA, US

Abstract

Drug addiction is a chronically relapsing disorder characterized by compulsive drug use and loss of control over intake. Currently, the available therapeutic approaches fail to completely treat and address the compulsive nature of drug seeking and drug taking associated with addiction. This chronic relapsing disorder is believed to result from neurodysregulation and long-lasting neuroadaptations induced by prolonged drug use. Over the past decade, metabotropic glutamate receptors (mGluRs) have received great interest in the addiction research field. Accumulating evidence indicates the potential importance and usefulness of agents that act on mGluR2/3 and mGluR5 receptors to prevent some of the behavioral effects of drugs of abuse, such as the reinforcing and addictive actions and withdrawal states. The negative modulation of mGluR5 or positive modulation of mGluR2/3 receptors is known to reduce drug self-administration, drug-seeking behavior, and drug withdrawal in animal models, with negligible effects when tested on conventional reinforcer-directed behavior. This chapter discusses several preclinical studies that suggest the key involvement of mGluR5 and mGluR2/3 receptors in drug addiction and how targeting these receptors may potentially reduce drug intake and withdrawal and prevent relapse.

Keywords: mGluR5, mGluR2/3, addiction, cocaine, nicotine, opioid, heroin, morphine, ethanol, alcohol

[*] Corresponding author: Dr. Alessandra Matzeu, Molecular and Cellular Neuroscience Department, The Scripps Research Institute, SP30-2120, 10550 North Torrey Pines Road, La Jolla, CA 92037 US. E-mail: amatzeu@scripps.edu.

Introduction

Drug addiction is a chronic relapsing disorder characterized by compulsive drug seeking and use [1-4]. One hypothesis that can explain the long-lasting nature of drug-seeking behavior is that the circuits that "normally" contribute to the adaptive and beneficial aspect of hedonia, learning, and memory, which are crucial for long-term survival, are usurped and more robustly activated by drugs of abuse, potentially creating new motivational states that normally govern responding for natural rewards toward drug-directed behavior. Therefore, in individuals with drug addiction, these neurocircuits are preferentially activated by drugs of abuse, creating pharmacological states of euphoria, pleasure, and reward [5, 6].

Addiction can be conceptualized as a disorder that involves elements of both impulsivity (predisposition toward rapid, unplanned reactions) and compulsivity (perseveration in responding in the face of adverse consequences) [7]. Genetic/epigenetic, cellular, and molecular neuroadaptations mediate the transition from occasional drug use to the loss of control over drug seeking and drug taking, with chronic relapse even after protracted periods of abstinence [8]. Indeed, repeated drug exposure produces neuronal and behavioral adaptations that, in turn, will lead to escalation and compulsive drug intake with protracted use that establish addiction and dependence and in the absence of drugs produce the emergence of negative emotional states that are characteristic of withdrawal [7].

The high relapse rate and long-lasting vulnerability to relapse in abstinent individuals is one of the challenges encountered for effective treatment of drug addiction. Unfortunately, the available therapeutic approaches fail to completely treat and address the compulsive nature of drug seeking and drug taking associated with addiction. Ideally, an efficient pharmacological approach for the prevention of relapse will be a compound that is universally efficacious across different classes of drugs of abuse and without nonspecific side effects, such as minimal/lack of action on natural reward-directed behavior. Traditionally, because of the common interactions between drugs of abuse and the mesolimbic system, addiction research has focused on dopaminergic mechanisms to find an effective pharmacotherapy to prevent relapse (e.g., [9-11]). However, more recently, the focus has shifted to glutamatergic neurotransmission because of its known implication in the development and maintenance of addiction (e.g., self-administration, reward-related processes, and relapse [12-14]). Over the past decade, metabotropic glutamate receptors (mGluRs) have received great interest because of their modulatory role in glutamatergic transmission (e.g., [15]). In particular, accumulating evidence indicates the potential beneficial effects of agents that target mGluR5 and mGluR2/3 receptors to help alleviate some of the behavioral effects of drugs of abuse, such as the reinforcing actions and withdrawal signs [16-25].

Eight receptor subtypes have been identified to classify the various mGluRs, with these eight subtypes grouped into three families. The Group I family of mGluRs consists of mGluR1 and mGluR5 receptors. The Group II family consists of mGluR2 and mGluR3 receptors. The Group III family consists of mGluR4, mGluR6, mGluR7, and mGluR8 receptors [26, 27]. mGluR5 receptors are part of the Group I mGluR family and activate G_q proteins that activate phospholipase C (PLC) and protein kinase C (PKC). mGluR5 receptors are primarily localized postsynaptically. Once they are activated, they positively modulate excitatory synaptic transmission [26, 27]. mGluR2/3 receptors belong to the Group II mGluR family and activate G_i/G_o proteins to negatively modulate adenylyl cyclase (AC), cyclic adenosine monophosphate (cAMP), and cAMP-dependent protein kinase A (PKA) activation. mGluR2/3 receptors are localized predominantly presynaptically, and their activation negatively modulates excitatory synaptic transmission [26, 27]. mGluR5 and mGluR2/3 receptors are highly present in brain regions that are influenced by drugs of abuse [28]. The negative modulation of mGluR5 or positive modulation of mGluR2/3 receptors is known to reduce drug (e.g., cocaine and alcohol) self-administration, drug-seeking behavior, and drug withdrawal in animal models, with negligible effects when tested on conventional reinforcer-directed behavior (e.g., [22, 29, 30]).

This chapter discusses several preclinical studies that suggest the key involvement of mGluR5 and mGluR2/3 receptors in cocaine, nicotine, opioid, and ethanol addiction and how pharmacological manipulations of these receptors may potentially reduce drug intake, mitigate negative withdrawal effects associated with dependence, and prevent relapse.

Cocaine

Cocaine addiction is a chronic relapsing disorder that affects more than 1.6 million Americans [31]. The 2012 National Survey on Drug Use and Health estimated that 13.6% of all individuals who received substance abuse treatment fell into the category of cocaine abuse, the fourth highest prevalence of all abused drugs [32].

Recent findings suggest that changes in synaptic plasticity due to cocaine exposure mediate the transition to cocaine addiction, particularly repeated cocaine-induced neuroplasticity that involves excitatory amino acid transmission [33-35]. Especially interesting in this regard are mGluR5 and mGluR2/3 receptors because of their abundance in brain regions that regulate stress and anxiety [36-40] or mediate drug reward and incentive motivation [39, 41, 42].

Growing evidence implicates mGluR5 and mGluR2/3 receptors in the modulation of several neurobehavioral effects of cocaine, including the drug's psychomotor stimulant and rewarding actions [16, 22, 29].

mGluR5 Receptors

Over the past decade, mGluR5 receptors have been shown to play a key role in some of the behavioral effects of cocaine use. For example, mGluR5 knockout mice do not self-administer cocaine and show no increased locomotor activity following cocaine treatment [16], whereas mGluR5 knockout mice present attenuated anhedonic and somatic effects of psychostimulant withdrawal [43].

Systemic manipulation of mGluR5 receptors. The negative pharmacological manipulation of mGluR5 receptors reduces cocaine reinforcement and reinstatement [28, 29, 44, 45]. For example, 2-methyl-6-(phenylethynyl)-pyridine (MPEP), a selective mGluR5 negative allosteric modulator, has been shown to reduce the acquisition of cocaine-induced conditioned place preference (CPP) [46] and decrease both cocaine self-administration and cocaine prime-induced reinstatement of cocaine seeking without any effects on food intake [47-50]. In addition, MPEP has a significantly greater efficacy in reducing the breakpoint on a progressive-ratio schedule of reinforcement (a measure of the motivational and rewarding action of a reinforcer) for cocaine than the breakpoint for food [51], implying a more specific role for mGluR5 receptors in the regulation of the reinforcing actions of cocaine than natural rewards. Both MPEP and 2-methyl-(1,3-thiazol-4-ylethynyl)-pyridine (MTEP), a more specific mGluR5 negative allosteric modulator, produce a reduction of the acute locomotor stimulant effects of cocaine and a reduction of the prime- and cue-induced reinstatement of cocaine seeking [29, 52-54]. In addition, antagonizing mGluR5 receptors protected against cocaine's toxic effects, in which MTEP attenuated lethality induced by acute cocaine overdose [55]. Moreover, MTEP prevented the expression of cocaine-induced locomotor sensitization [55] and attenuated the cue- and stress-induced reinstatement of cocaine seeking [29, 45], thus supporting the beneficial role of mGluR5 blockade against cocaine addiction.

Central manipulation of mGluR5 receptors. Injections of MTEP into discrete brain regions reveal that mGluR5 receptors mediate different aspects of the cocaine addiction process in a variety of distinct brain regions. For instance, when injected into the nucleus accumbens core, MTEP prevented both the context- and cue-induced reinstatement of cocaine seeking following extinction training [56]. Injections in the dorsolateral striatum attenuated extinction learning (higher amounts of lever presses during extinction) without interfering with cocaine seeking [56]. When injected in the medial prefrontal cortex, MTEP prevented the expression of cocaine sensitization [57]. MPEP injections in the nucleus accumbens shell attenuated cocaine prime induced reinstatement of drug seeking, while the mGluR1/5 agonist DHPG (3,5-dihydroxyphenylglycine) promotes cocaine-seeking behavior [58]. Thus, mGluR5 receptors regulate distinct aspects of cocaine addiction, depending on where it is injected in the brain.

Neuroadaptations of mGluR5 receptors induced by cocaine. mGluR5 receptors are highly expressed in the forebrain and mesolimbic system [59, 60]. Several studies have clarified the relationship between a history of cocaine exposure and the induced neuroadaptation/ dysregulated neurotransmission at the mGluR5 level. For example, in the dorsolateral striatum and nucleus accumbens shell, cocaine exposure increased mGluR5 mRNA [61], whereas the extinction of cocaine-seeking behavior decreased postsynaptic mGluR5 density in the dorsomedial prefrontal cortex [62]. In the hippocampus, cocaine experience increased the gene expression and protein synthesis of mGluR5 receptors, contributing to changes in synaptic efficacy, such as long-term potentiation, associated with cocaine administration [63]. Other evidence of functional changes in the function of mGluR5 receptors was documented in a recent study [24] that used an animal model of cocaine dependence. Following extended cocaine access (6 h/day for 22 days), reduced levels of mGluR5 receptors were found in the medial prefrontal cortex, and reduced mGluR5 function was observed in the nucleus accumbens [24]. These results suggest that cocaine induces both the downregulation and hypofunction of mGluR5 receptors, and that it could be one of the underlying mechanisms responsible for the compulsive nature of cocaine seeking. Altogether, these studies indicate that repeated cocaine exposure produces long-term changes in mGluR5 receptors in regions of the brain associated with the enduring behavioral effects of cocaine use.

mGluR2/3 Receptors

The beneficial effects of targeting mGluR2/3 receptors to prevent some of the behavioral actions of cocaine have been documented in rodents and nonhuman primates.

Systemic manipulation of mGluR2/3 receptors. (–)-2-Oxa-4-aminobicylco[3.1.0]hexane-4,6-dicarboxylic acid (LY379268), an mGluR2/3 agonist, has been shown to reduce cocaine prime-induced dopamine release in the rat nucleus accumbens, attenuate cocaine self-administration and intracranial self-stimulation [22, 64], block cocaine prime-induced relapse [64, 65], and reduce cue- and stress-induced reinstatement of cocaine-seeking behavior [22, 45]. This compound has valid potential as a candidate for cocaine treatment at doses that allow for mGluR2/3 selectivity. At lower doses, LY379268 specifically blocks cocaine-seeking behavior but prevents food (food pellet or sweetened condensed milk)-seeking behavior only at a higher dose [22, 65].

Several studies have attempted to identify the specific participation of mGluR2 *vs.* mGluR3 in the behavioral effects of cocaine. For example, mGluR2 knockout mice show higher cocaine-induced locomotor sensitization, stronger CPP, and increased dopamine and glutamate release in the nucleus accumbens in response to a cocaine injection, suggesting that LY379268's effects are the result of mGluR2 activation [66]. In contrast, no difference was observed between mGluR3 knockout mice and their littermate controls in the cue-induced reinstatement of cocaine seeking [67], suggesting a prominent role for mGluR2 *vs.* mGluR3 receptors in cocaine addiction. Recent studies reported that newly developed selective mGluR2 positive allosteric modulators (PAMs), such as BINA (potassium 3'-([(2-cyclopentyl-6-7-dimethyl-1-oxo-2,3-dihydro-1H-inden-5-yl)oxy]methyl)biphenyl-4-carboxylate), attenuated the reward-enhancing effects of cocaine (i.e., increased hypothalamic intracranial self-stimulation threshold) and decreased cue-induced reinstatement of cocaine-seeking behavior, without affecting food-directed behavior [68].

Central manipulation of mGluR2/3 receptors. Cerebral injections of LY379268 have also efficiently blocked some of cocaine's behavioral effects. Microinjections into the nucleus accumbens core and ventral tegmental area blocked the cocaine prime-induced reinstatement of cocaine seeking [65, 69], and infusions into the central nucleus of the amygdala reduced the incubation of cocaine craving and blocked conditioned reinstatement [70]. Considering the important role of the ventral tegmental area-nucleus accumbens circuit in the reinforcing effect of cocaine [71, 72], these data suggest that mGluR2/3 receptors in both the nucleus accumbens and ventral tegmental area are involved in regulating the rewarding properties of cocaine, whereas mGluR2/3 receptors in the central nucleus of the amygdala are critical for the expression of craving incubation.

Neuroadaptation of mGluR2/3 receptors induced by cocaine. Evidence of the cocaine-induced neuroadaptations of mGluR2/3 receptors comes from studies that combined behavioral models of drug abuse and molecular pharmacological approaches. Acute withdrawal from cocaine (24 h after the last cocaine injection) reduced mGluR2/3 function in the central nucleus of the amygdala. After prolonged withdrawal (e.g., several weeks), mGluR2/3 function was reduced in the prefrontal cortex and nucleus accumbens [73-77]. Interestingly, animal models of cocaine non-dependence with short daily access showed down-regulation of mGlu2R/3 receptors, whereas studies that used a model of cocaine dependence demonstrated an opposite trend. For instance, the efficacy of LY379268 was improved following prolonged cocaine exposure [24, 78]. In a model of extended access to cocaine in self-administering rats (6 h/day), the efficacy of LY379268 in diminishing cocaine-maintained progressive ratio performance increased compared with animals that only had short access (1 h/day) to cocaine self-administration [24]. Extended-access animals also exhibited an increased sensitivity to the anxiolytic-like effect of LY379268 when tested in the defensive burying paradigm [78]. These early behavioral pharmacological studies were complemented by molecular approaches that attempted to characterize functional changes in mGluR2/3 receptors that result from chronic cocaine exposure. Changes in the pharmacological profile of LY379268's were associated with enhanced functional mGluR2/3 activity in rats that had extended cocaine access [24]. Additionally, mGluR2/3 density in the dorsal striatum has been shown to be elevated following chronic cocaine intake [79]. In summary, these results implicate mGluR2/3-mediated glutamate transmission in the regulation of cocaine reinforcement and reinstatement and reveal an upregulation of mGluR2/3 function with prolonged cocaine use.

Such results implicate mGluR5- and mGlu2R2/3-mediated glutamate transmission in regulating the positive and negative aspects of cocaine addiction, respectively. Exploiting this dichotomy by decreasing the efficacy of mGluR5 receptors and increasing the efficacy of mGluR2/3 receptors may represent a promising strategy against cocaine addiction.

Nicotine

Tobacco use remains the single largest preventable cause of death and disease in the United States. According to the 2010 US Surgeon General's report, approximately 443,000 US adults die from smoking-related illnesses each year [80]. Therapeutic approaches to treat tobacco dependence are predicated on the assumption that nicotine is the primary reinforcer derived from tobacco. Although bupropion (an atypical antidepressant) and nicotine replacement are currently the most commonly used cessation aids, relapse rates during the first year remain very high (80%), demonstrating the urgent need for alternative treatments for nicotine dependence [81, 82].

Nicotine increases glutamate release through an agonist action at excitatory presynaptic nicotinic acetylcholine receptors (nAchRs) on glutamate terminals in various brain sites (for review [83, 84]). It is hypothesized that both the reinforcing and motivational effects of nicotine and stimuli previously associated with nicotine intake are at least partially mediated by nicotine-induced increases in glutamate that acts at postsynaptic mGluR5 receptors [85]. Conversely, the depression-like aspects of early nicotine withdrawal states are hypothesized to be associated with decreased glutamate transmission mediated by the increased activity of presynaptic inhibitory mGluR2/3 receptors [85]. Thus, these metabotropic receptors may be valuable targets for the development of new therapeutics that may assist people in abstaining and maintaining abstinence from tobacco smoking.

mGluR5 Receptors

The blockade of mGluR5 receptors has been shown to produce several beneficial effects in preventing nicotine reinforcement and reinstatement.

Systemic manipulation of mGluR5 receptors. MPEP and MTEP decrease nicotine self-administration and the motivation to self-administer nicotine in a progressive-ratio procedure [17, 47, 48, 51, 86-88]. The administration of MPEP decreases the cue- and prime-induced reinstatement of nicotine-seeking behavior and reduces both the acute locomotor stimulant effect of nicotine and expression of locomotor sensitization induced by repeated nicotine administration [48, 87, 89]. It is important to note that the doses of MPEP that successfully block nicotine self-administration and reinstatement had no effect on food self-administration or reinstatement [17, 87, 89]. Additionally, when tested in the progressive-ratio schedule, the effect of MPEP on breakpoints for nicotine tended to be larger than those for food [51]. This selective effectiveness of MPEP for nicotine versus food suggests that the negative modulation of mGluR5 receptors with this compound can be useful for nicotine treatment and is almost devoid (at certain doses) of nonspecific side effects related to food reinforcement and food reinstatement. Research has been conducted to determine the cerebral sites of action of nicotine and interaction between mGluR5 receptors and nicotinic acetylcholine receptors (nAChRs). Pharmacological studies have demonstrated MPEP's ability to inhibit dopamine overflow induced by nicotine in the nucleus accumbens core and shell [90], suggesting that mGluR5 receptors are also implicated in the primary reinforcing properties of nicotine, which depend on increased dopamine overflow in the nucleus accumbens.

Despite the ability of negative modulation of mGluR5 receptors to ameliorate different aspects of nicotine dependence, MPEP also increased the somatic signs of nicotine withdrawal and elevated intracranial self-stimulation thresholds associated with nicotine withdrawal [20]. This is important when considering targeting mGluR5 receptors to treat nicotine dependence during abstinence.

Central manipulation of mGluR5 receptors. Site-specific injections of MPEP have been tested in the nucleus accumbens shell and ventral tegmental area, two brain areas importantly involved in mediating the reinforcing effects of nicotine [91-96]. Administered directly in the nucleus accumbens shell, MPEP dose-dependently decreased nicotine self-administration without affecting food self-administration. Injections in the ventral tegmental area decreased responding for both nicotine and food at lower doses [97].Thus, while mGluR5 receptors in the nucleus accumbens shell specifically mediate the reinforcing effect of nicotine, mGluR5 receptors in the ventral tegmental area possibly mediate a more general action in the reinforcing effects of both nicotine and food.

Neuroadaptation of mGluR5 receptors induced by nicotine. Of particular interest, a recent human study described the down-regulation of mGluR5 receptors in dependent smokers and ex-smokers (abstinent for at least 3 months), which affected all brain regions except the brainstem [98]. Interestingly, ex-smokers exhibited less mGluR5 down-regulation compared with smokers [98]. This decrease in mGluR5 receptors may be an adaptation to chronic increases in glutamate induced by chronic nicotine administration, and the attenuated down-regulation seen in ex-smokers may be attributable to incomplete recovery of the receptors [98].

The data support the conclusion that targeting mGluR5 receptors may be a valuable strategy against nicotine dependence. Nevertheless, mGluR5 antagonists may exacerbate the consequences of nicotine withdrawal [20], which needs to be considered in future studies.

mGluR2/3 Receptors

mGluR2/3 antagonism may be beneficial to treat withdrawal symptoms, whereas positive modulation may be beneficial for counteracting nicotine self-administration and the cue-induced reinstatement of nicotine seeking.

Systemic manipulation of mGluR2/3 receptors. Pretreatment with the mGluR2/3 agonist LY379268 in nicotine-experienced rats in the presence of a context associated with nicotine self-administration completely blocked the nicotine-induced increase in dopamine in the nucleus accumbens shell [99]. Moreover, LY379268 decreased both nicotine self-administration and cue-induced reinstatement of nicotine-seeking behavior, with no effect on food self-administration; it affected food-seeking behavior only at the highest dose tested [95]. Altogether, these results suggest a key role for mGluR2/3 receptors in the reinforcing effects of nicotine and motivational influence of environmental stimuli previously associated with nicotine self-administration. While facilitating the activation of mGluR2/3 receptors was beneficial against nicotine self-administration and reinstatement, mGluR2/3 antagonism appears to be beneficial against withdrawal. Indeed, the Group II metabotropic agonist LY314582 [a racemic mixture of LY354740 ([+]-2-aminobicyclo[3.1.0]hexane-2,6-dicarboxylic acid])] precipitated withdrawal-like threshold elevations (measured by intracranial self-stimulation reward thresholds in nicotine-dependent rats). Systemic injection of the antagonist LY341495 (2*S*-2-amino-2-[1*S*,2*S*-2-carboxycyclopropan-1-yl]-3-[xanth-9-yl]propionic acid) reversed the threshold elevations observed in rats undergoing spontaneous nicotine withdrawal [23]. This strongly suggests an important role for these receptors in generating the reward deficits associated with nicotine withdrawal.

Central manipulation of mGluR2/3 receptors. As mentioned above regarding systemic injection, intra-ventral tegmental area administration of the mGluR2/3 agonist LY314582 precipitated withdrawal-like intracranial self-stimulation threshold elevations [23]. LY379268 injections in the ventral tegmental area or nucleus accumbens shell reduced nicotine self-administration and cue-induced reinstatement of nicotine-seeking behavior, with no effect on food self-administration. Food-seeking behavior was affected at the highest dose tested [95]. Such studies suggest a critical role for mGluR2/3 receptors in the ventral tegmental area and nucleus accumbens shell in the primary rewarding effects of nicotine, motivational impact of environmental stimuli previously associated with nicotine self-administration, and reward deficits experienced during withdrawal.

Neuroadaptation of mGluR2/3 receptors induced by nicotine. It can be hypothesized that adaptations develop to counteract the effects of the persistent increase in glutamate induced by repeated nicotine administration [85]. Indeed, nicotine self-administration down-regulates mGluR2/3 function in corticolimbic rat brain sites involved in reward and motivational processes, such as the prefrontal cortex, nucleus accumbens, ventral tegmental area, amygdala, hippocampus, and hypothalamus [95]. The adaptations that occur in the function of mGluR2/3 receptors with chronic nicotine exposure are likely to contribute to the maintenance of nicotine dependence and reinstatement of nicotine-seeking behavior. Moreover, other studies have demonstrated that nicotine withdrawal precipitated a deficit in brain reward function, reflected by elevations in intracranial self-stimulation reward thresholds, similar to those observed in rats that undergo withdrawal from other major drugs of abuse [100]. As mentioned above, increased mGluR2/3 activity has been implicated in mediating the depression-like aspect of the early nicotine withdrawal state [85]. Thus, during nicotine withdrawal, when the stimulatory effects of nicotine on excitatory glutamate transmission are no longer present, increased mGluR2/3 function would be expected to decrease glutamate transmission, thereby causing the reward deficits associated with nicotine withdrawal.

Opioids

The cycle of opioid dependence is punctuated by periods of abstinence, drug treatment, and relapse [101]. Dependent opioid users often surface through either their enrollment in treatment programs or entry into the legal system by incarceration. Despite efforts to intervene or the consequences of incarceration, dependent users continue their drug use over the span of decades [102-105]. Within 1 year after completing a given episode of drug treatment, the majority of users relapse to opioid use [106].

The currently approved medications for the rehabilitation of opioid dependence are methadone and buprenorphine. They act as replacement agents for maintenance and/or detoxification of opioid-addicted individuals [for review, see [107]].

Opioids are the best resource for analgesia and are invaluable for the task of pain management in medical care [108]. However, pain alleviation through opioid use quite often paradoxically results in more sensitivity to nociceptive stimuli. This is known as opioid-induced hyperalgesia [109].

Rodent studies of heroin exposure have revealed spontaneous reductions of mechanical nociceptive thresholds [110], further supporting the paradoxical relationship between opioids and opioid-induced hyperalgesia and suggesting a recruitment/sensitization of pro-nociceptive systems [111, 112]. The negative effects of opioid withdrawal may be augmented by the resulting opioid-induced hyperalgesia from chronic opioid use, acting as a strong negative reinforcer for further opioid consumption. Recent studies have implicated certain mGluR receptors (mGluR5, mGluR2/3, and mGluR7) in nociceptive transmission through their presence in regions of the central nervous system that are important for the transmission of nociceptive pain [113]. mGluR5 receptors can be found in nociceptive pathways, including the dorsal root ganglion and spinal dorsal horn neurons [114, 115].Their localization in nociceptive pathways suggests their involvement in pain. Their involvement in nociceptive transmission may suggest possible targets that may be beneficial in attenuating opioid-induced hyperalgesia, with a focus on diminishing the negative affect of withdrawal associated with opioid addiction. Here, the focus will remain on heroin and morphine.

mGluR5 Receptors

Compared with the evidence mentioned above for cocaine and nicotine, the involvement of mGluR5 receptors in opioid addiction is not as extensive.

Systemic manipulation of mGluR5 receptors. Van der Kam et al. demonstrated that the mGluR5s antagonist MPEP is able to modestly reduce heroin self-administration only at a high dose (20 mg/kg, i.p.), with no effect on food self-administration [116]. These results indicate a specific effect of this compound on heroin *vs.* non-drug self-administration.

However, despite not showing any reward-like properties using a CPP paradigm [117], MPEP is readily self-administered by rats, suggesting that MPEP has reinforcing actions of its own [118]. With MPEP substitution (1 mg/kg/infusion), rats that previously self-administered heroin (0.05 mg/kg/infusion) sustained stable MPEP self-administration behavior [118]. These findings indicate that MPEP substitute for heroin. Popik and Wrobel demonstrate MPEP's ability to attenuate both acquisition and expression of conditioned morphine reward in mice with no previous self-administration training [119]. The attenuating effect of MPEP on intravenous heroin self-administration appears to be attributable to an increase rather than a decrease on the rewarding/reinforcing effects of heroin [117]. Although the negative modulation of mGluR5 receptors may have therapeutic potential for the treatment of heroin abuse, with an ability to reduce heroin self-administration, the lack of efficacy at lower MPEP doses on heroin self-administration and its ability to potentiate the reinforcing effect of heroin appear to exclude MPEP as a possible candidate for heroin addiction treatment.

mGluR2/3 Receptors

Exposure to environmental stimuli previously associated with opiate intake can provoke drug relapse in humans and animals [120-123].

Systemic manipulation of mGluR2/3 receptors. The role of mGluR2/3 receptors in heroin addiction has been primarily studied with a focus on the ability of environmental stimuli previously associated with heroin to induce the reinstatement of heroin-seeking behavior. Indeed, exposure to a heroin-associated context (i.e., a context where rats were trained to self-administered heroin but different from the extinction context) induced robust reinstatement of drug seeking, an effect that was attenuated by a systemic injection of the positive allosteric modulator LY379268 [124]. Moreover, LY379268 was able to attenuate the discrete cue-induced reinstatement of heroin seeking without affecting heroin self-administration [125]. One study of the atypical stimulant modafinil suggested the involvement of mGluR2/3 receptors in the reinstatement of morphine-induced CPP. Acute modafinil pretreatment blocked the reinstatement of extinguished morphine-induced CPP in an animal model of relapse to opiate abuse at a dose of 300 mg/kg. Additionally, this anti-reinstatement effect was fully suppressed by the mGluR2/3 receptors antagonist LY341495, further revealing the required stimulation of mGluR2/3 receptors to block the reinstatement of morphine-induced CPP [126].

Central Manipulation of mGluR2/3 receptors. Intracerebral injections of LY379268 in the ventral tegmental area and nucleus accumbens shell dose-dependently attenuated context-induced reinstatement of heroin seeking, whereas injections into the nucleus accumbens core were effective only at higher doses, with no effect when injected into the caudate putamen [124, 127].

Neuroadaptation of mGluR2/3 receptors induced by opioids. The aforementioned data indicate the important involvement of mGluR2/3 receptors in the context- and cue-induced reinstatement of heroin-seeking behavior and the important involvement of the ventral tegmental area–nucleus accumbens (particularly the shell portion) in the brain mechanisms that regulate heroin reinstatement behavior. Current evidence suggests the Group II metabotropic glutamate receptors are potential therapeutic targets for preventing heroin relapse.

Ethanol

Alcoholism is a major public health problem. A 2012 survey revealed current alcohol drinkers account for slightly more than half of Americans over the age of 12, with 23% of these drinkers falling into the category of binge drinkers and 6.5% of the population qualifying as heavy drinkers [32]. Initial exposure to alcohol (or ethanol) triggers the brain's reward systems, and this rewarding effect mediates the positive reinforcement associated with further alcohol intake. Over time and as a result of chronic exposure to alcohol, long-term neuroadaptive changes occur, with the recruitment of distinct negative reinforcing mechanisms that further drive excessive alcohol intake [128]. These are the characteristics of a prolonged history of dependence. Currently, the approved medications for alcohol cessation and rehabilitation are the aldehyde dehydrogenase blocker disulfiram, opioid receptor antagonist naltrexone, and functional glutamate receptor antagonist acamprosate (for review, see [129]). However, to develop better pharmacotherapies to treat alcohol addiction, the effects of other drugs that hold promise for treating alcoholism may be assessed using animal models. Strong contenders include modulators of mGluR5 and mGluR2/3 receptors [129].

The neurobiological and molecular bases of relapse to ethanol seeking are still not fully understood. However, preclinical and clinical data imply that relapse is mediated by different neurotransmitter systems [130-132]. Among these, excitatory glutamatergic transmission in the brain is known to be affected by the use and abuse of ethanol, such that acute alcohol intake reduces excitatory glutamatergic synaptic neurotransmission [133-136]. Chronic alcohol intake leads to compensatory changes that result in hyper-responsive glutamatergic transmission that occurs specifically during abstinence [135-138]. These observations suggest that lessening glutamatergic system hyperexcitability through pharmacological manipulations of mGluR5 or mGluR2/3 receptors could be beneficial for the treatment of abstinent alcoholics.

mGluR5 Receptors

mGluR5 receptors have received great attention in the field of alcohol addiction because of overwhelming evidence that implicates these receptors in both the reinforcing and addictive effects of alcohol and the ability of mGluR5 antagonists to treat alcohol dependence.

Systemic manipulation of mGluR5 receptors. Significant numbers of studies have shown that antagonizing mGluR5 receptors with MPEP or MTEP reduced the discriminative effect of ethanol in a drug-discrimination paradigm [139, 140], decreased cue-induced reinstatement and ethanol consumption after an ethanol-deprivation period [141, 142], significantly attenuated the expression of CPP [143], reduced ethanol intake without affecting plain water or 10% sugar water intake [144], and reduced both ethanol self-administration and the reinstatement of ethanol seeking induced by footshock stress [19]. Sidhpura et al. reported no difference in the inhibition profile of ethanol self-administration and stress-induced reinstatement induced by MTEP between non-dependent and post-dependent rats (an animal model of alcohol addiction; [19], implying that ethanol exposure at dependence-inducing levels does not alter mGluR5 sensitivity.

In alcohol-preferring rats, a genetic model of high alcohol drinking, MPEP reduced the breakpoint for ethanol [145] and decreased operant ethanol self-administration during maintenance and after repeated alcohol deprivation [146]. These findings indicate that glutamatergic activity at mGluR5 receptors regulates the motivation that drives ethanol self-administration, the maintenance of ethanol self-administration, and relapse. mGluR5 knockout mice consumed less ethanol in a two-bottle free-choice procedure compared with wildtype mice [147]. The knockout mice also exhibited less severity of alcohol-induced withdrawal and more sensitivity to the sedative effect of ethanol [148]. Unfortunately, while the usefulness of mGluR5 antagonism in self-administration and relapse was encouraging, its utilization fell short in the treatment of hyperexcitability during alcohol withdrawal assessed by the handling-induced convulsion paradigm [149].

Central manipulation of mGluR5 receptors. The reduction of glutamatergic hyperexcitability during alcohol abstinence could be best achieved by site-specifically targeting postsynaptic mGluR5s because of their abundance in brain regions known to be involved in drug reinforcement [59, 60, 150, 151]. Indeed, in alcohol-preferring rats, MPEP injections in the nucleus accumbens reduced ethanol but not sucrose self-administration [152]. This suggests that mGluR5 activity in the nucleus accumbens is required for the expression of ethanol's reinforcing effects in individuals with a genetic predisposition for heavy alcohol drinking [152]. Moreover, MTEP injections in the basolateral amygdala or nucleus accumbens core [153] reduced cue-induced reinstatement of ethanol-seeking behavior but not sucrose-seeking behavior, further implicating these two brain regions in the suppression of the cue-induced reinstatement of ethanol-seeking behavior. Furthermore, the blockade of mGluR5 receptors in the nucleus accumbens shell reduced ethanol reinforcement [154], providing a possible neuroanatomical area where these receptors may regulate ethanol self-administration.

Neuroadaptation of mGluR5 receptors induced by ethanol. Chronic alcohol intake leads to compensatory changes in the mGluR5 system that complements the effects of glutamate. In rodents, these changes were particularly evident in binge models of alcohol consumption that have been shown to increase extracellular glutamate levels in the nucleus accumbens [155]. Prenatal ethanol exposure reduced mGluR5 expression in the dentate gyrus of the hippocampus in adult rat offspring [156]. In humans, changes in mGluR5 function have been observed, specifically in the hippocampus. A post-mortem study in human alcoholics reported an increased density of mGluR1/5 receptors in the hippocampal CA2 region [157].

mGluR2/3 Receptors

Group II mGluR ligands offer another approach to reduce excessive glutamatergic neurotransmission and have characteristics that recommend exploring their use for treating alcohol dependence.

Systemic manipulation of mGluR2/3 receptors. mGluR2/3 receptors are expressed largely in mesocorticolimbic brain regions implicated in alcohol self-administration and dependence [158]. Numerous examples suggest significant potential targeting these receptors in alcohol dependence. Indeed, systemic administration of LY379268 inhibits the discriminative stimulus effect of alcohol [159], attenuates alcohol self-administration [19, 160], decreases cue-induced reinstatement [30, 160], and decreases stress-induced reinstatement [19, 30]. The inhibitory profile of the effects of LY379268 on ethanol self-administration, stress-induced reinstatement, and cue-induced reinstatement reveals a significantly greater reduction of these behaviors at the lowest dose in post-dependent *vs.* non-dependent rats [19, 25].

Central manipulation of mGluR2/3 receptors. The activation of mGluR2/3 receptors by LY329768 was targeted in both the amygdala and nucleus accumbens because of their involvement in regulating the discriminative stimulus effects of ethanol [161-164]. When injected into the amygdala but not nucleus accumbens, LY329768 inhibited the discriminative stimulus effects of ethanol [159]. These findings suggest an unconventional role for the amygdala in the modulation of the discriminative stimulus properties of ethanol.

Neuroadaptation of mGluR2/3 receptors induced by ethanol. The increased efficacy of LY379268 in post-dependent rats is likely the result of changes in the functional activity of mGluR2/3 receptors that reduces hyperglutamatergic tone associated with ethanol dependence [14, 136]. Footshock and ethanol-associated cues activate the central nucleus of the amygdala and laterodorsal bed nucleus of the stria terminalis [30]. Interestingly, these same brain regions exhibit increased functional activity of mGluR2/3 receptors in post-dependent *vs.* non-dependent rats [25]. The stimulation of mGluR2/3 receptors reduces presynaptic glutamate release and exerts inhibitory control over postsynaptic excitability [40, 158]. Therefore, mGluR2/3 stimulation may exert a greater dampening effect on bed nucleus of the stria terminalis- and central nucleus of the amygdala-mediated reinforcement in post-dependent rats than in their non-dependent counterparts because of the increased agonist sensitivity of mGluR2/3 receptors in these sites [25]. Thus, a history of ethanol dependence produces neuroadaptive changes that may explain the increased sensitivity to the effects of LY379268. Additionally, the medial prefrontal cortex in post-dependent rats and corresponding anterior cingulate cortex in human alcoholics exhibited robust down-regulation of mGluR2 receptors [165]. Therefore, the normalization of mGluR2 function in these brain circuits may have therapeutic value.

Such findings reveal that facilitation of the action of mGluR2/3 receptors was able to prevent ethanol self-administration and protect against relapse elicited by common triggering factors, including alcohol-related stimuli and stress.

Conclusion

Drug addiction is a disorder of the central nervous system characterized by compulsive drug intake and a high risk of relapse despite negative consequences. Presently, the existing effective therapies fail to entirely treat and address the compulsive nature of drug seeking and drug taking that leads to relapse behavior associated with addiction. Animal models that simulate drug reinforcement and relapse suggest that the negative modulation of mGluR5 receptors and positive modulation of mGluR2/3 receptors may have potential as pharmacotherapies for treating addiction to cocaine, nicotine, opioids, and ethanol, while interfering minimally with conventional reinforcer-directed behavior.

One challenge in addiction research is relating aberrant molecular and cellular adaptations to alterations in the function of the neural systems that underlie compulsive drug-seeking behavior. Chronic repeated drug use leads to the formation of powerful and lasting associations between the drug's effects, drug-specific withdrawal symptoms, and environmental cues and contexts that are present at the time when drug-related effects are experienced. Chronic drug use causes neuroadaptations that affect mGluR5 and mGluR2/3 receptors, revealing the key involvement of these receptors in the pathophysiology of addiction and strongly suggest the treatment potential of targeting these metabotropic glutamate receptors.

Acknowledgments

This is publication number 25000 from The Scripps Research Institute. Research support was provided by NIH/NIDA DA033344 (R.M-F.).

We thank M. Arends for editorial assistance.

Alessandra Matzeu and Eva Rebecca Zamora-Martinez participated equally in the writing of this chapter.

References

[1] O'Brien, C. P., McLellan, A. T. (1996) Myths about the treatment of addiction. *Lancet* 347:237-240.

[2] Leshner, A. I. (1997) Addiction is a brain disease, and it matters. *Science* 278:45-47.

[3] O'Brien, C. P., Childress, A. R., Ehrman, R., Robbins, S. J. (1998) Conditioning factors in drug abuse: can they explain compulsion? *J. Psychopharmacol.* 12:15-22.

[4] McLellan, A. T., Lewis, D. C., O'Brien, C. P., Kleber, H. D. (2000) Drug dependence, a chronic medical illness: implications for treatment, insurance, and outcomes evaluation. *JAMA* 284:1689-1695.

[5] Kelley, A. E., Berridge, K. C. (2002) The neuroscience of natural rewards: relevance to addictive drugs. *The Journal of neuroscience: the official journal of the Society for Neuroscience* 22:3306-3311.

[6] Kalivas, P. W., O'Brien, C. (2008) Drug addiction as a pathology of staged neuroplasticity. *Neuropsychopharmacology* 33:166-180.

[7] Koob, G. F., Volkow, N. D. (2010) Neurocircuitry of addiction. *Neuropsychopharmacology* 35:217-238.

[8] Koob, G. F., Zorrilla, E. P. (2010) Neurobiological mechanisms of addiction: focus on corticotropin-releasing factor. *Curr. Opin. Invest. Drugs* 11:63-71.

[9] Spanagel, R., Weiss, F. (1999) The dopamine hypothesis of reward: past and current status. *Trends Neurosci.* 22:521-527.

[10] Self, D. W., Nestler, E. J. (1998) Relapse to drug-seeking: neural and molecular mechanisms. *Drug and alcohol dependence* 51:49-60.

[11] Koob, G. F., Le Moal, M. (1997) Drug abuse: hedonic homeostatic dysregulation. *Science* 278:52-58.

[12] Tzschentke, T. M., Schmidt, W. J. (2003) Glutamatergic mechanisms in addiction. *Mol. Psychiatry* 8:373-382.

[13] Kalivas, P. W. (2004) Glutamate systems in cocaine addiction. *Curr. Opin. Pharmacol.* 4:23-29.

[14] Gass, J. T., Olive, M. F. (2008) Glutamatergic substrates of drug addiction and alcoholism. *Biochem. Pharmacol.* 75:218-265.

[15] Pilc, A., Chaki, S., Nowak, G., Witkin, J. M. (2008) Mood disorders: regulation by metabotropic glutamate receptors. *Biochem. Pharmacol.* 75:997-1006.

[16] Chiamuler, C., Epping-Jordan, M. P., Zocchi, A., Marcon, C., Cottiny, C., Tacconi, S., Corsi, M., Orzi, F., Conquet, F. (2001) Reinforcing and locomotor stimulant effects of cocaine are absent in mGluR5 null mutant mice. *Nature Neurosci.* 4:873-874.

[17] Paterson, N. E., Semenova, S., Gasparini, F., Markou, A. (2003) The mGluR5 antagonist MPEP decreased nicotine self-administration in rats and mice. *Psychopharmacology* 167:257-264.

[18] Rasmussen, K., Martin, H., Berger, J. E., Seager, M. A. (2005) The mGlu5 receptor antagonists MPEP and MTEP attenuate behavioral signs of morphine withdrawal and morphine-withdrawal-induced activation of locus coeruleus neurons in rats. *Neuropharmacology* 48:173-180.

[19] Sidhpura, N., Weiss, F., Martin-Fardon, R. (2010) Effects of the mGlu2/3 agonist LY379268 and the mGlu5 antagonist MTEP on ethanol seeking and reinforcement are differentially altered in rats with a history of ethanol dependence. *Biol. Psychiatry* 67: 804-811.

[20] Liechti, M. E., Markou, A. (2007) Interactive effects of the mGlu5 receptor antagonist MPEP and the mGlu2/3 receptor antagonist LY341495 on nicotine self-administration and reward deficits associated with nicotine withdrawal in rats. *European journal of pharmacology* 554:164-174.

[21] Vandergriff, J., Rasmussen, K. (1999) The selective mGlu2/3 receptor agonist LY 354740 attenuates morphine-withdrawal-induced activation of locus coeruleus neurons and behavioral signs of morphine withdrawal. *Neuropharmacology* 38:217-222.

[22] Baptista, M. A., Martin-Fardon, R., Weiss, F. (2004) Preferential effects of the metabotropic glutamate 2/3 receptor agonist LY379268 on conditioned reinstatement versus primary reinforcement: comparison between cocaine and a potent conventional reinforcer. *J. Neurosci.* 24:4723-4727.

[23] Kenny, P. J., Gasparini, F., Markou, A. (2003) Group II metabotropic and alpha-amino-3-hydroxy-5-methyl-4-isoxazole propionate (AMPA)/kainate glutamate receptors regulate the deficit in brain reward function associated with nicotine withdrawal in rats. *J. Pharmacol. Exp. Ther.* 306:1068-1076.

[24] Hao, Y., Martin-Fardon, R., Weiss, F. (2010) Behavioral and functional evidence of metabotropic glutamate receptor 2/3 and metabotropic glutamate receptor 5 dysregulation in cocaine-escalated rats: factor in the transition to dependence. *Biol. Psychiatry* 68:240-248.

[25] Kufahl, P. R., Martin-Fardon, R., Weiss, F. (2011) Enhanced sensitivity to attenuation of conditioned reinstatement by the mGluR 2/3 agonist LY379268 and increased functional activity of mGluR 2/3 in rats with a history of ethanol dependence. *Neuropsychopharmacology* 36:2762-2773.

[26] Conn, P. J., Pin, J. P. (1997) Pharmacology and functions of metabotropic glutamate receptors. *Annu. Rev. Pharmacol.* 37:205-237.

[27] Pin, J. P., Duvoisin, R. (1995) The metabotropic glutamate receptors - structure and functions. *Neuropharmacology* 34:1-26.

[28] Olive, M. F. (2009) Metabotropic glutamate receptor ligands as potential therapeutics for addiction. *Curr. Drug Abuse Rev.* 2:83-98.

[29] Martin-Fardon, R., Baptista, M. A., Dayas, C. V., Weiss, F. (2009) Dissociation of the effects of MTEP [3-[(2-methyl-1,3-thiazol-4-yl)ethynyl]piperidine] on conditioned reinstatement and reinforcement: comparison between cocaine and a conventional reinforcer. *J. Pharmacol. Exp. Ther.* 329:1084-1090.

[30] Zhao, Y., Dayas, C. V., Aujla, H., Baptista, M. A., Martin-Fardon, R., Weiss, F. (2006) Activation of group II metabotropic glutamate receptors attenuates both stress and cue-induced ethanol-seeking and modulates c-fos expression in the hippocampus and amygdala. *J. Neurosci.* 26:9967-9974.

[31] Dackis, C. A., O'Brien, C. P. (2001) Cocaine dependence: a disease of the brain's reward centers. *J. Subst. Abuse Treat.* 21:111-117.

[32] Substance Abuse and Mental Health Services Administration (2013) Results from the 2012 National Survey on Drug Use and Health: Summary of National Findings. Rockville, MD.

[33] Berke, J. D., Hyman, S. E. (2000) Addiction, dopamine, and the molecular mechanisms of memory. *Neuron* 25:515-532.

[34] Ungless, M. A., Whistler, J. L., Malenka, R. C., Bonci, A. (2001) Single cocaine exposure in vivo induces long-term potentiation in dopamine neurons. *Nature* 411:583-587.

[35] Wolf, M. E. (1998) The role of excitatory amino acids in behavioral sensitization to psychomotor stimulants. *Prog. Neurobiol.* 54:679-720.

[36] Spooren, W. P., Vassout, A., Neijt, H. C., Kuhn, R., Gasparini, F., Roux, S., Porsolt, R. D., Gentsch, C. (2000) Anxiolytic-like effects of the prototypical metabotropic glutamate receptor 5 antagonist 2-methyl-6-(phenylethynyl)pyridine in rodents. *J. Pharmacol. Exp. Ther.* 295:1267-1275.

[37] Spooren, W. P., Schoeffter, P., Gasparini, F., Kuhn, R., Gentsch, C. (2002) Pharmacological and endocrinological characterisation of stress-induced hyperthermia in singly housed mice using classical and candidate anxiolytics (LY314582, MPEP and NKP 608). *Eur. J. Pharmacol.* 435:161-170.

[38] Chojnacka-Wojcik, E., Klodzinska, A., Pilc, A. (2001) Glutamate receptor ligands as anxiolytics. *Curr. Opin. Invest. Drugs* 2:1112-1119.

[39] Kenny, P. J., Markou, A. (2004) The ups and downs of addiction: role of metabotropic glutamate receptors. *Trends Pharmacol. Sci.* 25:265-272.

[40] Swanson, C. J., Bures, M., Johnson, M. P., Linden, A. M., Monn, J. A., Schoepp, D. D. (2005) Metabotropic glutamate receptors as novel targets for anxiety and stress disorders. *Nature Rev. Drug Discov.* 4:131-144.

[41] Ohishi, H., Shigemoto, R., Nakanishi, S., Mizuno, N. (1993) Distribution of the messenger RNA for a metabotropic glutamate receptor, mGluR2, in the central nervous system of the rat. *Neuroscience* 53:1009-1018.

[42] Ohishi, H., Shigemoto, R., Nakanishi, S., Mizuno, N. (1993) Distribution of the mRNA for a metabotropic glutamate receptor (mGluR3) in the rat brain: an in situ hybridization study. *J. Comp. Neurol.* 335:252-266.

[43] Stoker, A. K., Olivier, B., Markou, A. (2012) Involvement of metabotropic glutamate receptor 5 in brain reward deficits associated with cocaine and nicotine withdrawal and somatic signs of nicotine withdrawal. *Psychopharmacology* 221:317-327.

[44] Olive, M. F. (2010) Cognitive effects of Group I metabotropic glutamate receptor ligands in the context of drug addiction. *Eur. J. Pharmacol.* 639:47-58.

[45] Martin-Fardon, R., Weiss, F. (2012) (-)-2-oxa-4-aminobicylco[3.1.0]hexane-4,6-dicarboxylic acid (LY379268) and 3-[(2-methyl-1,3-thiazol-4-yl)ethynyl]piperidine (MTEP) similarly attenuate stress-induced reinstatement of cocaine seeking. *Addict. Biol.* 17:557-564.

[46] McGeehan, A. J., Olive, M. F. (2003) The mGluR5 antagonist MPEP reduces the conditioned rewarding effects of cocaine but not other drugs of abuse. *Synapse* 47:240-242.

[47] Kenny, P. J., Paterson, N. E., Boutrel, B., Semenova, S., Harrison, A. A., Gasparini, F., Koob, G. F., Skoubis, P. D., Markou, A. (2003) Metabotropic glutamate 5 receptor antagonist MPEP decreased nicotine and cocaine self-administration but not nicotine and cocaine-induced facilitation of brain reward function in rats. *Ann. NY Acad. Sci.* 1003:415-418.

[48] Tessari, M., Pilla, M., Andreoli, M., Hutcheson, D. M., Heidbreder, C. A. (2004) Antagonism at metabotropic glutamate 5 receptors inhibits nicotine- and cocaine-taking behaviours and prevents nicotine-triggered relapse to nicotine-seeking. *Eur. J. Pharmacol.* 499:121-133.

[49] Lee, B., Platt, D. M., Rowlett, J. K., Adewale, A. S., Spealman, R. D. (2005) Attenuation of behavioral effects of cocaine by the metabotropic glutamate receptor 5 antagonist 2-methyl-6-(phenylethynyl)-pyridine in squirrel monkeys: comparison with dizocilpine. *J. Pharmacol. Exp. Ther.* 312:1232-1240.

[50] Platt, D. M., Rowlett, J. K., Spealman, R. D. (2008) Attenuation of cocaine self-administration in squirrel monkeys following repeated administration of the mGluR5 antagonist MPEP: comparison with dizocilpine. *Psychopharmacology* 200:167-176.

[51] Paterson, N. E., Markou, A. (2005) The metabotropic glutamate receptor 5 antagonist MPEP decreased break points for nicotine, cocaine and food in rats. *Psychopharmacology* 179:255-261.

[52] Mcgeehan, A. J., Janak, P. H., Olive, M. F. (2004) Effect of the mGluR5 antagonist 6-methyl-2-(phenylethynyl)pyridine (MPEP) on the acute locomotor stimulant properties of cocaine, D-amphetamine, and the dopamine reuptake inhibitor GBR12909 in mice. *Psychopharmacology* 174:266-273.

[53] Dravolina, O. A., Danysz, W., Bespalov, A. Y. (2006) Effects of group I metabotropic glutamate receptor antagonists on the behavioral sensitization to motor effects of cocaine in rats. *Psychopharmacology* 187:397-404.

[54] Kumaresan, V., Yuan, M., Yee, J., Famous, K. R., Anderson, S. M., Schmidt, H. D., Pierce, R. C. (2009) Metabotropic glutamate receptor 5 (mGluR5) antagonists attenuate cocaine priming- and cue-induced reinstatement of cocaine seeking. *Behav. Brain Res.* 202:238-244.

[55] Kotlinska, J., Bochenski, M. (2011) Pretreatment with group I metabotropic glutamate receptors antagonists attenuates lethality induced by acute cocaine overdose and expression of sensitization to hyperlocomotor effect of cocaine in mice. *Neurotox. Res.* 19:23-30.

[56] Knackstedt, L. A., Trantham-Davidson, H. L., Schwendt, M. (2013) The role of ventral and dorsal striatum mGluR5 in relapse to cocaine-seeking and extinction learning. *Addict. Biol.*, in press.

[57] Timmer, K. M., Steketee, J. D. (2012) Examination of a role for metabotropic glutamate receptor 5 in the medial prefrontal cortex in cocaine sensitization in rats. *Psychopharmacology* 221:91-100.

[58] Schmidt, H. D., Schassburger, R. L., Guercio, L. A., Pierce, R. C. (2013) Stimulation of mGluR5 in the accumbens shell promotes cocaine seeking by activating PKC gamma. *J. Neurosci.* 33:14160-14169.

[59] Romano, C., Sesma, M. A., McDonald, C. T., O'Malley, K., Van den Pol, A. N., Olney, J. W. (1995) Distribution of metabotropic glutamate receptor mGluR5 immunoreactivity in rat brain. *J. Comp. Neurol.* 355:455-469.

[60] Shigemoto, R., Nomura, S., Ohishi, H., Sugihara, H., Nakanishi, S., Mizuno, N. (1993) Immunohistochemical localization of a metabotropic glutamate receptor, mGluR5, in the rat brain. *Neurosci. Lett.* 163:53-57.

[61] Ghasemzadeh, M. B., Nelson, L. C., Lu, X. Y., Kalivas, P. W. (1999) Neuroadaptations in ionotropic and metabotropic glutamate receptor mRNA produced by cocaine treatment. *J. Neurochem.* 72:157-165.

[62] Ghasemzadeh, M. B., Vasudevan, P., Giles, C., Purgianto, A., Seubert, C., Mantsch, J. R. (2011) Glutamatergic plasticity in medial prefrontal cortex and ventral tegmental area following extended-access cocaine self-administration. *Brain Res.* 1413:60-71.

[63] Freeman, W. M., Brebner, K., Lynch, W. J., Robertson, D. J., Roberts, D. C., Vrana, K. E. (2001) Cocaine-responsive gene expression changes in rat hippocampus. *Neuroscience* 108:371-380.

[64] Adewale, A. S., Platt, D. M., Spealman, R. D. (2006) Pharmacological stimulation of group ii metabotropic glutamate receptors reduces cocaine self-administration and cocaine-induced reinstatement of drug seeking in squirrel monkeys. *J. Pharmacol. Exp. Ther.* 318:922-931.

[65] Peters, J., Kalivas, P. W. (2006) The group II metabotropic glutamate receptor agonist, LY379268, inhibits both cocaine- and food-seeking behavior in rats. *Psychopharmacology* 186:143-149.

[66] Morishima, Y., Miyakawa, T., Furuyashiki, T., Tanaka, Y., Mizuma, H., Nakanishi, S. (2005) Enhanced cocaine responsiveness and impaired motor coordination in metabotropic glutamate receptor subtype 2 knockout mice. *Proc. Nat. Acad. Sci. US* 102: 4170-4175.

[67] Cannella, N., Halbout, B., Uhrig, S., Evrard, L., Corsi, M., Corti, C., Deroche-Gamonet, V., Hansson, A. C., Spanagel, R. (2013) The mGluR2/3 agonist LY379268 induced anti-reinstatement effects in rats exhibiting addiction-like behavior. *Neuropsychopharmacology* 38:2048-56.

[68] Jin, X., Semenova, S., Yang, L., Ardecky, R., Sheffler, D. J., Dahl, R., Conn, P. J., Cosford, N. D., Markou, A. (2010) The mGluR2 positive allosteric modulator BINA decreases cocaine self-administration and cue-induced cocaine-seeking and counteracts cocaine-induced enhancement of brain reward function in rats. *Neuropsychopharmacology* 35:2021-2036.

[69] Lu, L., Xue, Y., Steketee, J. D., Rebec, G. V., Sun, W. (2012) Regulation of cocaine-induced reinstatement by group II metabotropic glutamate receptors in the ventral tegmental area. *Psychopharmacology* 220:75-85.

[70] Lu, L., Uejima, J. L., Gray, S. M., Bossert, J. M., Shaham, Y. (2007) Systemic and central amygdala injections of the mGluR(2/3) agonist LY379268 attenuate the expression of incubation of cocaine craving. *Biol. Psychiatry* 61:591-598.

[71] Koob, G. F. (1998) Circuits, drugs, and drug addiction. *Adv. Pharmacol.* 42:978-982.

[72] Wise, R. A. (2002) Brain reward circuitry: insights from unsensed incentives. *Neuron* 36: 229-240.

[73] Neugebauer, V., Zinebi, F., Russell, R., Gallagher, J. P., Shinnick-Gallagher, P. (2000) Cocaine and kindling alter the sensitivity of group II and III metabotropic glutamate receptors in the central amygdala. *J. Neurophysiol.* 84:759-770.

[74] Ghasemzadeh, M. B., Mueller, C., Vasudevan, P. (2009) Behavioral sensitization to cocaine is associated with increased glutamate receptor trafficking to the postsynaptic density after extended withdrawal period. *Neuroscience* 159:414-426.

[75] Xi, Z. X., Ramamoorthy, S., Baker, D. A., Shen, H., Samuvel, D. J., Kalivas, P. W. (2002) Modulation of group II metabotropic glutamate receptor signaling by chronic cocaine. *J. Pharmacol. Exp. Ther.* 303:608-615.

[76] Xie, X., Steketee, J. D. (2008) Repeated exposure to cocaine alters the modulation of mesocorticolimbic glutamate transmission by medial prefrontal cortex Group II metabotropic glutamate receptors. *J. Neurochem.* 107:186-196.

[77] Xie, X., Steketee, J. D. (2009) Effects of repeated exposure to cocaine on group II metabotropic glutamate receptor function in the rat medial prefrontal cortex: behavioral and neurochemical studies. *Psychopharmacology* 203:501-510.

[78] Aujla, H., Martin-Fardon, R., Weiss, F. (2008) Rats with extended access to cocaine exhibit increased stress reactivity and sensitivity to the anxiolytic-like effects of the mGluR2/3 agonist LY379268 during abstinence. *Neuropsychopharmacology* 33:1818-1826.

[79] Beveridge, T. J., Smith, H. R., Nader, M. A., Porrino, L. J. (2011) Group II metabotropic glutamate receptors in the striatum of non-human primates: dysregulation following chronic cocaine self-administration. *Neurosci. Lett.* 496:15-19.

[80] United States, Public Health Service, Office of the Surgeon General (2010) How tobacco smoke causes disease: the biology and behavioral basis for smoking-attributable disease. A report of the Surgeon General, US Dept. of Health and Human Services, Public Health Service.

[81] Jorenby, D. E., Leischow, S. J., Nides, M. A., Rennard, S. I., Johnston, J. A., Hughes, A. R., Smith, S. S., Muramoto, M. L., Daughton, D. M., Doan, K., Fiore, M. C., Baker, T. B. (1999) A controlled trial of sustained-release bupropion, a nicotine patch, or both for smoking cessation. *N. Eng. J. Med.* 340:685-691.

[82] Croghan, G. A., Sloan, J. A., Croghan, I. T., Novotny, P., Hurt, R. D., DeKrey, W. L., Mailliard, J. A., Ebbert, L. P., Swan, D. K., Walsh, D. J., Wiesenfeld, M., Levitt, R., Stella, P., Johnson, P. A., Tschetter, L. K., Loprinzi, C. (2003) Comparison of nicotine patch alone versus nicotine nasal spray alone versus a combination for treating smokers: a minimal intervention, randomized multicenter trial in a nonspecialized setting. *Nicotine Tobacco Res.* 5:181-187.

[83] Mansvelder, H. D., McGehee, D. S. (2000) Long-term potentiation of excitatory inputs to brain reward areas by nicotine. *Neuron* 27:349-357.

[84] Markou, A. (2004) Nicotine dependence perpetuating tobacco smoking may be treatable by drugs acting at glutamate receptors. *Discov. Med.* 4:246-254.

[85] Markou, A. (2007) Metabotropic glutamate receptor antagonists: novel therapeutics for nicotine dependence and depression? *Biol. Psychiatry* 61:17-22.

[86] Palmatier, M. I., Liu, X., Donny, E. C., Caggiula, A. R., Sved, A. F. (2008) Metabotropic glutamate 5 receptor (mGluR5) antagonists decrease nicotine seeking, but do not affect the reinforcement enhancing effects of nicotine. *Neuropsychopharmacology* 33: 2139-2147.

[87] Tronci, V., Vronskaya, S., Montgomery, N., Mura, D., Balfour, D. J. (2010) The effects of the mGluR5 receptor antagonist 6-methyl-2-(phenylethynyl)-pyridine (MPEP) on behavioural responses to nicotine. *Psychopharmacology* 211:33-42.

[88] Markou, A., Paterson, N. E., Semenova, S. (2004) Role of gamma-aminobutyric acid (GABA) and metabotropic glutamate receptors in nicotine reinforcement: potential pharmacotherapies for smoking cessation. *Ann. NY Acad. Sci.* 1025:491-503.

[89] Bespalov, A. Y., Dravolina, O. A., Sukhanov, I., Zakharova, E., Blokhina, E., Zvartau, E., Danysz, W., van Heeke, G., Markou, A. (2005) Metabotropic glutamate receptor (mGluR5) antagonist MPEP attenuated cue- and schedule-induced reinstatement of nicotine self-administration behavior in rats. *Neuropharmacology* 49 Suppl. 1:167-178.

[90] Tronci, V., Balfour, D. J. (2011) The effects of the mGluR5 receptor antagonist 6-methyl-2-(phenylethynyl)-pyridine (MPEP) on the stimulation of dopamine release evoked by nicotine in the rat brain. *Behav. Brain Res.* 219:354-357.

[91] Corrigall, W. A., Franklin, K. B., Coen, K. M., Clarke, P. B. (1992) The mesolimbic dopaminergic system is implicated in the reinforcing effects of nicotine. *Psychopharmacology* 107:285-289.

[92] Corrigall, W. A., Coen, K. M., Adamson, K. L. (1994) Self-administered nicotine activates the mesolimbic dopamine system through the ventral tegmental area. *Brain Res.* 653:278-284.

[93] Ikemoto, S., Qin, M., Liu, Z. H. (2006) Primary reinforcing effects of nicotine are triggered from multiple regions both inside and outside the ventral tegmental area. *J. Neurosci.* 26:723-730.

[94] Kenny, P. J., Chartoff, E., Roberto, M., Carlezon, W. A. Jr, Markou, A. (2009) NMDA receptors regulate nicotine-enhanced brain reward function and intravenous nicotine self-administration: role of the ventral tegmental area and central nucleus of the amygdala. *Neuropsychopharmacology* 34:266-281.

[95] Liechti, M. E., Lhuillier, L., Kaupmann, K., Markou, A. (2007) Metabotropic glutamate 2/3 receptors in the ventral tegmental area and the nucleus accumbens shell are involved in behaviors relating to nicotine dependence. *J. Neurosci.* 27:9077-9085.

[96] Wang, F., Chen, H., Steketee, J. D., Sharp, B. M. (2007) Upregulation of ionotropic glutamate receptor subunits within specific mesocorticolimbic regions during chronic nicotine self-administration. *Neuropsychopharmacology* 32:103-109.

[97] D'Souza, M. S., Markou, A. (2011) Metabotropic glutamate receptor 5 antagonist 2-methyl-6-(phenylethynyl)pyridine (MPEP) microinfusions into the nucleus accumbens shell or ventral tegmental area attenuate the reinforcing effects of nicotine in rats. *Neuropharmacology* 61:1399-1405.

[98] Akkus, F., Ametamey, S. M., Treyer, V., Burger, C., Johayem, A., Umbricht, D., Gomez Mancilla, B., Sovago, J., Buck, A., Hasler, G. (2013) Marked global reduction in mGluR5 receptor binding in smokers and ex-smokers determined by [11C]ABP688 positron emission tomography. *Proc. Nat. Acad. Sci. US* 110:737-742.

[99] D'Souza, M. S., Liechti, M. E., Ramirez-Nino, A. M., Kuczenski, R., Markou, A. (2011) The metabotropic glutamate 2/3 receptor agonist LY379268 blocked nicotine-induced increases in nucleus accumbens shell dopamine only in the presence of a nicotine-associated context in rats. *Neuropsychopharmacology* 36:2111-2124.

[100] Epping-Jordan, M. P., Watkins, S. S., Koob, G. F., Markou, A. (1998) Dramatic decreases in brain reward function during nicotine withdrawal. *Nature* 393:76-79.

[101] Termorshuizen, F., Krol, A., Prins, M., van Ameijden, E. J. (2005) Long-term outcome of chronic drug use: the Amsterdam Cohort Study among Drug Users. *Am. J. Epidemiol.* 161:271-279.

[102] Goldstein, A., Herrera, J. (1995) Heroin addicts and methadone treatment in Albuquerque: a 22-year follow-up. *Drug Alcohol Depend.* 40:139-150.

[103] Hser, Y. I., Hoffman, V., Grella, C. E., Anglin, M. D. (2001) A 33-year follow-up of narcotics addicts. *Arch. Gen. Psychiatry* 58:503-508.

[104] Oppenheimer, E., Tobutt, C., Taylor, C., Andrew, T. (1994) Death and survival in a cohort of heroin addicts from London clinics: a 22-year follow-up study. *Addiction* 89: 1299-1308.

[105] Hser, Y. I. (2007) Predicting long-term stable recovery from heroin addiction: findings from a 33-year follow-up study. *J. Addict. Dis.* 26:51-60.

[106] Gerstein, D. R., Lewin, L. S. (1990) Treating drug problems. *N. Eng. J. Med.* 323:844-848.

[107] Bonhomme, J., Shim, R. S., Gooden, R., Tyus, D., Rust, G. (2012) Opioid addiction and abuse in primary care practice: a comparison of methadone and buprenorphine as treatment options. *JAMA* 104:342-350.

[108] Koob, G. F., Le Moal, M. (2006) *Neurobiology of Addiction*. Elsevier Academic Press, Amsterdam.

[109] Angst, M. S., Clark, J. D. (2006) Opioid-induced hyperalgesia: a qualitative systematic review. *Anesthesiology* 104:570-587.

[110] Simonnet, G., Rivat, C. (2003) Opioid-induced hyperalgesia: abnormal or normal pain? *Neuroreport* 14:1-7.

[111] Celerier, E., Laulin, J. P., Corcuff, J. B., Le Moal, M., Simonnet, G. (2001) Progressive enhancement of delayed hyperalgesia induced by repeated heroin administration: a sensitization process. *J. Neurosci.* 21:4074-4080.

[112] Edwards, S., Vendruscolo, L. F., Schlosburg, J. E., Misra, K. K., Wee, S., Park, P. E., Schulteis, G., Koob, G. F. (2012) Development of mechanical hypersensitivity in rats during heroin and ethanol dependence: alleviation by CRF_1 receptor antagonism. *Neuropharmacology* 62:1142-1151.

[113] Osikowicz, M., Mika, J., Makuch, W., Przewlocka, B. (2008) Glutamate receptor ligands attenuate allodynia and hyperalgesia and potentiate morphine effects in a mouse model of neuropathic pain. *Pain* 139:117-126.

[114] Berthele, A., Platzer, S., Laurie, D. J., Weis, S., Sommer, B., Zieglgansberger, W., Conrad, B., Tolle, T. R. (1999) Expression of metabotropic glutamate receptor subtype mRNA (mGluR1-8) in human cerebellum. *Neuroreport* 10:3861-3867.

[115] Bhave, G., Karim, F., Carlton, S. M., Gereau, R. Wt. (2001) Peripheral group I metabotropic glutamate receptors modulate nociception in mice. *Nature Neurosci.* 4: 417-423.

[116] Van der Kam, E. L., de Vry, J., Tzschentke, T. M. (2007) Effect of 2-methyl-6-(phenyl-ethynyl) pyridine on intravenous self-administration of ketamine and heroin in the rat. *Behav. Pharmacol.* 18:717-724.

[117] Van der Kam, E. L., De Vry, J., Tzschentke, T. M. (2009) 2-methyl-6-(phenylethynyl)-pyridine (MPEP) potentiates ketamine and heroin reward as assessed by acquisition, extinction, and reinstatement of conditioned place preference in the rat. *Eur. J. Pharmacol.* 606:94-101.

[118] Van der Kam, E. L., De Vry, J., Tzschentke, T. M. (2009) The mGlu5 receptor antagonist 2-methyl-6-(phenylethynyl)pyridine (MPEP) supports intravenous self-administration and induces conditioned place preference in the rat. *Eur. J. Pharmacol.* 607:114-120.

[119] Popik, P., Wrobel, M. (2002) Morphine conditioned reward is inhibited by MPEP, the mGluR5 antagonist. *Neuropharmacology* 43:1210-1217.

[120] O'Brien, C. P., Childress, A. R., McLellan, A. T., Ehrman, R. (1992) Classical conditioning in drug-dependent humans. *Ann. NY Acad. Sci.* 654:400-415.

[121] Fuchs, R. A., See, R. E. (2002) Basolateral amygdala inactivation abolishes conditioned stimulus- and heroin-induced reinstatement of extinguished heroin-seeking behavior in rats. *Psychopharmacology* 160:425-433.

[122] McFarland, K., Ettenberg, A. (1997) Reinstatement of drug-seeking behavior produced by heroin-predictive environmental stimuli. *Psychopharmacology* 131:86-92.

[123] Shalev, U., Grimm, J. W., Shaham, Y. (2002) Neurobiology of relapse to heroin and cocaine seeking: a review. *Pharmacol. Rev.* 54:1-42.

[124] Bossert, J. M., Liu, S. Y., Lu, L., Shaham, Y. (2004) A role of ventral tegmental area glutamate in contextual cue-induced relapse to heroin seeking. *J. Neurosci.* 24:10726-10730.

[125] Bossert, J. M., Busch, R. F., Gray, S. M. (2005) The novel mGluR2/3 agonist LY379268 attenuates cue-induced reinstatement of heroin seeking. *Neuroreport* 16: 1013-1016.

[126] Tahsili-Fahadan, P., Carr, G. V., Harris, G. C., Aston-Jones, G. (2010) Modafinil blocks reinstatement of extinguished opiate-seeking in rats: mediation by a glutamate mechanism. *Neuropsychopharmacology* 35:2203-2210.

[127] Bossert, J. M., Gray, S. M., Lu, L., Shaham, Y. (2006) Activation of group II metabotropic glutamate receptors in the nucleus accumbens shell attenuates context-induced relapse to heroin seeking. *Neuropsychopharmacology* 31:2197-2209.

[128] Koob, G. F. (2013) Addiction is a reward deficit and stress surfeit disorder. *Front Psychiatry* 4:72.

[129] Heilig, M., Egli, M. (2006) Pharmacological treatment of alcohol dependence: target symptoms and target mechanisms. *Pharmacol. Ther.* 111:855-876.

[130] Spanagel, R., Zieglgansberger, W. (1997) Anti-craving compounds for ethanol: new pharmacological tools to study addictive processes. *Trends Pharmacol. Sci.* 18:54-59.

[131] Le, A., Shaham, Y. (2002) Neurobiology of relapse to alcohol in rats. *Pharmacol. Ther.* 94:137-156.

[132] Weiss, F., Porrino, L. J. (2002) Behavioral neurobiology of alcohol addiction: recent advances and challenges. *J. Neurosci.* 22:3332-3337.

[133] Lovinger, D. M., White, G., Weight, F. F. (1989) Ethanol inhibits NMDA-activated ion current in hippocampal neurons. *Science* 243:1721-1724.

[134] Lima-Landman, M. T., Albuquerque, E. X. (1989) Ethanol potentiates and blocks NMDA-activated single-channel currents in rat hippocampal pyramidal cells. *FEBS Lett.* 247:61-67.

[135] Weight, F. F., Peoples, R. W., Wright, J. M., Lovinger, D. M., White, G. (1993) Ethanol action on excitatory amino acid activated ion channels. *Alcohol Alcohol Suppl.* 2: 353-358.

[136] Tsai, G., Coyle, J. T. (1998) The role of glutamatergic neurotransmission in the pathophysiology of alcoholism. *Ann. Rev. Med.* 49:173-184.

[137] Hoffman, P. L. (1995) Glutamate receptors in alcohol withdrawal-induced neuro-toxicity. *Metabol. Brain Dis.* 10:73-79.

[138] Hoffman, P. L., Tabakoff, B. (1994) The role of the NMDA receptor in ethanol withdrawal. *Exs* 71:61-70.

[139] Besheer, J., Hodge, C. W. (2005) Pharmacological and anatomical evidence for an interaction between mGluR5- and $GABA_A$ α1-containing receptors in the discriminative stimulus effects of ethanol. *Neuropsychopharmacology* 30:747-757.

[140] Besheer, J., Stevenson, R. A., Hodge, C. W. (2006) mGlu5 receptors are involved in the discriminative stimulus effects of self-administered ethanol in rats. *Eur. J. Pharmacol.* 551:71-75.

[141] Backstrom, P., Bachteler, D., Koch, S., Hyytia, P., Spanagel, R. (2004) mGluR5 antagonist MPEP reduces ethanol-seeking and relapse behavior. *Neuropsychopharma-cology* 29:921-928.

[142] Schroeder, J. P., Spanos, M., Stevenson, J. R., Besheer, J., Salling, M., Hodge, C. W. (2008) Cue-induced reinstatement of alcohol-seeking behavior is associated with increased ERK1/2 phosphorylation in specific limbic brain regions: blockade by the mGluR5 antagonist MPEP. *Neuropharmacology* 55:546-554.

[143] Kotlinska, J. H., Bochenski, M., Danysz, W. (2011) The role of group I mGlu receptors in the expression of ethanol-induced conditioned place preference and ethanol withdrawal seizures in rats. *Eur. J. Pharmacol.* 670:154-161.

[144] Gupta, T., Syed, Y. M., Revis, A. A., Miller, S. A., Martinez, M., Cohn, K. A., Demeyer, M. R., Patel, K. Y., Brzezinska, W. J., Rhodes, J. S. (2008) Acute effects of acamprosate and MPEP on ethanol Drinking-in-the-Dark in male C57BL/6J mice. *Alcohol Clin. Exp. Res.* 32:1992-1998.

[145] Besheer, J., Faccidomo, S., Grondin, J. J., Hodge, C. W. (2008) Regulation of motivation to self-administer ethanol by mGluR5 in alcohol-preferring (P) rats. *Alcohol Clin. Exp. Res.* 32:209-221.

[146] Schroeder, J. P., Overstreet, D. H., Hodge, C. W. (2005) The mGluR5 antagonist MPEP decreases operant ethanol self-administration during maintenance and after repeated alcohol deprivations in alcohol-preferring (P) rats. *Psychopharmacology* 179:262-270.

[147] Bird, M. K., Kirchhoff, J., Djouma, E., Lawrence, A. J. (2008) Metabotropic glutamate 5 receptors regulate sensitivity to ethanol in mice. *Int. J. Neuropsychopharmacol.* 11: 765-774.

[148] Blednov, Y. A., Harris, R. A. (2008) Metabotropic glutamate receptor 5 (mGluR5) regulation of ethanol sedation, dependence and consumption: relationship to acamprosate actions. *Int. J. Neuropsychopharmacol.* 11:775-793.

[149] Olive, M. F., Becker, H. C. (2008) Effects of the mGluR2/3 agonist LY379268 and the mGluR5 antagonist MPEP on handling-induced convulsions during ethanol withdrawal in mice. *Alcohol* 42:191-197.

[150] Lu, X. Y., Ghasemzadeh, M. B., Kalivas, P. W. (1999) Expression of glutamate receptor subunit/subtype messenger RNAS for NMDAR1, GLuR1, GLuR2 and mGLuR5 by accumbal projection neurons. *Brain Res. Mol. Brain Res.* 63:287-296.

[151] Volkow, N. D., Fowler, J. S., Wang, G. J. (2003) The addicted human brain: insights from imaging studies. *J. Clin. Invest.* 111:1444-1451.

[152] Besheer, J., Grondin, J. J., Cannady, R., Sharko, A. C., Faccidomo, S., Hodge, C. W. (2010) Metabotropic glutamate receptor 5 activity in the nucleus accumbens is required for the maintenance of ethanol self-administration in a rat genetic model of high alcohol intake. *Biol. Psychiatry* 67:812-822.

[153] Sinclair, C. M., Cleva, R. M., Hood, L. E., Olive, M. F., Gass, J. T. (2012) mGluR5 receptors in the basolateral amygdala and nucleus accumbens regulate cue-induced reinstatement of ethanol-seeking behavior. *Pharmacol. Biochem. Behav.* 101:329-335.

[154] Gass, J. T., Olive, M. F. (2009) Role of protein kinase C epsilon (PKCε) in the reduction of ethanol reinforcement due to mGluR5 antagonism in the nucleus accumbens shell. *Psychopharmacology* 204:587-597.

[155] Szumlinski, K. K., Diab, M. E., Friedman, R., Henze, L. M., Lominac, K. D., Bowers, M. S. (2007) Accumbens neurochemical adaptations produced by binge-like alcohol consumption. *Psychopharmacology* 190:415-431.

[156] Galindo, R., Frausto, S., Wolff, C., Caldwell, K. K., Perrone-Bizzozero, N. I., Savage, D. D. (2004) Prenatal ethanol exposure reduces mGluR5 receptor number and function in the dentate gyrus of adult offspring. *Alcohol Clin. Exp. Res.* 28:1587-1597.

[157] Kupila, J., Karkkainen, O., Laukkanen, V., Tupala, E., Tiihonen, J., Storvik, M. (2013) mGluR1/5 receptor densities in the brains of alcoholic subjects: a whole-hemisphere autoradiography study. *Psychiatry Res.* 212:245-250.

[158] Schoepp, D. D. (2001) Unveiling the functions of presynaptic metabotropic glutamate receptors in the central nervous system. *J. Pharmacol. Exp. Ther.* 299:12-20.

[159] Cannady, R., Grondin, J. J., Fisher, K. R., Hodge, C. W., Besheer, J. (2011) Activation of group II metabotropic glutamate receptors inhibits the discriminative stimulus effects of alcohol via selective activity within the amygdala. *Neuropsychopharmacology* 36: 2328-2338.

[160] Backstrom, P., Hyytia, P. (2005) Suppression of alcohol self-administration and cue-induced reinstatement of alcohol seeking by the mGlu2/3 receptor agonist LY379268 and the mGlu8 receptor agonist (S)-3,4-DCPG. *Eur. J. Pharmacol.* 528:110-118.

[161] Besheer, J., Cox, A. A., Hodge, C. W. (2003) Coregulation of ethanol discrimination by the nucleus accumbens and amygdala. *Alcohol Clin. Exp. Res.* 27:450-456.

[162] Hodge, C. W., Cox, A. A., Bratt, A. M., Camarini, R., Iller, K., Kelley, S. P., Mehmert, K. K., Nannini, M. A., Olive, M. F. (2001) The discriminative stimulus properties of self-administered ethanol are mediated by $GABA_A$ and NMDA receptors in rats. *Psychopharmacology* 154:13-22.

[163] Hodge, C. W., Cox, A. A. (1998) The discriminative stimulus effects of ethanol are mediated by NMDA and $GABA_A$ receptors in specific limbic brain regions. *Psychopharmacology* 139:95-107.
[164] Hodge, C. W., Alken, A. S. (1996) Discriminative stimulus function of ethanol: role of $GABA_A$ receptors in the nucleus accumbens. *Alcohol Clin. Exp. Res*. 20:1221-1228.
[165] Meinhardt, M. W., Hansson, A. C., Perreau-Lenz, S., Bauder-Wenz, C., Stahlin, O., Heilig, M., Harper, C., Drescher, K. U., Spanagel, R., Sommer, W. H. (2013) Rescue of infralimbic mGluR2 deficit restores control over drug-seeking behavior in alcohol dependence. *J. Neurosci*. 33:2794-2806.

In: Metabotropic Glutamate Receptors
Editor: M. Foster Olive

ISBN: 978-1-63117-451-3

Chapter 9

Therapeutic Potential of mGluR Agents on Substance Abuse and Mood Disorders: Role of Stress Reactivity during Abstinence and Withdrawal

Peter R. Kufahl [*] ***and Sara B. Taylor***
Department of Psychology, Arizona State University, AZ, US

Abstract

Pharmacological modulators of metabotropic glutamate receptor (mGluR) activity have been clinically investigated for treatment of various psychiatric disorders, including substance abuse, depression, and anxiety. However, the effectiveness of mGluR agents as anti-addiction therapies varies with different drugs of abuse and degree of chemical dependency as well as individual vulnerabilities to develop comorbid mood disorders. Interestingly, depression, anxiety, and substance abuse disorders are frequently characterized by maladaptive stress reactivity. Thus, one explanation for the clinical limitations of mGluR agents in the treatment of addiction is that the extent and nature of changes incurred by the stress-mediating systems from drug abuse depend on various developmental and environmental factors. This chapter reviews the evidence for mGluR agents as treatments for substance abuse disorders, and relates the co-occurrence of stress dysregulation and mood disorders with the advanced stages of drug addiction to potential interference with treatment strategies.

Keywords: Stress, relapse, reinstatement, anxiety, addiction, withdrawal, dependence

[*] Corresponding Author address: Dr. Peter R. Kufahl, Department of Psychology, Arizona State University, PO Box 871104, Tempe, AZ, 85287-1104, Email: pkufahl@asu.edu.

Introduction

Drug addiction is a chronic, relapsing disorder characterized by habitual pursuit and intake of psychoactive substances, even in the face of harmful consequences [1]. Addiction is known as a progressive disease; initial experimentation and consumption of drugs provide highly reinforcing experiences [2], but in many individuals drug use eventually becomes a means of self-medication against chronic negative mood states, including symptoms of depression and anxiety [3]. This negative affect, frequently brought on by physiological drug withdrawal [4,5], is often accompanied by a persistent desire to procure the drug often referred to as craving [6,7]. Whether it is characterized as an explicit feeling or as an unconscious memory trace, craving leads to relapse behaviors of drug use, even after significant periods of abstinence [8,9]. Relapses can occur spontaneously or following external triggers in the environment [10]. Such triggers include sensory cues that activate drug-related memories [11], intake of other psychoactive substances [12] and emotionally stressful events [13]. While the occurrence of craving sensations is strongly predictive of relapse to drug taking [11], these environmental triggers also provoke autonomic responses that can bring about relapse without first recalling drug-related memories or eliciting feelings of craving [14-16]. Therefore, maintenance of extended abstinence from drug use and reduction of self-reports of craving are both goals of addiction treatment strategies [17].

The standard treatment strategy for addiction is the combination of a therapeutic drug regimen, where available, with behavioral and psychosocial treatment [18, 19]. The presently available pharmacological therapies for addiction have had limited success in maintaining abstinence over an extended period in patients, and the variety of side effects associated with these drugs has hindered the rate of compliance [20-22]. Additionally, many forms of chemical addiction, including methamphetamine and cocaine dependence continue to lack approved pharmacotherapies of any kind. Contemporary research into new treatments has therefore focused on drug targets that offer efficacy in reducing intake and craving while maintaining a widely acceptable profile of side effects. At the same time, several clinically approved therapeutics of other disorders have been investigated as possible treatments of addictive disorders [23]. These drugs in particular are of interest for treatment of patients with addiction and comorbid psychiatric disorders, whose effective treatment represents an additional significant public health challenge [24].

Many agents being investigated as potential treatments for addictive disorders target various aspects of the glutamatergic neurotransmission system. Glutamate is the predominant excitatory neurotransmitter [25] and its function is widely implicated in the various reinforcement learning, cue memory, and cognitive control aspects of addiction [26,27]. The system of glutamatergic projections from the medial prefrontal cortex (mPFC) to the nucleus accumbens (NAc) of the ventral striatum, hypothesized to be an important link between the processing of environmental cues and goal-directed behavior, is a critical substrate of relapse to drug-seeking [28-30]. Because chronic exposure to drugs of abuse has been shown to induce long-lasting alterations in this pathway [31,32], compounds have been developed to target metabotropic glutamate receptors (mGluRs), which modulate glutamate production, transmission and uptake [33,34].

Group I and Group II mGluRs are thought to act in tandem to manage the production, release, and uptake of synaptic glutamate when stimulated by glutamate in the extracellular space [30]. While glutamate is the most abundant excitatory amino acid in the brain and most neurons and glial cells express glutamate receptors [35], its extracellular and intracellular concentrations are starkly different to avoid neurotoxic effects [36]. Disruption of the mechanisms, including the glutamate-cysteine antiporter xc- and glutamate transporter 1 (GLT1) that maintain this concentration gradient by chronic drug abuse is thought to result in suboptimal functioning of Group I and Group II mGluRs [30]. Recovery of normal function of xc- and GLT1 by administration of their substrates N-acetylcysteine and ceftriaxone, respectively, has shown promise as a potential therapeutic strategy against addictive behaviors [37,38].

Many studies were performed to evaluate the potential for mGluR ligands to exert control over drug self-administration, withdrawal behaviors, and cue- or drug-triggered drug-seeking [39]. However, the results of these investigations have been mixed, with glutamatergic agents showing promise as potent and specific anti-addictive treatments in some circumstances but failing in others. Despite the well-characterized roles identified for mGluRs in stress and anxiety behaviors, the influence of the stress-mediating brain systems on the effectiveness of mGluR treatments in addiction models remains largely unexplored. This chapter provides an overview of mGluR contributions to behaviors associated with the progressive stages of addiction, details the involvement of mGluRs in stress-related brain systems, and proposes a direction of future research aimed at revealing the circumstances in which mGluR-based treatments would be specific and reliable interventions against drug-seeking and drug-taking behaviors.

The Roles of mGluRs in the Progressive Stages of Addiction

Due to their dual roles in mediating glutamatergic synaptic plasticity and control of synaptic glutamate release, mGluRs have emerged as therapeutic targets of interest in the study of drug addiction [27]. Many studies have shown that mGluRs change in population, distribution, and function in response to acute drug exposure, chronic drug exposure, and both immediate and protracted drug withdrawal [27,40,41]. There exist widely varying conceptualizations of addiction, but many of them generally agree that the disease involves a progression from experimentation and highly reinforcing drug experiences, to a habitual self-treatment of chronic withdrawal and negative affect, to a state of persistent and encompassing craving and preoccupation for drugs [42]. The ubiquity of glutamate neurotransmission is reflected in the evidence for the involvement of mGluR function in all of these behavioral stages, for several drugs of abuse.

Glutamatergic Influences on Brain Reward Activity

While the reinforcing and stimulatory effects of addictive drugs have long been associated with neuroadaptations of dopamine receptor expression and function within the mesolimbic dopamine system [43], recent research has also implicated important changes in the glutamatergic system. Cocaine administration has been shown to induce a long-lasting increase in the α-amino-3-hydroxy-5-methyl-4-isoxazolepropionic acid (AMPA) receptor-mediated excitatory synaptic strength of mesolimbic dopamine neurons [44]. Initially associated with the development of locomotor sensitization to passively administered psychostimulants [45], potentiated glutamate release in the mesolimbic neurons has also been tied to the development of drug self-administration [46]. This long-term potentiation (LTP) effect was found to occur over several sessions of training that paired sensory cues to subsequent reward [47], making the potentially critical association between reward learning and the glutamate system. *Ex vivo* stimulation of mGluR1 reverses drug-induced LTP in mesolimbic neurons [48]. The mesolimbic pathway is therefore a potentially important substrate for the effects of mGluR ligands, in addition to the prefrontal cortical projections to the NAc mentioned earlier [41].

Stage 1: Experimentation and Intoxication

The initial stage of drug addiction is characterized by episodes of experimental and typically highly mood-enhancing drug experiences, which engage a learning process known as positive reinforcement [49-51]. These processes are modeled in animals by drug self-administration experiments, where operant responding for drug delivery is learned over several daily sessions [52]. Drugs of abuse also facilitate the rewarding properties of subsequent experiences, modeled in animals with intracranial self-stimulation experiments (ICSS). Animals readily acquire operant behavior that results in electrical stimulation, as long as the frequency and magnitude of the electrical current is sufficient to support this activity [53]. When pretreated with a mood-enhancing drug of abuse, animals require a smaller current threshold to continue responding for intracranial self-stimulation [54-56]. Both the directly rewarding and mood-enhancing effects of drug administration have been confirmed in functional magnetic resonance imaging (fMRI) experiments of addicted humans, where metabolic activation of reward-sensitive brain centers coincided with subject reports of drug rush and elevated mood [57,58]. For many individuals, these rewarding experiences result in a positive craving for further drug experiences, despite knowledge of the costs and dangers of drug abuse [17]. Behavioral pharmacology experimentation has established roles for mGluRs in the reward-enhancing and reinforcing properties of several drugs of abuse.

Psychostimulants

Group I mGluRs (mGluR1 and mGluR5), particularly mGluR5, have been associated with the primary reinforcing effects of drugs of abuse. A pioneering study demonstrated that knockout mice lacking mGluR5, a subtype of mGluR normally located throughout the forebrain and limbic system, failed to exhibit an enhanced locomotive response to cocaine and did not self-administer cocaine to the levels of wild-type mice [59]. These findings instigated a series of studies across several laboratories linking mGluR5 expression and function with the direct effects of psychostimulants [60-63].

The negative allosteric modulator (NAM) of mGluR5 2-methyl-6-(phenylethynyl) pyridine hydrochloride (MPEP) has been shown to reduce self-administration for cocaine in various animal models [64-66]; self-administration performance was gauged as daily consumption rate, or performance during progressive ratio testing. In the latter case, the minimum threshold number of operant responses per reinforcement was determined by stepping through increasing number of responses. MPEP was also found to degrade reward function as measured by increased ICSS threshold, indicating nonspecific psychological effects underpinning its effectiveness against cocaine self-administration [67]. The more selective mGluR5 NAM 3-[(2-methyl-1,3-thiazol-4-yl)ethynyl]-pyridine (MTEP) was also found to attenuate self-administration of cocaine [68], and both self-administration responding and breakpoint performance for methamphetamine [62]. However, the short half-life and tolerance characteristics of MPEP and MTEP preclude their use as clinical treatments, and alternative mGluR5 agents have been sought for similar behavioral effects [34]. The anxiolytic drug N-(3-chlorophenyl)-N′-(4,5-dihydro-1-methyl-4-oxo-1H-imidazole-2-yl)urea (fenobam), currently investigated as a possible treatment for Fragile-X syndrome, was recently discovered to be a selective mGluR5 NAM [69]. Fenobam treatment resulted in reduced cocaine self-administration, but also attenuated responding for sucrose pellets, indicating nonspecific suppression of reward-directed behaviors [70].

Group II mGluRs (mGluR2 and mGluR3) are predominantly presynaptically and perisynaptically located throughout the cortex and limbic brain areas, and generally act as inhibitory modulators over synaptic glutamate release. Stimulation of Group II mGluRs has been investigated as a means of reducing drug self-administration, most frequently with the selective agonist (−)-2-oxa-4-aminobicyclo[3.1.0]hexane-4,6-dicarboxylate (LY379268). Treatment with LY379268 effectively reduced cocaine self-administration in monkeys [71] but not as successfully in rats [72]. A different study found that LY379268 non-selectively reduced cocaine and food self-administration [73]. However, the selective mGluR2 positive allosteric modulator (PAM) 4-[3-[(2-cyclopentyl-6,7-dimethyl-1-oxo-2,3-dihydroinden-5-yl)oxymethyl]phenyl]benzoic acid (BINA) effectively reduced cocaine self-administration without changing responding for sucrose [73]. Therefore, limited evidence suggests that selective stimulation of mGluR2 could be a potential strategy for reducing the psychostimulant use in active drug abusers.

The Group III mGluR (mGluR4, mGluR6, mGluR7, and mGluR8) subfamily has also been investigated for its role on cocaine-induced behaviors, following the development of the systemically active mGluR7 PAM N,N'-dibenzhydrylethane-1,2-diamine dihydrochloride (AMN082) [74]. Treatment with AMN082 dose-dependently reversed cocaine-lowered ICSS thresholds and diminished cocaine self-administration, which was in turn restored with an injection of a mGluR7 antagonist into the NAc [75,76]. However, the presence of baseline locomotor effects imposed by AMN082 on rodents has been a matter of some disagreement [76,77,78].

Nicotine

Tobacco smoking, a major driver of preventable morbidity and mortality, is heavily associated with dependence on nicotine, a stimulant that directly activates nicotinic acetylcholine receptors (nAchRs) throughout the brain [79]. Stimulation of nAchRs on mesolimbic dopaminergic neurons results in downstream effects throughout the complex network of dopamine, glutamate, and γ-aminobutyric acid (GABA) systems that mediate the activity of the mesolimbic reward pathway [80,81]. Nicotine also indirectly stimulates mesolimbic dopamine neurons by acting on nAchRs on glutamatergic afferents [82]. These circuits undergo adaptive changes when chronically stimulated by nicotine, notably by long-lasting changes in the expression of NMDA and AMPA receptors [83]. Nicotine use also affects densities of mGluR5 in the brain. In a recent study utilizing ^{11}C-ABP688 positron emission tomography, mGluR5 levels were significantly (about 20%) reduced in the caudate, putamen, and insula cortex of smokers, with extended abstinence from smoking associated with the recovery of these levels [84]. The insula has been previously associated with the urge to smoke in a well-known clinical study of recovering stroke patients [85].

Preclinical models of nicotine abuse have conclusively linked the transmission of glutamate at the midbrain with the reinforcing properties of nicotine [86]. When administered into the midbrain or systemically, compounds that block the function of postsynaptic N-methyl-d-aspartic acid (NMDA) receptors reduced nicotine reward and self-administration [87-89]. Likewise, interference of mGluR5 receptors with MPEP reduced nicotine self-administration [64,90,91], the progressive schedule breakpoint for nicotine reinforcement [65] and the ability of nicotine to enhance the rewarding attributes of other stimuli [92]. Unfortunately, MPEP also reduced the progressive schedule breakpoint for food reinforcement, indicating a nonspecific effect on motivation [65]. MPEP treatment reduced the ability of nicotine to lower the threshold current for electrical self-stimulation reinforcement [but see 93,94], but also increased this threshold in drug-naïve animals [93]. MTEP was found to decrease self-administration of nicotine as well, by counteracting the primary reinforcement, but not the reward-enhancing, property of nicotine [93,95]. When infused into the mesolimbic dopamine system, the mGluR2/3 agonist LY379268 has also been shown to reduce nicotine self-administration in rats, though context affected its potency [96,97].

Alcohol

Alcohol exerts its psychoactive effects through various neurotransmitter systems, including direct antagonism of NMDA receptors in the mesolimbic dopamine system [98]. In studies of isolated cells and *Xenopus* oocytes, alcohol has been shown to suppress mGluR5 activity via the protein kinase C molecular pathway [99,100]. The compound acamprosate, one of three clinically available treatments for alcoholism, exerts some of its therapeutic effects by acting as an antagonist of mGluR5 [101,102]. Treatment with MPEP or MTEP interfered with the onset and continued maintenance of alcohol self-administration in mice [103-105]. MPEP and MTEP also reduced alcohol drinking behavior in rats bred for a high alcohol preference [106-108]. MPEP treatment also largely cancelled the tendency for alcohol self-administering animals to exhibit increased drinking following a period of forced abstinence, a phenomenon known as the alcohol deprivation effect [109, 110]. However, the potency of MTEP in attenuating alcohol consumption in outbred rats was shown to be subject to the alcohol exposure history of the animals [111]. Additionally, studies of genetic knockout mouse models have linked the development of alcohol drinking behavior with a dysfunction of mGluR7 [112]. Stimulation of presynaptic mGluR7 with AMN082 was shown to diminish alcohol drinking in mice, but this treatment suppressed sucrose self-administration as well [113].

Opiates

Although MTEP was found to attenuate self-administration of morphine in mice [114], administration of MPEP failed to significantly affect responding for heroin in the rat model [115,116]. MPEP reduced the expression of morphine conditioned place preference [117], but facilitated the development of ketamine and heroin conditioned place preference [116].

Therefore, the evidence supporting the potential of mGluR ligands as a strategy to reduce opiate consumption remains inconsistent.

Summary

The primary brain circuits engaged by acute drug abuse in the early stage of addiction, even bingeing, are those mediating reward and incentive salience [118]. All drugs of abuse directly or indirectly activate the mesolimbic dopamine system, and presumably exert effects over the glutamatergic synapses that connect it to the prefrontal and limbic brain areas. However, the primary substrates of the acute effects of drugs vary by drug type: cocaine and other psychostimulants target dopamine and other monoamines, nicotine directly affects acetylcholine neurotransmission, and the reinforcing effects of alcohol arise from engagement of GABA and opioid peptides. This variation helps explain the different degrees of efficacy of mGluR ligands against drug consumption and drug-elicited sensitization.

Inconsistent effects on opiate self-administration and reward aside, the mGluR5 modulators MPEP and MTEP inhibit reinforcement-supported behavior across a broad range of drugs of abuse [64,105,106]. Additionally, mGluR5 antagonism with MTEP interferes with the expression of conditioned food reward, indicating a role for this receptor in the acquisition of incentive motivation, regardless of reinforcement strategy [119]. While these broad effects on appetitive behavior are undesirable in therapeutic interventions against drug abuse, they may encourage the investigation of mGluR5 agents for the facilitation of learning the incentive properties of alternative non-drug rewards [120].

Stage 2: Withdrawal and Negative Affect

The hypothesized stage of addiction at which neuroadaptations from chronic drug use manifest as decreased function of the reward system and facilitation of an anti-reward system, already in place to limit reward, is known as the *withdrawal/negative affect* stage [51,121]. This stage is marked by the presence of chronic irritability, physical discomfort and emotional pain, loss of motivation for non-drug rewards and the use of drugs as a means to combat these persistent negative feelings [122-124]. Additionally, drugs are often consumed to mentally escape negative situations brought about by past drug use [125]. Acute chemical withdrawal from the chronic intake of any recognized drug of abuse results in elevated ICSS thresholds, which describes a compromised reward system [126]. When drugs are self-administered to stave off the physical and emotional symptoms of chemical withdrawal, their use becomes more likely as a result and engages a process of negative reinforcement [121]. Given their existing indications as candidate anxiety and depression treatments [127], compounds that target mGluRs have shown promise in alleviating withdrawal symptoms associated with several drugs of abuse.

Psychostimulants

Chronic psychostimulant abuse is accompanied by intermittent withdrawal periods that can bear a strong resemblance to clinical depression [128]. Though many aspects of depression and negative affect can be difficult to model in rodents, rats have been shown to develop conditioned place aversion to an environment paired with cocaine withdrawal [129]. Withdrawal from amphetamine resulted in a decreased motivation for sucrose reinforcement, an example of reduced desire for natural rewards [130]. However, the degree to which anhedonia and other depressive symptoms contribute to the overall experience of psychostimulant withdrawal is not entirely consistent within the clinical evidence, particularly in the case of methamphetamine withdrawal [131].

The anhedonia that marks psychostimulant withdrawal periods serves to motivate increased drug intake [4,132], as modeled by the escalation model of self-administration. The self-administration of cocaine or methamphetamine does not typically result in uncontrolled use in animal models, as long as daily sessions are restricted to a relatively short time (60 min to 2 hr, known as short access). However, when subjected to self-administration sessions lasting several hours (long access sessions), animals frequently escalate their drug intake toward very high and unstable levels [133,134]. Typically, exposure to long access sessions results in an increased motivation to self-administer psychostimulants, as reflected in higher breakpoint thresholds when tested in a session governed by a progressive ratio schedule of reinforcement [4,135-137]. While the ICSS thresholds tend to remain unchanged during short-access cocaine self-administration training, increased ICSS thresholds between training sessions accompany long-access cocaine self-administration [133]. The presence of a similar state of negative effect has also been confirmed in rats subjected to extended sessions of methamphetamine self-administration [138].

The escalation model has been recently utilized to distinguish circumstances under which treatment by glutamatergic compounds have a greater effect on self-administration [139]. MTEP pretreatment was not as effective in reducing the self-administration breakpoints in rats with a history of escalated cocaine intake, versus animals with a restricted drug history [137]. However, LY379268 treatment reduced cocaine breakpoints in post-dependent rats at smaller doses than in nondependent cocaine-trained rats [137]. LY379268 was also more efficient in reducing anxiety-like defensive burying behavior in post-dependent rats [140]. The specific mGluR2 PAM BINA was equally effective in reducing cocaine self-administration in rats trained in short and prolonged sessions, although this report also demonstrated an equivalent effect of LY379268 [73]. Together, these reports could suggest a promising role for Group II mGluR PAMs as treatments against drug-taking in addicted individuals with a history of high drug intake. However, the capabilities of mGluR5 and mGluR7 drugs in this role are still largely untested.

Recent studies in preclinical depression models have shown that antagonists of mGluR5 may also be an important emerging target for the treatment of depressive symptoms [141]. The Flinders Sensitive Line (FSL) of rats, a strain of Sprague-Dawley rats that reliably exhibit depressive symptoms, have been found to possess lower densities of mGluR5 throughout the brain, relative to controls [142]. This finding was in accordance with positron emission tomography (PET) reports of lower levels of mGluR5 in the brains of human patients of major depressive disorder [143,144]. FSL rats have also been found to have deficient expression of mGluR2/3 in the hippocampus, and pretreatment with LY379268 alleviated some of their performance deficiencies in a forced-swim test [145]. Therefore, both Group I and Group II families of mGluRs have potential as targets for treatment against the emotional symptoms of psychostimulant withdrawal. On the other hand, the role of Group III mGluRs in psychostimulant withdrawal is largely unknown, aside from the observation of elevated mGluR8 expression following cessation from chronic amphetamine treatment [146].

Nicotine

Withdrawal from nicotine is marked by anhedonia, lack of motivation and other signs of distress similar to withdrawal from other psychostimulants [147,148]. In animal models, nicotine withdrawal is characterized by elevated ICSS thresholds [149,150] and sensitivity to environmental stressors [151]. Rats subjected to withdrawal from chronic nicotine exposure also exhibit an enhanced startle response to light [152] as well as other somatic behaviors not regularly observed in models of withdrawal from cocaine [153]. Similarly, human subjects experiencing nicotine withdrawal have demonstrated increased startle response when tested with unpredictable, but not cue-predicted, mild electric shock, implicating heightened anxiety but not fear [154].

As discussed previously, ligands of mGluRs have shown promising effects in reducing the behavioral symptoms of anxiety and depression in animal models. Group II mGluRs have been shown over multiple studies to mediate some of the affective consequences of nicotine withdrawal [150], beginning with the report that the mGluR2/3 agonist (+)-2-aminobicyclo[3.1.0]hexane-2,6-dicarboxylate (LY354740) diminished withdrawal-enhanced startle response [155]. However, the mGluR2/3 antagonist bicyclo(3.1.0)hexane-2,6-dicarboxylic acid (LY314582) reversed the withdrawal-elicited increase in ICSS thresholds [150], indicating a complex role for these receptors in numerous aspects of withdrawal [39].

Alcohol

An important role for glutamate in the physiological consequences of acute alcohol withdrawal has been established in numerous studies [156]. Alcohol intoxication is a state of depression of the central nervous system, characterized in part by attenuated glutamate neurotransmission [157]. After the cessation of heavy alcohol consumption, glutamatergic activity increases toward abnormally high and even neurotoxic levels [158]. *In vivo* electrophysiological studies have observed that alcohol withdrawal is marked by increased NMDA receptor-mediated activity and long-lasting changes in glutamate receptor expression [159,160]. Ethanol withdrawal symptoms including seizures have been blocked by treatment with the NMDA blocker MK-801 [161].

Notably, the potentially harmful consequences of alcohol on the glutamate system are exacerbated by chronic episodes of alcohol intoxication and withdrawal. Such experiences are modeled in animals by overnight housing in the presence of vaporized ethanol (or, alternatively, forced ingestion of ethanol via gavage or liquid diet), which induces sufficient intoxication to result in a state of chemical dependency [162-164]. Mice subjected to repeated intervals of withdrawal developed a vulnerability to seizures, known as the *kindling* effect [165,166]. Group I, II, and III mGluRs have been shown to be significant treatment targets for novel anticonvulsant drugs [167], and were anticipated to be effective treatment targets for physical symptoms of alcohol withdrawal. However, systemic pretreatment with the mGluR2/3 agonist LY379268 or mGluR5 NAM MTEP had negligible effects on handling-induced convulsions in mice undergoing acute alcohol withdrawal [168]. Also, the pathogenesis of withdrawal-induced seizures and *delirium tremens* did not appear to be associated with the currently known mGluR7 and mGluR8 polymorphisms in human subjects [169]. Alcohol withdrawal, especially when experienced in several episodes, also appears to increase anxiety symptoms [170]. Rats given five cycles of intoxication and withdrawal demonstrated reduced performance in a social interaction test, a symptom of heightened anxiety [171]. In several pharmacological investigations targeting neurotransmitters associated with stress, many anxiety behaviors that resulted from alcohol withdrawal, including reduced social interaction, increased freezing in a radial arm maze, and heightened acoustic startle, were at least partially chemically reversible [172-174]. Therefore, the therapeutic potential for mGluR compounds during alcohol withdrawal may be found in the treatment of the psychological aspects of stress reactivity and anxiety, rather than the prevention of seizures and tremors.

As with psychostimulants, models of extended alcohol exposure have been used to test the potency of mGluR compounds on self-administration behaviors. "Post-dependent" rodents with a history of chemical dependence typically exhibit a pattern of excessive alcohol drinking that last several subsequent sessions [163,175]. In mice, this increased drinking was accompanied by a nearly twofold increase in extracellular glutamate release in the NAc [176]. Pretreatment with MTEP was significantly less effective in reducing drinking post-dependent rats than rats with histories of restricted alcohol exposure [111].

Opiates

Opiate withdrawal is characterized in humans and animals by distinctive and often severe physical symptoms, including shaking, teeth chattering and diarrhea [177]. Opiate withdrawal, similar to withdrawal from other drugs of abuse, is accompanied by elevated ICSS thresholds, demonstrating a deficit in the mesolimbic reward system [178]. However, different substrates appear to govern the physical dependence and motivational effects of opiate withdrawal [179].

Opiate withdrawal differs significantly from psychostimulant withdrawal in that it lacks the hypothesized long-lasting suppression of glutamate neurotransmission. However, glutamate transmission in limbic structures was shown to be important for the expression of some of the aversive effects of morphine withdrawal [180]. Furthermore, an experiment using (S)-4-carboxyphenylglycine ((S)-4CPG), a nonspecific blocker of mGluRs, demonstrated involvement of mGluRs in the development of chemical morphine dependence [181]. Further studies determined that the Group I and Group II mGluRs and mediated the development of morphine dependence and were therefore treatment targets for the management of withdrawal symptoms [182-185].

Summary

The motivational and somatic features of drug withdrawal depend on the class of drug no longer consumed, and the presence or absence of physiological symptoms may determine whether abstinent drug addicts will maintain treatment with therapeutic compounds. Several experimental drugs that target Group I and Group II mGluRs have anxiolytic and antidepressant properties, and hold potential as a strategy to alleviate the negative affect associated with drug withdrawal. This strategy could improve patient compliance, an aspect of pharmacological treatment that has been particularly problematic for the treatment of substance abuse [186].

Stage 3: Preoccupation and Craving

The final stage of the addiction cycle is one of persistent feelings of craving for the drug, leading to chronic relapse behaviors. Relapse often occurs with the individual's apparent understanding of the negative consequences of continued drug use, even following several years of successful abstinence. Importantly, environmental cues that invoke memories of past drug use or encounters with stressful stimuli trigger the feelings of craving that often precede relapse [187,188]. Addicts with a history of severe intoxication and bingeing tend to experience greater feelings of craving in response to these cues, as well as to stressful events [189-191]. The ability of these triggers to elicit craving is notoriously resistant to the efforts of counseling and pharmacological treatment, and this persistence is analogous to the long-lasting sensitivity to cues in previously trained animals [192].

Craving and relapse are frequently modeled in animals with the *extinction/reinstatement* procedure. Drug self-administration training sessions are followed by forced abstinence or a series of *extinction* sessions, during which the formerly drug-reinforced operant response results in no reward. The animals are then tested using cues, stress, or experimenter-administered drug primes that elicit the formerly extinguished self-administration behavior [193]. The goal of most studies utilizing the extinction/reinstatement model is to identify a pharmacological treatment that significantly and dose-dependently attenuates operant responding during the *reinstatement* test sessions, without the development of tolerance or the exhibition of nonspecific effects on locomotor behavior or the pursuit of natural rewards (i.e., food or sucrose pellets).

Loss of control over craving feelings is associated with a compromised link between the cognitive centers of the brain within the mPFC and the substrates of motivational drive, which are centered on the mesolimbic dopaminergic neurons that terminate in subcortical areas, particularly the NAc. The terminals of this projection are glutamatergic, and typically populated by presynaptic and perisynaptic mGluR2 complexes and postsynaptic mGluR5 complexes [194]. In contrast, cortical and NAc mGluR3 is primarily concentrated on glia [195]. Glutamate overflow at these synapses, which can be inhibited by stimulation of mGluR2/3 activity [196], has been repeatedly shown to rise in sufficient amounts to disperse beyond the synaptic cleft and into the extrasynaptic space when an animal is seeking drug reward [32,197]. Hence, two strategies have emerged for targeting mGluRs to inhibit drug relapse: targeting mGluRs with specific ligands, and indirectly managing mGluR activity by targeting the proteins that tightly manage extracellular glutamate concentrations, a process known as glutamate homeostasis [30].

Much of the pursuit of the glutamate homeostasis approach involves the commercially available compound N-acetylcysteine, a substrate of the cysteine-glutamate antiporter (xc-), a glial membrane protein that exchanges extracellular cysteine for intracellular glutamate [198, 30]. Well-tolerated in animals and humans and widely used against the effects of acetaminophen overdose [199], n-acetylcysteine has been evaluated as an anti-craving agent in clinical trials and preclinical experiments, yielding mixed results [200]. The antibiotic ceftriaxone, commonly used in medicine to treat pneumonia infection, is known to stimulate the glutamate transporter protein GLT1 [201], a primary means by which extracellular glutamate is reduced [36]. The activity level of xc- has been found to be directly related to the activity level of glutamate transport [202,203]. GLT1 and xc- coregulate extracellular glutamate, making N-acetylcysteine and ceftriaxone useful agents for the investigation of glutamate homeostasis as a treatment target against drug craving and relapse [38].

Psychostimulants

Systemically administered compounds blocking mGluR5 have been shown in several cases to attenuate reinstatement of cocaine seeking [204-206]. MTEP was effective in reducing cocaine-seeking behavior and cocaine self-administration, and also attenuated operant behavior for cues paired with sweetened milk reinforcement [68]. Blocking mGluR5 function with systemic MTEP pretreatment has been shown to effectively reduce cue- and drug-triggered reinstatement of methamphctamine [62]. The mGluR5 NAM fenobam also attenuated cue- and drug-induced methamphetamine seeking, but its range of effective doses overlapped with the doses sufficient to reduce cue-induced sucrose-seeking in stimulant-naïve rats [207]. A study of fenobam sulfate (fenobam reformulated to have a longer half-life) treatment on cocaine-trained rats reported similar attenuation of cocaine-seeking and sucrose-seeking behavior [70]. These secondary effects are consistent with the observation that mGluR5 mediates appetitive behaviors [208], but the recently developed mGluR5 NAM 3-fluoro-5-[(6-methylpyridin-2-yl)ethynyl]benzonitrile (MFZ 10-7) has shown greater potency against cocaine reinstatement than sucrose reinstatement [209]. The role of mGluR5 activity on psychostimulant reinstatement has been intensively studied, and has been at least partially localized to the corticofugal projections terminating in the NAc core [210,211]. However, the strategy of interfering with mGluR5 function in order to prevent stimulant relapse without significant motivational side effects has not been realized in clinically viable formulations.

Compounds that stimulate the activity of Group II mGluRs, especially the specific mGluR2/3 agonist LY379268, have been shown to inhibit a broad variety of drug-seeking behaviors [212]. LY379268 pretreatment attenuated both cocaine-seeking behavior in rats, and also reduced cue-elicited food seeking behavior at a dose three times higher [213]. In another study utilizing discriminative cues indicating drug availability or absence in separate test conditions, LY379268 pretreatment was more affective at reducing cocaine-seeking behavior in the presence of drug-conditioned cues than reducing cocaine intake during self-administration sessions [72]. Given the role mGluR2/3 appear to have in the working memory of rodents [214], it is tempting to explain that LY379268 has been more effective in tests involving memory-driven behaviors. However, methamphetamine-induced memory deficits were found to be more closely associated with *ex vivo* mGluR5 expression than mGluR2/3 expression in the hippocampus [215].

Following the glutamate homeostasis hypothesis of addiction described earlier [216], and promising long-lasting effects against cocaine reinstatement [217,218], N-acetylcysteine has been evaluated in some studies as a clinical treatment against cocaine craving. Orally administered N-acetylcysteine reduced reports of cocaine rush but not cocaine-elicited feelings of craving [219], but significantly attenuated the desire to procure cocaine after viewing drug-related videos [220]. Another pilot study demonstrated potential of N-acetylcysteine to facilitate cocaine abstinence and reduced cravings [221]. These reports were followed by a double-blind placebo-controlled study with over 100 cocaine users: N-acetylcysteine extended days of abstinence from cocaine, but only in users who had already achieved abstinence before starting treatment [222]. This result offered some promise for N-acetylcysteine as a well-tolerated maintenance treatment, but the conclusion of the overall study was negative [222]. Similarly, a double-blind study of 31 methamphetamine addicts failed to differentiate the effects of a combined N-acetylcysteine and naltrexone (an opioid antagonist) from placebo [223]. A strategy of combining N-acetylcysteine with a mGluR5

NAM to reliably reduce drug craving and relapse has been suggested, but has so far not been tested [41].

The antibiotic ceftriaxone has also demonstrated effectiveness in reducing cue- and drug-primed cocaine reinstatement in trained rats [38,224]. Ceftriaxone also reduced reinstatement of food-seeking activity, which unfortunately lacked a control for the reputed effects of antibiotic treatments on appetite [224]. Ceftriaxone injections produced an attenuation of cue-induced reinstatement of cocaine seeking two weeks after cessation of treatment [225], doing so by renormalizing levels of extracellular glutamate after exposure to cocaine self-administration training [226]. Analogous effects were found in a model of methamphetamine reward, although ceftriaxone was used in this case to attenuate the expression of conditioned place preference and not operant behavior [227].

The escalation model of cocaine and methamphetamine self-administration, described earlier, has been used to demonstrate the contribution of withdrawal-related anti-reward processes to drug- and cue-primed reinstatement behavior [228]. Rats trained to self-administer cocaine in prolonged sessions have subsequently exhibited greater levels of drug seeking during drug-primed reinstatement tests [229-231]. This observation has also been extended to drug-primed reinstatement of methamphetamine seeking [232] and stress-induced reinstatement of cocaine seeking [233]. Interestingly, the level of operant activity during cue-elicited reinstatement does not appear to be different for animals with histories of prolonged versus restricted self-administration sessions [232,234].

Multiple recent studies have shown that the potential of MTEP and LY379268 as treatments against addictive behavior depends on the animal's history of psychostimulant exposure [139]. Although effective at reducing reinstatement of cocaine seeking and methamphetamine seeking in previously self-administering rats after extinction training [62,235], the capability of MTEP to attenuate self-administration was compromised in rats with a history of extensive cocaine intake [137]. LY379268, on the other hand, prevented the expression of drug-induced anxiety in rats with a history of prolonged access to cocaine, versus rats that had only restricted access [140]. LY379268 was also effective at smaller doses in rats with a history of high methamphetamine intake against cue-elicited reinstatement of methamphetamine seeking, but not drug-primed reinstatement [234].

Another possible disadvantage of blocking mGluR5 arises from the reputed effects mGluR5 NAMs have on learning and memory processes [236,237]. This observation has inspired the investigation of synthesized positive allosteric modulators (PAMs) of mGluR5 for cognitive-enhancing effects [238]. Recently developed systemically active mGluR5 PAMs include 3-cyano-N-(2,5-diphenylpyrazol-3-yl)benzamide (CDPPB) [239], (4-fluorophenyl)-[(3S)-3-[3-(4-fluorophenyl)-1,2,4-oxadiazol-5-yl] piperidin-1-yl]methanone (ADX47273) [240], 2-[4-[2-(benzyloxy)acetyl]piperazin-1-yl]benzonitrile (VU0364289), and 1-(4-(2,4-difluorophenyl)piperazin-1-yl)-2-((4-fluorobenzyl)oxy)ethanone (DPFE) [241]. Cognitive enhancement with chronic mGluR5 PAM treatment has been tested as a method to improve the rate of extinction learning, as well as subsequent resistance to known triggers of reinstatement [242,243]. CDPPB treatment accelerated extinction after cocaine self-administration training [244], but required a high dose seen later to possibly have overt locomotor effects [245]. CDPPB treatment improved the rate of extinction learning after methamphetamine self-administration at a smaller dose, but had a negligible effect on subsequent cue-induced reinstatement of drug seeking [246]. CDPPB treatment was also ineffective at preventing reinstatement in the presence of a methamphetamine-paired context

[247]. Reinstatement is a well-recognized consequence of extinction training for all psychostimulants [248, 249], and the attempts thus far at preventing all varieties of triggered reinstatement by pharmacologically augmenting extinction learning have not been successful. However, recent reports have claimed success at reducing reinstatement through timing manipulations during the initiation of extinction training [250,251], and these novel techniques may lead to improved results with mGluR-based cognitive enhancers [252].

The highly selective mGluR7 PAM AMN082 was shown to convincingly diminish drug-primed reinstatement of cocaine-seeking behavior [253]. Together with its reported anxiolytic and anti-depressant effects, AMN082 has been discussed as a promising avenue for anti-relapse agents for a wider variety of drugs and circumstances [253]. However, AMN082 was shown to be rapidly metabolized into a compound having 30-fold less affinity for mGluR7, but with significant affinity for the dopamine and norepinephrine transporter proteins and a very high affinity for the serotonin transporter [254]. Any lasting beneficial effects of AMN082 against the anti-reward system, negative affect, or other drivers of drug craving in advanced addiction may involve its metabolites acting as selective serotonin reuptake inhibitor (SSRI)-type of antidepressant, rather than its regulation of mGluR7 [254]. SSRIs are common treatments of comorbid mood and anxiety disorders in addicts, but their lasting effectiveness against substance abuse has proven to be limited and highly variable [255].

Nicotine

Like animal models for the other drugs of abuse, rodents trained to self-administer intravenous nicotine and then have that behavior extinguished in a series of extinction sessions are liable to reinstate operant behavior when reintroduced to drug-paired cues [94,256], given a priming nicotine injection [257,258] or subjected to unpredicted shock [259,260]. Administration of mGluR1 antagonist (3-ethyl-2-methylquinolin-6-yl)-(4-methoxycyclohexyl)methanone (EMQMCM) attenuated cue- and nicotine-induced nicotine behavior at doses smaller than the dose affecting food-seeking behavior [256]. MPEP treatment attenuated the reinstatement of nicotine seeking in animals that were reintroduced to nicotine-paired cues [94] or given a nicotine priming injection [90]. Group II mGluRs are also significant treatment targets for nicotine relapse, as shown by an experiment using systemic LY379268 administration to reduce cue-triggered nicotine seeking [96]. This study also found LY379268 to attenuate cue-induced food seeking, albeit at a threefold higher pretreatment dose than what was effective against nicotine seeking [96].

Correction of extracellular glutamate concentrations with N-acetylcysteine was tested as a strategy for nicotine craving in humans and animals. The dose of N-acetylcysteine required to significantly attenuate nicotine-induced reinstatement of nicotine seeking in rats overlapped with the doses sufficient to reduce feeding behavior [261]. N-acetylcysteine was tested for smoking cessation in nicotine-dependent smokers: while the subjects reported using fewer cigarettes, the physiological measurements suggested that the treated individuals compromised by taking deeper drags per cigarette [262]. Another small study in abstinent smokers found that N-acetylcysteine had negligible effects on nicotine withdrawal but made the first cigarette after abstinence feel less rewarding [263]. Ceftriaxone was also shown to effectively reduce the effects of nicotine withdrawal on glutamate transmission and prevent reinstatement of nicotine conditioned place preference [264]. Therefore, despite the limited clinical outcomes to date, correction of extracellular glutamate release remains an intriguing target for prevention of nicotine relapse.

Unlike other stimulants and virtually every other drug of abuse, extended access to nicotine reinforcement does not result in escalated self-administration behavior in trained rats [265]. The reasons for this lack of escalated self-administration are not completely understood but most likely related to the aversive effects of nicotine, mediated by a subclass of nAchRs in the medial habenula [266]. However, significant escalation of nicotine intake was obtained in animal models where extended access sessions were separated by intermittent periods of abstinence [267]. Nightly exposure (12h/day) to intoxicating levels of vaporized nicotine also resulted in escalated self-administration [268]. Chronic withdrawal experiences may therefore be a critical component of the drug intake behavior leading to nicotine dependence [269]. Withdrawal after extended exposure to nicotine was shown to result in reduced xc- functioning, and therefore compromised Group I and Group II mGluR activity in a manner analogous to the xc- effects of cocaine withdrawal [262].

Alcohol

One of the limitations of current alcoholism treatments is their lack of ability to prevent feelings of craving in severely dependent individuals, particularly when they encounter stimuli that trigger strong memories of past alcohol use [187]. Due partly to its influence on cognitive control and memory systems, and also to its role in the physiological symptoms of alcohol withdrawal, glutamate neurotransmission has been intensively studied for novel treatments of alcoholism [27]. Of particular importance has been the Group I and Group II mGluRs, for their capability to manage synaptic glutamate release and intrinsic role in drug-induced neuroplasticity. A gene variation in mGluR5 has been associated with an increased risk for the development of alcoholism in a genetic study of over 3000 subjects [270]. Inhibiting mGluR5 with central injections to the NAc and basolateral amygdala has been shown to interfere with cue reinstatement of alcohol seeking [108, 271,272].

Treatment with LY379268 attenuated footshock-induced and cue-induced reinstatement of ethanol-seeking [273], and did not reduce sucrose self-administration [274]. However, the dose required to attenuate alcohol consumption in rats has been associated with nonspecific locomotor effects [275]. This finding agreed with a report that the mGluR2/3 agonist (1R,4S,5S,6S)-4-amino-2,2-dioxo-2λ6-thiabicyclo[3.1.0]hexane-4,6-dicarboxylic acid (LY404039) reduced alcohol seeking but not self-administration in P-rats, which were animals selectively bred for high alcohol consumption [276]. Group II mGluRs may therefore be more appropriate treatment targets against alcohol craving than consumption, with beneficial effects in managing stressful stimuli as well as memories of past alcohol abuse.

Past research in mouse models has unveiled that alcohol-related changes in the glutamate system grow more severe with increased exposure to intoxicating levels of ethanol and subsequent withdrawal episodes [165,166]. In alcohol-trained rats, various mGluR5 NAM compounds attenuated reinstatement to alcohol seeking triggered by stress or cues [109,277,278]. However, the potency of MTEP in reducing reinstatement behaviors as well as drinking was compromised in post-dependent animals [111]. In contrast, the mGluR2/3 agonist LY379268 was increasingly effective in controlling alcohol seeking in rats with a history of ethanol dependence, relative to rats that had been trained to self-administer ethanol but not made dependent [279]. The range of doses effective against alcohol-seeking without locomotor effects is narrow (threefold) for LY379268 [213,275], but wider (tenfold) for post-dependent rats [279]. Later research has found that mGluR2 receptor function was compromised in key prefrontal cortical areas associated with relapse behaviors in post-dependent animals [280]. However, previous findings demonstrated *increased* functional mGluR2/3 activity in the stress-mediating basal forebrain and central nucleus of the amygdala [279]. These results implicate mGluR2 expression and function in alcohol seeking, but the long-term effects of alcohol dependence on these receptors may be heterogeneous and influenced by other experimental conditions, such as the length of time the animals spent inside the intoxication chambers [280].

Acamprosate is a clinically approved treatment for alcoholism, and although its mechanisms of action are currently a topic of active research [281], one of its purported therapeutic mechanisms is as an mGluR5 antagonist [102]. Acamprosate has a history of clinical trials establishing its effectiveness in reducing alcohol consumption and alcohol craving [281-284], but the recent large-scale clinical COMBINE study failed to find a statistical difference in the primary outcomes of acamprosate versus placebo [285]. One possible reason for this failure was acamprosate was administered to subjects that were actively drinking, but acamprosate may be most effective against alcohol cravings when administered following detoxification [286]. The very poor bioavailability of acamprosate has not only proven a limit to preclinical addiction models, but also could have provided a significant obstacle to treatment compliance [23].

Treatment with the widely available cysteine prodrug N-acetylcysteine has been investigated for benefits in models of alcoholism, with encouraging effects against oxidative stress [287,288]. It is the compound's ability to restabilize extracellular glutamate levels and reinstate the normal function of mGluR2/3 and mGluR5 that is of interest to studies of alcohol craving [200], but thus far no results have been published. In a recent study, N-acetylcysteine treatment reduced alcohol seeking in both nondependent and post-dependent alcohol-trained rats (P. Kufahl, unpublished observations). Ceftriaxone has also been tested in alcohol-preferring strains of rats, and while demonstrating reduced alcohol intake with or without an imposed multiday abstinence, these experiments did not fully account for the anti-feeding effects of antibiotics [289,290].

Opiates

Craving for heroin and other opiates have been modeled in animals using a contextual reinstatement procedure. Rats are first trained to self-administer the opiate in one environment, subjected to repeated sessions of forced abstinence in a second environment and then returned to the first environment and tested for the number of unreinforced operant responses. A series of microdialysis experiments has shown conclusive evidence that glutamate signaling in the NAc is necessary for the expression of heroin seeking in this model [291]. Direct infusion of the mGluR2/3 agonist LY379268 into the NAc resulted in attenuated reinstatement of heroin seeking inside the self-administration environment [292,293]. It also provided evidence that the same prefrontal projection to the NAc was critical for heroin and cocaine seeking [291], suggesting a common treatment strategy for these classes of drugs, despite the differences in ideal training methods [30].

Chronic treatment with N-acetylcysteine was also effective in reducing heroin-seeking behavior, presumably through its ability to increase mGluR2/3 function by renormalizing extracellular glutamate concentrations [294]. This result was particularly important for its lasting effect on heroin motivation: rats treated with N-acetylcysteine prior to extinction sessions demonstrated reduced operant behavior in subsequent tests for cue-elicited reinstatement [295].

Summary

Both Group I and Group II mGluR allosteric modulators are effective in attenuating drug-seeking behaviors in animal models of craving. Although the research of these receptors has been dominated by experiments utilizing compounds that will not enter clinical trials, later-generation drugs are being developed that exhibit greater half-life and reduced off-target effects [209,212]. The mGluR5 NAM Methyl-(3aR,4S,7aR)-4-hydroxy-4-[(3-methylphenyl)ethynyl]octahydro-1H-indole-1-carboxylate (AFQ056) has been investigated as a potential treatment of Fragile-X syndrome [296], and is also the subject of an ongoing Phase II clinical trial for obsessive-compulsive disorder [297]. The mGluR5 NAM 2-[(3-fluorophenyl)ethynyl]-4,6-dimethyl-3-pyridinamine hydrochloride (ADX10059) has been investigated in a Phase II clinical trial of patients with gastrointestinal reflux disorder [298]. An analogue of LY379268, the mGluR2/3 PAM LY354740 has been under investigation for beneficial effects against anxiety and panic disorders [299, 300]. Its limited bioavailability has motivated the development of a prodrug form: (1S,2S,5R,6R)-2-[[(2S)-2-aminopropanoyl] amino]bicyclo[3.1.0]hexane-2,6-dicarboxylic acid (LY544344) [301, 302]. Additionally, the mGluR2/3 antagonist 2-amino-3-(3,4-dichlorobenzyloxy)-6-fluorobicyclo(3.1.0)hexane-2,6-dicarboxylic acid (BCI-632), the mGluR2/3 NAMs RO499819 and RO4432717, and mGluR5 NAMs AZD2066 and RO4917523 are all currently under investigation for the treatment of major depression [303].

The mGluR5 NAM fenobam is currently being evaluated for the treatment of Fragile-X syndrome [304] and also has demonstrated promise as treatment for dyskinesia [305]. An ongoing Phase I clinical trial is being conducted to determine its pharmacokinetics and side effects in healthy adults [306]. As described previously, fenobam has shown the potential to reduce methamphetamine-seeking and cocaine-seeking behaviors, but nonspecific motivational deficits accompany these effects [70,207].

In studies of cocaine, methamphetamine and alcohol, Group II mGluRs have been found to be at least as effective in animals with histories of prolonged drug exposure than animals with restricted drug exposure, suggesting that this receptor subfamily could be particularly important in treating the craving feelings of severely addicted individuals [111,234,279]. However, the lack of specific potency exhibited by mGluRs against consumptive behaviors may indicate that mGluR5 agents may be the more appropriate class of compounds to develop for actively abusing addicts, and perhaps those individuals experiencing a transition from one from of substance abuse to another.

While drugs directly targeting mGluRs have yet to demonstrate their effectiveness in large-scale controlled clinical trials, there has been some success in identifying agents whose beneficial actions are at least partially mediated by glutamate modulation. Unfortunately, large-scale tests of acamprosate and N-acetylcysteine have not met the statistical benchmarks for successful treatment against relapsing episodes. By themselves, these treatment strategies may be too sensitive to the intersubject variability in the underlying causes of craving and relapse, including drug exposure history, comorbid affective disorders, and sensitivity to aversive or stressful events.

Moreover, the experience of multiple episodes of intoxication and withdrawal has been linked to a growing importance of stress as a primary contributor to addictive behaviors in treatment-resistant addicts [151,307]. The indications for mGluR-targeting compounds for symptoms of depression and anxiety, and the relationship between these symptoms and stress-induced drug use, suggest that mGluRs could be instrumental targets for individualized treatment regimens. The following section presents evidence that stress impacts mGluR function, and that mGluR agents have shown effectiveness in treating stress-induced anxiety behaviors, as well as withdrawal-induced behaviors.

Physiology of Stress Systems

Stress is most simply defined as the adaptive response to a threat (i.e., the stressor) against homeostasis, which is the tendency of the body to maintain stable internal conditions. When a stressor is encountered, several central nervous system structures are activated to initiate the stress response, including the brainstem and limbic forebrain structures [308]. These structures then recruit neural and neuroendocrine systems to initiate a stress response, which is mediated by the neuroendocrine hypothalamic-pituitary-adrenal (HPA) axis and extra hypothalamic brain stress systems.

Stressors can be categorized on the primary means of perception. Physical (systemic) stressors are predominantly identified through involuntary (autonomic) mechanisms via brainstem structures and circumventricular organs. Psychological (or anticipatory) stressors are perceived threats against homeostasis or well-being, based on previous experiences or instinctual responses, and are primarily under the control of limbic brain regions such as the hippocampus, amygdala, and prefrontal cortex [308]. Important qualities of psychological stressors include loss of control, unpredictability and/or novelty. Abstinence from drugs of abuse is likely associated with simultaneous physical (e.g., physiological withdrawal) and psychological (e.g., loss of sense of control) stressors. Psychological stressors are intimately related to anxiety disorders due to their ability to induce maladaptive emotional states, especially in response to the cues associated with an aversive event.

The Hypothalamic-Pituitary-Adrenal Axis

When a stressor is detected the stress response is activated, culminating in the release of glucocorticoid hormones (GCs). The secretion of the adrenal GCs (corticosterone in rodents, cortisol in humans) is mediated by the medial parvocellular neurons in the paraventricular nucleus (PVN) of the hypothalamus. These neurons secrete corticotropin releasing hormone (CRH) and arginine vasopressin (AVP) into the hypophyseal circulation and stimulate the corticotrophic cells of the anterior pituitary to release adrenocorticotrophic hormone (ACTH) into the peripheral circulation. ACTH activates the cortex of the adrenal gland to secrete GCs, which are carried to every organ via the circulation to allow for a coordinated and adaptive stress response [309,310]. Corticosterone (CORT), the primary GC in rodents, is the ligand for the glucocorticoid receptor (GR) and mineralocorticoid receptor (MR), steroid hormone receptors that function as transcription factors and regulate neuronal gene transcription. GRs have a key role in terminating GC release via negative feedback loops at the level of the pituitary and PVN, as well as the hippocampus and the mPFC [311].

Importantly, acute administration of most drugs of abuse activates the HPA axis and may serve to facilitate both drug reward and the acquisition of reward-seeking behavior [312-315]. However, repeated activation of the HPA axis response, such as would occur with chronic stress or drug use, is associated with marked dysregulation of HPA axis function. A number of characteristic effects are associated with chronic stress exposure, including high basal GC secretion, hypertrophy of the adrenal glands, atrophy of the thymus, down regulated GRs and reduced efficacy of negative feedback, enhanced GC response to novel stressors, and dysregulated CRF mediated responses to stress [310,316]. Furthermore, following chronic stress, the modulatory and regulatory roles of extrahypothalamic stress circuitries appear to be altered and likely contribute to the development of maladaptive behaviors such as increased depression- and anxiety-like phenotypes.

Extrahypothalamic Brain Stress Systems

An integrative limbic neurocircuitry, including the hippocampus, mPFC, and the amygdala regulates the stress response. These limbic regulatory regions are activated in a stressor specific manner. For the purposes of this chapter, the focus remains on psychological or anticipatory stressors. Importantly, these limbic brain regions have relatively few direct inputs to the PVN, but instead project to several relay nuclei, the most studied of which is the bed nucleus of the stria terminalis (BNST, also described as part of the extended amygdala). The neurons of the BNST are predominantly GABAergic and project directly to the PVN [317]. The hippocampal regulation of the HPA axis is mediated primarily through ventral subiculum (vSUB) outputs. Excitatory inputs from these regions are integrated through the BNST to exert inhibitory control over the PVN in response to acute anticipatory stressors, but do not appear to be required for HPA axis responses to chronic stress [318]. Similarly, excitatory outputs of the prelimbic (PL) subregion of the mPFC also exert inhibitory control over the HPA axis via the posterior BNST (as well as other PVN-projecting regions), primarily in response to psychological stressors [319,320]. The infralimbic (IL) region of the mPFC appears to be more critical for the activation of the HPA axis in response to psychological stressors and the inhibition of responses to physical stressors [321,322].

In general the amygdala is considered to contribute to the activation of the HPA axis, though the role of individual subregions varies considerably. The central amygdala (CeA) is more responsive to physical stressors while the medial amygdala (MeA) is more responsive to psychological stressors [318]. Direct projections from the CeA and MeA to the PVN are limited, but both regions project heavily to the BNST [323]. In this case, the output of both the CeA/MeA and BNST are predominantly GABAergic and thus the activating effects on the HPA axis occur as a result of disinhibition [309]. The basolateral amygdala (BLA) is strongly activated by psychological stressors, but does not appear necessary for acute HPA axis responses to these stressors [318,324,325]. The BLA primarily projects to the other subregions of the amygdala, and thus the excitatory output of the BLA modulates the inhibitory output of the CeA and MeA [309]. However, following chronic stress, the BLA appears to exert an inhibitory influence on the HPA axis in response to novel stressors while the roles of the CeA and MeA appear to be reduced [318,326].

Importantly, what is termed the extended amygdala (BNST, CeA, and the posterior medial shell of the NAc) plays a central role in integrating stress, anxiety, and addiction [124,126]. Both the CeA and BNST are activated during anxiety-related behaviors, and lesions of the BNST prevents mPFC from exerting regulatory control over the PVN [327]. Similarly, increased release of corticotropin releasing factor (CRF; same peptide as CRH but referred to as factor when acting as a neurotransmitter or localized outside of the HPA axis) in the CeA/BNST is hypothesized to underlie behavioral responses to stress, non-drug related anxiety-like behaviors, as well as abstinence induced anxiety-like behaviors [328-332]. However, enhanced CRF mRNA expression in the CeA or BNST is not found after all chronic stress paradigms (e.g., chronic variable stress), despite the presence of anxiety-like phenotypes [333]. A substantial amount of work has focused on the sensitizing effects of chronic drug use or dependence on the extrahypothalamic CRF system. Withdrawal from cocaine, opioids, ethanol, nicotine, and cannabis (Δ^9-tetrahydrocannabinol) all lead to increased CRF expression within the CeA as well as anxiety-like behaviors [329,334-338]. Thus, there exists a complex interplay between drug abuse and withdrawal and HPA axis activation, dysregulation, and anxiety-like states. Furthermore, significant co-morbidity is reported for drug dependence and anxiety or mood disorders in clinical literatures.

mGluRs and the Acute Neuroendocrine Stress Response

Glutamate plays a central role in regulating the stress response [339-341]. It is proposed that the primary actions of glutamate occur centrally. Glutamate infusion into the third ventricle or injection directly into the PVN results in increased ACTH release as well as direct release of CRH from hypothalamic slices [339,342,343]. Furthermore, blockade of the NMDA receptor with kynurenate in the PVN decreases the acute stress response measured by circulating CORT [344]. Evidence suggests that both ionotropic and metabotropic glutamate receptors mediate the actions of glutamate in the PVN [344-346]. Of relevance to this chapter, mGluRs are expressed the PVN, as well as other regions implicated in mediating the stress response such as the PFC, hippocampus, amygdala, and BNST [347-352] and central administration of mGluR agents significantly impacts HPA axis function.

For example, intracerebroventricular (i.c.v.) treatment with agonists or antagonists of mGlu1/5 receptors both resulted in increased circulating CORT [345]. This finding was corroborated by another study using systemic administration of the mGluR5 antagonist, MPEP, which dose dependently increased circulating CORT concentrations [353]. On the other hand, systemic administration of the mGluR2/3 antagonist, LY341495, but not the agonist, LY379268, resulted in increased circulating CORT [346]. In addition, this same study found that while mRNA for mGluRs 1,3,4,5,7 and 8 were expressed in the hypothalamus, only mGluR2/3 was detected as protein (but see also [354,355]). Furthermore, the mGluR2/3 antagonist LY341495 was found to increase CRH secretion from isolated hypothalami [346]. The authors concluded that mGluR2/3 exerts a tonic inhibition of the HPA axis at the level of hypothalamic CRH secretion.

Group III mGluRs have also been implicated in the regulation of the neuroendocrine response. Central administration of non-selective group III mGluR agonists also increases circulating CORT [345]. In addition, mice lacking the mGluR7 demonstrate dysregulated HPA axis function, such that GR-dependent negative feedback was enhanced and circulating CORT was slightly reduced [356]. Interestingly, mGluR7 null mice also demonstrated an overall reduction in depressive- and anxiety-like behaviors [356]. Further support for the role of mGluR7 in stress regulation comes from the systemic administration of the mGluR7 agonist, which results in increased circulating CORT [74].

Together, these studies indicate that activation of mGluR1/5 and mGluR7 (and potentially other group III receptors) increase HPA axis activation, while mGlu2/3 receptors normally exert a tonic inhibition on the HPA axis which can be disinhibited via administration of mGluR2/3 antagonists.

Chronic stress is associated with changes in mGluR expression changes as well. For example, nine weeks of chronic mild stress, which is typically associated with an anhedonic phenotype, results in increased mGluR5 protein expression in the CA1 and decreased mGluR5 expression in the CA3 region of the hippocampus [357]. In addition, 21 days of CORT treatment (40 mg/kg) results in decreased mGluR5 protein (but not mRNA expression) in the whole hippocampus. However, no changes in mGluR5 expression were identified in the frontal cortex and mGluR1 expression remained stable in both brain regions [358]. These findings support a potential role for mGluR5 in the dysregulation of stress systems, particularly with respect to hippocampal regulation of the stress response.

Importantly for the present discussion, chronic drug use is also associated with numerous alterations in mGluR expression and function. For example, following repeated intermittent cocaine exposure, the sensitivity of mGlu2/3 receptors to agonist drugs was reduced in the CeA amygdala [359]. Examination of radiolabeled agonist binding to target G protein coupled receptors in brain tissue (utilizing LY379268 in a ^{35}S-radiolabelled guanosine 5'-O-[gamma-thio]triphosphate, or [^{35}S]GTPγS, assay) collected from chemically post-dependent and nondependent rats trained to self-administer alcohol found enhanced mGluR2/3 function in the post-dependent rats, but this change was isolated to the BNST and CeA [279]. Similarly, elevated expression of mGluR2/3 function in BNST, CeA, hippocampus, and mPFC was revealed in brain tissue harvested from rats exposed to prolonged versus restricted cocaine self-administration sessions [137]. In contrast, mGluR5 functional levels (assessed by MTEP in a [^{35}S]GTPγS assay) demonstrated little variation as a consequence of escalated cocaine intake [137]. Escalated self-administration of methamphetamine was found to result in decreased mGluR2/3 expression in the mPFC, NAc and dorsal striatum, although these effects were largely reversed in the NAc and dorsal striatum by extinction training [360]. Fluctuating alterations in the expression and functional activity of mGluRs, and Group II mGluRs in particular, have been highlighted in numerous behavioral, physiological, and neurochemical studies mostly focused on the glutamatergic projections from the mPFC to the NAc and VTA [41]. However, mGluR2/3 changes also occur in the stress-mediating extended amygdala as a consequence of drug and alcohol exposure, perhaps (in the case of alcohol) requiring less time and exposure than analogous changes in the mPFC [280].

mGluR Agents and Stress- and Anxiety-Related Behaviors

mGluR agents have been extensively studied in both the prevention of relapse and the treatment of anxiety-like disorders. Given that chronic drug use leads to significant dysregulation of the stress response and potentiated anxiety-like behaviors in preclinical studies, a consideration of targeting withdrawal-induced anxiety-like states with mGluR treatments is warranted. Anxiety-like behaviors can be measured with a number of preclinical assays. Measures that tap into both dysregulated stress systems and associated anxiety-like states typically observe a reduction in exploration (i.e., avoidance) of open areas (open field test), open arms (elevated plus maze) or enhanced defensive responses (increased burying of the shock probe in the defensive burying test), punishment-induced reduction of exploration or responding for water, food in conflict tests (Four-plate test, Vogel test, Geller-Seifter test), increased startle to a loud stimulus, increased fear-potentiated startle (FPS) and increased freezing to a conditioned fear stimulus [127]. Many of these behaviors are directly reduced by administration of CRF antagonists (i.c.v. or intra-amygdala) and thus are implicated in the long-term outcomes of dysregulated stress systems [361,362]. mGluR agents have demonstrated efficacy in the reduction of anxiety- and stress-related behaviors. Specifically, group II (mGluR2/3) agonists/PAMs and group I (mGluR1/5) antagonists/NAMs have been extensively investigated. Group III mGluRs are less extensively studied, though knockout studies suggest that mGluR7 and mGluR8 play opposing roles in anxiety-like behavior. For a more in-depth coverage of this topic, readers are directed to a comprehensive review by Swanson and colleagues [127].

Group I mGluR Antagonists

Extensive evidence indicates anxiolytic effects following the blockade of mGluR1/5 receptors. MPEP, an mGluR5 NAM, is most frequently examined in this context. Systemic administration (i.p. and/or oral) in rodents (most frequently rats) blocks acquisition/expression of fear potentiated startle, increases open arm time in the elevated plus maze, increase social interaction and number of shocks or punished responses in conflict tests, and also reduces defensive marble burying and stress-induced hyperthermia [363-366]. Intra-amygdala administration of MPEP also reduces the acquisition of conditioned fear, though no effects were found on the expression of fear (i.e., freezing) following conditioning [367]. Interestingly, administration of a high dose of MPEP, other mGluR1/5 antagonists ((+)-2-methyl-4-carboxyphenylglycine (LY367685) and 3-[(2S)-2-amino-1-hydroxy-1-oxo-3-(9H-thioxanthen-9-yl)propan-2-yl]cyclobutane-1-carboxylic acid (LY393675)) as well as mGluR1/5 agonists (2-amino-2-(3,5-dihydroxyphenyl)acetic acid (DHPG), 2-amino-2-(2-chloro-5-hydroxyphenyl)acetic acid (CHPG), and (1S,3R)-1-aminocyclopentane-1,3-dicarboxylic acid (ACPD)) all result in increased plasma CORT [368]. These findings indicate potentially opposing roles of Group I mGluRs in the extrahypothalamic stress circuitry (e.g., the amygdala) and the HPA axis. To determine the role of mGluR1/5 receptors in the response to stress, additional studies investigating the behavioral and physiological response to stressors in the presence of mGluR1/5 antagonism.

Group II mGluR Agonists

mGluR2/3 agonists appear to play a more prominent role in alleviating anxiety-like states than other mGluR agents. In rodents, systemic administration of mGluR2/3 agonists (primarily LY367685) is effective at blocking expression of fear-potentiated startle in rodents, increasing open arm time in the elevated plus maze and increasing punished responding/shocks accepted in conflict tests in rodents, as well as blocking lactate-induced panic in the rat [369-376]. In addition, mGluR2/3 agonists can reduce stress-induced hyperthermia and the neurochemical response to some stressors [366,377]. While several compounds have been developed for both pre-clinical and clinical use, clinical trials of the most promising compound, LY354740 (eglumegad) appears to have stalled after Phase II clinical trials for anxiety disorders. However, it was found that LY354740 was able to block expression of fear-potentiated startle in the absence of sedating effects. In addition, the same compound was found to reduce some anxiety symptoms in patients with generalized anxiety disorder as well as prevent CO_2-induced anxiety in patients with panic attacks [376,378,379].

Group III mGluR Compounds and Knockout Models

The investigation of mGluR7/8 agents on anxiety-like behavior is less investigated and established, primarily due to the lack of specific pharmacological tools. Additionally, findings from both pharmacological studies and knockout animals are inconsistent. For example, intrahippocampal administration of the mGluR7/8 agonist (1S,3R,4S)-1-aminocyclopentane-1,2,4-tricarboxylic acid (L-SOP/ACPT-1) resulted in anxiolytic effects in the Vogel conflict test, while administration of the Group II/III antagonist (RS)-α-methylserine-O-phosphate (MSOP) leads to similar anxiolytic effects [365,380-382]. Furthermore, mGlu7 knockout animals exhibit an anxiolytic-like phenotype, while mGluR8 knockout animals exhibit increased anxiety-like behavior [383-385]. Thus, it is likely that mGluR7 and mGluR8 play opposing roles in anxiety-like states and more selective pharmacological tools will shed light on the relative involvement of these receptor subtypes.

Withdrawal-Induced Stress and Anxiety: Therapeutic Potential of mGluR Agents

Anxiety and dysregulated stress systems are strongly associated with withdrawal and abstinence following chronic drug use. In humans, acute withdrawal from numerous drugs is associated with increased CRF levels in cerebral spinal fluid (CSF), increased plasma ACTH, cortisol, norepinephrine, and epinephrine levels [151,386,387]. In addition, responsiveness to stressors is altered during early abstinence. For instance, chronic substance abusers exhibit a blunted ACTH and cortisol response to a stress or CRF challenge as well as altered autonomic nervous system functions [388-390]. Furthermore, negative emotional states including symptoms of irritability, anxiety, emotional distress, and aggressiveness are commonly found during early abstinence from alcohol, cocaine, opiates, nicotine, and marijuana [391-394]. Importantly, these abstinence-induced negative emotional states co-occur with stress dysregulation and it is reported that the severity of symptoms predicts treatment outcomes [395-398]. Thus, it is important to evaluate the potential pharmacological treatment of stress and anxiety-related behaviors during abstinence in pre-clinical models.

In rodents, a number of assays measure abstinence-induced stress or anxiety-like behaviors, including the elevated plus maze and defensive burying of a previously electrified probe. For example, increased defensive burying was reported during abstinence following two weeks of non-contingent cocaine administration (intraperitoneal injection) or three weeks of prolonged sessions of cocaine self-administration training [140,362,399]. However, three weeks of short-session cocaine self-administration training did not result in increased defensive probe burying compared to drug naïve rats [140]. In another study, two weeks of non-contingent cocaine lead to withdrawal-induced anxiety-like behavior as indicated by persistently increased freezing in the open arm of the elevated plus maze [400]. Others have found more traditional increases in anxiety-like behavior on the elevated plus maze (i.e., increased open arm time/entries) following withdrawal from two weeks non-contingent administration of cocaine [401,402]. Thus, it appears that withdrawal-induced anxiety-like behaviors result from occur after non-contingent administration or long access self-administration of cocaine, but not after short access self-administration.

The multifaceted overlap between withdrawal from chronic drug use and mGluR function, stress dysregulation, and anxiety-related behaviors and the effectiveness of mGluR agents at reducing anxiety-like states begs the question of whether mGluR agents might be effective treatments of withdrawal induced stress- and anxiety-related behaviors. Furthermore, if withdrawal-induced negative emotional and physiological states can be normalized, does the therapeutic efficacy of other relapse attenuating compounds change? At this time, one study has begun to address these questions. As mentioned previously, after long access self-administration of cocaine, rats show increased stress reactivity or anxiety-like behavior in the defensive burying test for up to six weeks of abstinence. Treatment with 1 mg/kg and 3 mg/kg of the mGluR2/3 agonist, LY379268, 30 minutes before the test prevented increased burying in abstinent rats, while only the 3 mg/kg LY379268 dose exhibited anxiolytic properties in non-dependent rats [140]. LY379268 pretreatment diminished stress-induced and cue-induced reinstatement to alcohol seeking, and subsequent immunohistochemistry analysis revealed a pattern of alterations in *c-fos* expression in the BNST, CeA, hippocampus, and PVT [273]. These results collectively point to Group II mGluRs as an intersection point between the neurocircuitry mediating stress-induced and cue-induced drug seeking [139].

Conclusion

Metabotropic glutamate receptors are the focus of a widespread and continuing research objective, due to their roles in drug- and alcohol-motivated behaviors in all purported stages of addiction. Many drugs that target these receptors have demonstrated promising effects in preclinical models of self-administration, withdrawal-induced anxiety, and reinstatement of drug-seeking behavior. The Group I and Group II mGluR subfamilies have been the most extensively studied for their effects on anxiety and stress-related behaviors, where their broad profile of beneficial effects may useful for addicted individuals driven to relapse while in variegated states of negative affect [23].

Allosteric modulators of mGluRs and compounds regulating extracellular glutamate are not completely proven as anti-addiction therapies, but they have a history of demonstrating encouraging effects against stress-related disorders in preclinical and small clinical trials. These therapeutic profiles may eventually be increasingly important in severely addicted individuals who are known to be driven to relapse by chronic experiences of stress, anxiety, and other varieties of negative affect. While most investigations have targeted Group I (particularly mGluR5) and Group II (both mGluR2 and mGluR3) mGluRs, Group III mGluRs have also been reported to mediate addictive behaviors in certain circumstances. However, the scope and totality of unretracted preclinical investigations into the lasting effects of mGluR7 ligands on drug-taking and seeking remain limited [253].

References

[1] Leshner AI (2000) The disease of addiction. *Lippincotts Prim Care Pract* 4:249-253.

[2] Pierce RC, Kumaresan V (2006) The mesolimbic dopamine system: the final common pathway for the reinforcing effect of drugs of abuse? *Neurosci Biobehav Rev* 30:215-238.

[3] Markou A, Kosten TR, Koob GF (1998) Neurobiological similarities in depression and drug dependence: a self-medication hypothesis. *Neuropsychopharmacology* 18:135-174.

[4] Ahmed SH, Koob GF (1998) Transition from moderate to excessive drug intake: change in hedonic set point. *Science* 282:298-300.

[5] Heilig M, Egli M, Crabbe JC, Becker HC (2010) Acute withdrawal, protracted abstinence and negative affect in alcoholism: are they linked? *Addict Biol* 15:169-184.

[6] Sinha R, O'Malley SS (1999) Craving for alcohol: findings from the clinic and the laboratory. *Alcohol Alcohol* 34:223-230.

[7] Sinha R, Fuse T, Aubin LR, O'Malley SS (2000) Psychological stress, drug-related cues and cocaine craving. *Psychopharmacology* (Berl) 152:140-148.

[8] Jaffe JH, Cascella NG, Kumor KM, Sherer MA (1989) Cocaine-induced cocaine craving. *Psychopharmacology* (Berl) 97:59-64.

[9] Gossop M, Stewart D, Browne N, Marsden J (2002) Factors associated with abstinence, lapse or relapse to heroin use after residential treatment: protective effect of coping responses. *Addiction* 97:1259-1267.

[10] O'Brien C (2001) Drug addiction and drug abuse. In: *The Pharmacological Basis of Therapeutics* (Hardman JG, Limbird LE, Gilman AG, eds), pp 621-642. New York: McGraw-Hill.

[11] Childress AR, McLellan AT, Ehrman R, O'Brien CP (1988) Classically conditioned responses in opioid and cocaine dependence: a role in relapse? *NIDA Res Monogr* 84:25-43.

[12] Volkow ND, Fowler JS, Wang GJ (2003) The addicted human brain: insights from imaging studies. *J Clin Invest* 111:1444-1451.

[13] Sinha R (2001) How does stress increase risk of drug abuse and relapse? *Psychopharmacology* (Berl) 158:343-359.

[14] Miller NS, Gold MS (1994) Dissociation of "conscious desire" (craving) from and relapse in alcohol and cocaine dependence. *Ann Clin Psychiatry* 6:99-106.

[15] Tiffany ST, Carter BL (1998) Is craving the source of compulsive drug use? *J Psychopharmacol* 12:23-30.

[16] Ingjaldsson JT, Thayer JF, Laberg JC (2003) Craving for alcohol and pre-attentive processing of alcohol stimuli. *Int J Psychophysiol* 49:29-39.

[17] Drummond DC (2001) Theories of drug craving, ancient and modern. *Addiction* 96:33-46.

[18] McLellan AT, Lewis DC, O'Brien CP, Kleber HD (2000) Drug dependence, a chronic medical illness: implications for treatment, insurance, and outcomes evaluation. *JAMA* 284:1689-1695.

[19] O'Brien CP (2008) Review. Evidence-based treatments of addiction. *Philos Trans R Soc Lond B Biol Sci* 363:3277-3286.

[20] Montoya ID, Vocci F (2008) Novel medications to treat addictive disorders. *Curr Psychiatry Rep* 10:392-398.

[21] Koob GF, Kenneth Lloyd G, Mason BJ (2009) Development of pharmacotherapies for drug addiction: a Rosetta stone approach. *Nat Rev Drug Discov* 8:500-515.

[22] Ross S, Peselow E (2009) Pharmacotherapy of addictive disorders. *Clin Neuropharmacol* 32:277-289.

[23] Olive MF, Cleva RM, Kalivas PW, Malcolm RJ (2012) Glutamatergic medications for the treatment of drug and behavioral addictions. *Pharmacol Biochem Behav* 100:801-810.

[24] Rachbeisel J, Scott J, Dixon L (1999) Co-occurring severe mental illness and substance use disorders: a review of recent research. *Psychiatr Serv* 50:1427-1434.

[25] Del Arco A, Mora F (2008) Prefrontal cortex-nucleus accumbens interaction: in vivo modulation by dopamine and glutamate in the prefrontal cortex. *Pharmacol Biochem Behav* 90:226-235.

[26] Kalivas PW (2004) Glutamate systems in cocaine addiction. *Curr Opin Pharmacol* 4:23-29.

[27] Gass JT, Olive MF (2008) Glutamatergic substrates of drug addiction and alcoholism. *Biochem Pharmacol* 75:218-265.

[28] Cornish JL, Kalivas PW (2000) Glutamate transmission in the nucleus accumbens mediates relapse in cocaine addiction. *J Neurosci* 20:RC89.

[29] Hyman SE, Malenka RC, Nestler EJ (2006) Neural mechanisms of addiction: the role of reward-related learning and memory. *Annu Rev Neurosci* 29:565-598.

[30] Knackstedt LA, Kalivas PW (2009) Glutamate and reinstatement. *Curr Opin Pharmacol* 9:59-64.

[31] McFarland K, Kalivas PW (2001) The circuitry mediating cocaine-induced reinstatement of drug-seeking behavior. *J Neurosci* 21:8655-8663.

[32] McFarland K, Lapish CC, Kalivas PW (2003) Prefrontal glutamate release into the core of the nucleus accumbens mediates cocaine-induced reinstatement of drug-seeking behavior. *J Neurosci* 23:3531-3537.

[33] Conn PJ, Pin JP (1997) Pharmacology and functions of metabotropic glutamate receptors. *Annu Rev Pharmacol Toxicol* 37:205-237.

[34] Nicoletti F, Bockaert J, Collingridge GL, Conn PJ, Ferraguti F, Schoepp DD, Wroblewski JT, Pin JP (2011) Metabotropic glutamate receptors: from the workbench to the bedside. *Neuropharmacology* 60:1017-1041.

[35] Schousboe A, Hertz L (1981) Role of astroglial cells in glutamate homeostasis. *Adv Biochem Psychopharmacol* 27:103-113.

[36] Danbolt NC (2001) Glutamate uptake. *Prog Neurobiol* 65:1-105.

[37] Moran MM, Melendez R, Baker D, Kalivas PW, Seamans JK (2003) Cystine/glutamate antiporter regulation of vesicular glutamate release. *Ann N Y Acad Sci* 1003:445-447.

[38] Knackstedt LA, Melendez RI, Kalivas PW (2010) Ceftriaxone restores glutamate homeostasis and prevents relapse to cocaine seeking. *Biol Psychiatry* 67:81-84.

[39] Kenny PJ, Markou A (2004) The ups and downs of addiction: role of metabotropic glutamate receptors. *Trends Pharmacol Sci* 25:265-272.

[40] Kalivas PW, Lalumiere RT, Knackstedt L, Shen H (2009) Glutamate transmission in addiction. *Neuropharmacology* 56 Suppl 1:169-173.

[41] Kalivas PW, Volkow ND (2011) New medications for drug addiction hiding in glutamatergic neuroplasticity. *Mol Psychiatry* 16:974-986.

[42] Koob GF, Le Moal M (2006) *Neurobiology of Addiction.* San Diego, CA: Academic Press.

[43] Wise RA (1996) Neurobiology of addiction. *Curr Opin Neurobiol* 6:243-251.

[44] Ungless MA, Whistler JL, Malenka RC, Bonci A (2001) Single cocaine exposure in vivo induces long-term potentiation in dopamine neurons. *Nature* 411:583-587.

[45] Kalivas PW, Duffy P (1995) D1 receptors modulate glutamate transmission in the ventral tegmental area. *J Neurosci* 15:5379-5388.

[46] Jones S, Bonci A (2005) Synaptic plasticity and drug addiction. *Curr Opin Pharmacol* 5:20-25.

[47] Stuber GD, Klanker M, de Ridder B, Bowers MS, Joosten RN, Feenstra MG, Bonci A (2008) Reward-predictive cues enhance excitatory synaptic strength onto midbrain dopamine neurons. *Science* 321:1690-1692.

[48] Bellone C, Luscher C (2006) Cocaine triggered AMPA receptor redistribution is reversed in vivo by mGluR-dependent long-term depression. *Nat Neurosci* 9:636-641.

[49] Stewart J, de Wit H, Eikelboom R (1984) Role of unconditioned and conditioned drug effects in the self-administration of opiates and stimulants. *Psychol Rev* 91:251-268.

[50] Wise RA (1988) The neurobiology of craving: implications for the understanding and treatment of addiction. *J Abnorm Psychol* 97:118-132.

[51] Koob GF, Le Moal M (1997) Drug abuse: hedonic homeostatic dysregulation. *Science* 278:52-58.

[52] Caine S, Lintz R, Koob G (1993) Intravenous drug self-administration techniques in animals. In: *Behavioural Neuroscience: A Practical Approach* (Shagal A, ed), pp 117-143. Oxford: IRL Press.

[53] Kornetsky C, Esposito RU, McLean S, Jacobson JO (1979) Intracranial self-stimulation thresholds: a model for the hedonic effects of drugs of abuse. *Arch Gen Psychiatry* 36:289-292.

[54] Esposito RU, Motola AH, Kornetsky C (1978) Cocaine: acute effects on reinforcement thresholds for self-stimulation behavior to the medial forebrain bundle. *Pharmacol Biochem Behav* 8:437-439.

[55] Esposito RU, McLean S, Kornetsky C (1979) Effects of morphine on intracranial self-stimulation to various brain stem loci. *Brain Res* 168:425-429.

[56] Kornetsky C, Esposito RU (1981) Reward and detection thresholds for brain stimulation: dissociative effects of cocaine. *Brain Res* 209:496-500.

[57] Breiter HC, Gollub RL, Weisskoff RM, Kennedy DN, Makris N, Berke JD, Goodman JM, Kantor HL, Gastfriend DR, Riorden JP, Mathew RT, Rosen BR, Hyman SE (1997) Acute effects of cocaine on human brain activity and emotion. *Neuron* 19:591-611.

[58] Kufahl PR, Li Z, Risinger RC, Rainey CJ, Wu G, Bloom AS, Li SJ (2005) Neural responses to acute cocaine administration in the human brain detected by fMRI. *Neuroimage* 28:904-914.

[59] Chiamulera C, Epping-Jordan MP, Zocchi A, Marcon C, Cottiny C, Tacconi S, Corsi M, Orzi F, Conquet F (2001) Reinforcing and locomotor stimulant effects of cocaine are absent in mGluR5 null mutant mice. *Nat Neurosci* 4:873-874.

[60] Kim JH, Vezina P (1998) Metabotropic glutamate receptors are necessary for sensitization by amphetamine. *Neuroreport* 9:403-406.

[61] Kim JH, Vezina P (2002) The mGlu2/3 receptor agonist LY379268 blocks the expression of locomotor sensitization by amphetamine. *Pharmacol Biochem Behav* 73:333-337.

[62] Gass JT, Osborne MP, Watson NL, Brown JL, Olive MF (2009) mGluR5 antagonism attenuates methamphetamine reinforcement and prevents reinstatement of methamphetamine-seeking behavior in rats. *Neuropsychopharmacology* 34:820-833.

[63] Simmons DL, Mandt BH, Ng CM, Richards TL, Yamamoto DJ, Zahniser NR, Allen RM (2013) Low- and high-cocaine locomotor responding rats differ in reinstatement of cocaine seeking and striatal mGluR5 protein expression. *Neuropharmacology* 75:347-355.

[64] [Kenny PJ, Paterson NE, Boutrel B, Semenova S, Harrison AA, Gasparini F, Koob GF, Skoubis PD, Markou A (2003) Metabotropic glutamate 5 receptor antagonist MPEP decreased nicotine and cocaine self-administration but not nicotine and cocaine-induced facilitation of brain reward function in rats. *Ann N Y Acad Sci* 1003:415-418.

[65] Paterson NE, Markou A (2005) The metabotropic glutamate receptor 5 antagonist MPEP decreased break points for nicotine, cocaine and food in rats. *Psychopharmacology* (Berl) 179:255-261.

[66] Platt DM, Rowlett JK, Spealman RD (2008) Attenuation of cocaine self-administration in squirrel monkeys following repeated administration of the mGluR5 antagonist MPEP: comparison with dizocilpine. *Psychopharmacology* (Berl) 200:167-176.

[67] Kenny PJ, Boutrel B, Gasparini F, Koob GF, Markou A (2005) Metabotropic glutamate 5 receptor blockade may attenuate cocaine self-administration by decreasing brain reward function in rats. *Psychopharmacology* (Berl) 179:247-254.

[68] Martin-Fardon R, Baptista MA, Dayas CV, Weiss F (2009) Dissociation of the effects of MTEP [3-[(2-methyl-1,3-thiazol-4-yl)ethynyl]piperidine] on conditioned reinstatement and reinforcement: comparison between cocaine and a conventional reinforcer. *J Pharmacol Exp Ther* 329:1084-1090.

[69] Porter RH, Jaeschke G, Spooren W, Ballard TM, Buttelmann B, Kolczewski S, Peters JU, Prinssen E, Wichmann J, Vieira E, Muhlemann A, Gatti S, Mutel V, Malherbe P (2005) Fenobam: a clinically validated nonbenzodiazepine anxiolytic is a potent, selective, and noncompetitive mGlu5 receptor antagonist with inverse agonist activity. *J Pharmacol Exp Ther* 315:711-721.

[70] Keck TM, Yang HJ, Bi GH, Huang Y, Zhang HY, Srivastava R, Gardner EL, Newman AH, Xi ZX (2013) Fenobam sulfate inhibits cocaine-taking and cocaine-seeking behavior in rats: implications for addiction treatment in humans. *Psychopharmacology* (Berl) 229:253-265.

[71] Adewale AS, Platt DM, Spealman RD (2006) Pharmacological stimulation of group ii metabotropic glutamate receptors reduces cocaine self-administration and cocaine-induced reinstatement of drug seeking in squirrel monkeys. *J Pharmacol Exp* Ther 318:922-931.

[72] Baptista MA, Martin-Fardon R, Weiss F (2004) Preferential effects of the metabotropic glutamate 2/3 receptor agonist LY379268 on conditioned reinstatement versus primary reinforcement: comparison between cocaine and a potent conventional reinforcer. *J Neurosci* 24:4723-4727.

[73] Jin X, Semenova S, Yang L, Ardecky R, Sheffler DJ, Dahl R, Conn PJ, Cosford ND, Markou A (2010) The mGluR2 positive allosteric modulator BINA decreases cocaine self-administration and cue-induced cocaine-seeking and counteracts cocaine-induced enhancement of brain reward function in rats. *Neuropsychopharmacology* 35:2021-2036.

[74] Mitsukawa K, Yamamoto R, Ofner S, Nozulak J, Pescott O, Lukic S, Stoehr N, Mombereau C, Kuhn R, McAllister KH, van der Putten H, Cryan JF, Flor PJ (2005) A selective metabotropic glutamate receptor 7 agonist: activation of receptor signaling via an allosteric site modulates stress parameters in vivo. *Proc Natl Acad Sci USA* 102:18712-18717.

[75] Li X, Gardner EL, Xi ZX (2008) The metabotropic glutamate receptor 7 (mGluR7) allosteric agonist AMN082 modulates nucleus accumbens GABA and glutamate, but not dopamine, in rats. *Neuropharmacology* 54:542-551.

[76] Li X, Li J, Peng XQ, Spiller K, Gardner EL, Xi ZX (2009) Metabotropic glutamate receptor 7 modulates the rewarding effects of cocaine in rats: involvement of a ventral pallidal GABAergic mechanism. Neuropsychopharmacology 34:1783-1796.

[77] Palucha A, Klak K, Branski P, van der Putten H, Flor PJ, Pilc A (2007) Activation of the mGlu7 receptor elicits antidepressant-like effects in mice. *Psychopharmacology* (Berl) 194:555-562.

[78] Palucha-Poniewiera A, Branski P, Lenda T, Pilc A (2010) The antidepressant-like action of metabotropic glutamate 7 receptor agonist N,N'-bis(diphenylmethyl)-1,2-ethanediamine (AMN082) is serotonin-dependent. *J Pharmacol Exp Ther* 334:1066-1074.

[79] Stolerman IP, Jarvis MJ (1995) The scientific case that nicotine is addictive. *Psychopharmacology* (Berl) 117:2-10.

[80] Picciotto MR, Corrigall WA (2002) Neuronal systems underlying behaviors related to nicotine addiction: neural circuits and molecular genetics. *J Neurosci* 22:3338-3341.

[81] Mansvelder HD, De Rover M, McGehee DS, Brussaard AB (2003) Cholinergic modulation of dopaminergic reward areas: upstream and downstream targets of nicotine addiction. *Eur J Pharmacol* 480:117-123.

[82] Mansvelder HD, Keath JR, McGehee DS (2002) Synaptic mechanisms underlie nicotine-induced excitability of brain reward areas. *Neuron* 33:905-919.

[83] Mao D, Gallagher K, McGehee DS (2011) Nicotine potentiation of excitatory inputs to ventral tegmental area dopamine neurons. *J Neurosci* 31:6710-6720.

[84] Hulka LM, Treyer V, Scheidegger M, Preller KH, Vonmoos M, Baumgartner MR, Johayem A, Ametamey SM, Buck A, Seifritz E, Quednow BB (2013) Smoking but not cocaine use is associated with lower cerebral metabotropic glutamate receptor 5 density in humans. *Mol Psychiatry*, in press.

[85] Naqvi NH, Rudrauf D, Damasio H, Bechara A (2007) Damage to the insula disrupts addiction to cigarette smoking. *Science* 315:531-534.

[86] Liechti ME, Markou A (2008) Role of the glutamatergic system in nicotine dependence : implications for the discovery and development of new pharmacological smoking cessation therapies. *CNS Drugs* 22:705-724.

[87] Schilstrom B, Nomikos GG, Nisell M, Hertel P, Svensson TH (1998) N-methyl-D-aspartate receptor antagonism in the ventral tegmental area diminishes the systemic nicotine-induced dopamine release in the nucleus accumbens. *Neuroscience* 82:781-789.

[88] Fu Y, Matta SG, Gao W, Brower VG, Sharp BM (2000) Systemic nicotine stimulates dopamine release in nucleus accumbens: re-evaluation of the role of N-methyl-D-aspartate receptors in the ventral tegmental area. *J Pharmacol Exp Ther* 294:458-465.

[89] Kosowski AR, Liljequist S (2004) The NR2B-selective N-methyl-D-aspartate receptor antagonist Ro 25-6981 [(±)-(R*,S*)-alpha-(4-hydroxyphenyl)-beta-methyl-4-(phenylmethyl)-1-piperidine propanol] potentiates the effect of nicotine on locomotor activity and dopamine release in the nucleus accumbens. *J Pharmacol Exp Ther* 311:560-567.

[90] Tessari M, Pilla M, Andreoli M, Hutcheson DM, Heidbreder CA (2004) Antagonism at metabotropic glutamate 5 receptors inhibits nicotine- and cocaine-taking behaviours and prevents nicotine-triggered relapse to nicotine-seeking. *Eur J Pharmacol* 499:121-133.

[91] Paterson NE, Bruijnzeel AW, Kenny PJ, Wright CD, Froestl W, Markou A (2005) Prolonged nicotine exposure does not alter $GABA_B$ receptor-mediated regulation of brain reward function. *Neuropharmacology* 49:953-962.

[92] Tronci V, Vronskaya S, Montgomery N, Mura D, Balfour DJ (2010) The effects of the mGluR5 receptor antagonist 6-methyl-2-(phenylethynyl)-pyridine (MPEP) on behavioural responses to nicotine. *Psychopharmacology* (Berl) 211:33-42.

[93] Harrison AA, Gasparini F, Markou A (2002) Nicotine potentiation of brain stimulation reward reversed by DHβE and SCH 23390, but not by eticlopride, LY 314582 or MPEP in rats. *Psychopharmacology* (Berl) 160:56-66.

[94] Bespalov AY, Dravolina OA, Sukhanov I, Zakharova E, Blokhina E, Zvartau E, Danysz W, van Heeke G, Markou A (2005) Metabotropic glutamate receptor (mGluR5) antagonist MPEP attenuated cue- and schedule-induced reinstatement of nicotine self-administration behavior in rats. *Neuropharmacology* 49 Suppl 1:167-178.

[95] Palmatier MI, Liu X, Donny EC, Caggiula AR, Sved AF (2008) Metabotropic glutamate 5 receptor (mGluR5) antagonists decrease nicotine seeking, but do not affect the reinforcement enhancing effects of nicotine. *Neuropsychopharmacology* 33:2139-2147.

[96] Liechti ME, Lhuillier L, Kaupmann K, Markou A (2007) Metabotropic glutamate 2/3 receptors in the ventral tegmental area and the nucleus accumbens shell are involved in behaviors relating to nicotine dependence. *J Neurosci* 27:9077-9085.

[97] D'Souza MS, Liechti ME, Ramirez-Nino AM, Kuczenski R, Markou A (2011) The metabotropic glutamate 2/3 receptor agonist LY379268 blocked nicotine-induced increases in nucleus accumbens shell dopamine only in the presence of a nicotine-associated context in rats. *Neuropsychopharmacology* 36:2111-2124.

[98] Woodward JJ (1999) Overview of the effects of alcohol on the cerebral nervous system. *Neurochem Int* 35:93-94.

[99] Netzeband JG, Gruol DL (1995) Modulatory effects of acute ethanol on metabotropic glutamate responses in cultured Purkinje neurons. *Brain Res* 688:105-113.

[100] Minami K, Gereau RWt, Minami M, Heinemann SF, Harris RA (1998) Effects of ethanol and anesthetics on type 1 and 5 metabotropic glutamate receptors expressed in Xenopus laevis oocytes. *Mol Pharmacol* 53:148-156.

[101] Harris BR, Prendergast MA, Gibson DA, Rogers DT, Blanchard JA, Holley RC, Fu MC, Hart SR, Pedigo NW, Littleton JM (2002) Acamprosate inhibits the binding and neurotoxic effects of trans-ACPD, suggesting a novel site of action at metabotropic glutamate receptors. *Alcohol Clin Exp Res* 26:1779-1793.

[102] Mann K, Kiefer F, Spanagel R, Littleton J (2008) Acamprosate: recent findings and future research directions. *Alcohol Clin Exp Res* 32:1105-1110.

[103] Olive MF, McGeehan AJ, Kinder JR, McMahon T, Hodge CW, Janak PH, Messing RO (2005) The mGluR5 antagonist 6-methyl-2-(phenylethynyl)pyridine decreases ethanol consumption via a protein kinase C epsilon-dependent mechanism. *Mol Pharmacol* 67:349-355.

[104] Hodge CW, Miles MF, Sharko AC, Stevenson RA, Hillmann JR, Lepoutre V, Besheer J, Schroeder JP (2006) The mGluR5 antagonist MPEP selectively inhibits the onset and maintenance of ethanol self-administration in C57BL/6J mice. *Psychopharmacology* (Berl) 183:429-438.

[105] Cowen MS, Krstew E, Lawrence AJ (2007) Assessing appetitive and consummatory phases of ethanol self-administration in C57BL/6J mice under operant conditions: regulation by mGlu5 receptor antagonism. *Psychopharmacology* (Berl) 190:21-29.

[106] Cowen MS, Djouma E, Lawrence AJ (2005) The metabotropic glutamate 5 receptor antagonist 3-[(2-methyl-1,3-thiazol-4-yl)ethynyl]-pyridine reduces ethanol self-administration in multiple strains of alcohol-preferring rats and regulates olfactory glutamatergic systems. *J Pharmacol Exp Ther* 315:590-600.

[107] Besheer J, Faccidomo S, Grondin JJ, Hodge CW (2008) Regulation of motivation to self-administer ethanol by mGluR5 in alcohol-preferring (P) rats. *Alcohol Clin Exp Res* 32:209-221.

[108] Besheer J, Grondin JJ, Cannady R, Sharko AC, Faccidomo S, Hodge CW (2010) Metabotropic glutamate receptor 5 activity in the nucleus accumbens is required for the maintenance of ethanol self-administration in a rat genetic model of high alcohol intake. *Biol Psychiatry* 67:812-822.

[109] Backstrom P, Bachteler D, Koch S, Hyytia P, Spanagel R (2004) mGluR5 antagonist MPEP reduces ethanol-seeking and relapse behavior. *Neuropsychopharmacology* 29:921-928.

[110] Schroeder JP, Overstreet DH, Hodge CW (2005) The mGluR5 antagonist MPEP decreases operant ethanol self-administration during maintenance and after repeated alcohol deprivations in alcohol-preferring (P) rats. *Psychopharmacology* (Berl) 179:262-270.

[111] Sidhpura N, Weiss F, Martin-Fardon R (2010) Effects of the mGlu2/3 agonist LY379268 and the mGlu5 antagonist MTEP on ethanol seeking and reinforcement are differentially altered in rats with a history of ethanol dependence. *Biol Psychiatry* 67:804-811.

[112] Gyetvai B, Simonyi A, Oros M, Saito M, Smiley J, Vadasz C (2011) mGluR7 genetics and alcohol: intersection yields clues for addiction. *Neurochem Res* 36:1087-1100.

[113] Salling MC, Faccidomo S, Hodge CW (2008) Nonselective suppression of operant ethanol and sucrose self-administration by the mGluR7 positive allosteric modulator AMN082. *Pharmacol Biochem Behav* 91:14-20.

[114] Brown RM, Stagnitti MR, Duncan JR, Lawrence AJ (2012) The mGlu5 receptor antagonist MTEP attenuates opiate self-administration and cue-induced opiate-seeking behaviour in mice. *Drug Alcohol Depend* 123:264-268.

[115] van der Kam EL, de Vry J, Tzschentke TM (2007) Effect of 2-methyl-6-(phenylethynyl) pyridine on intravenous self-administration of ketamine and heroin in the rat. *Behav Pharmacol* 18:717-724.

[116] van der Kam EL, De Vry J, Tzschentke TM (2009) 2-methyl-6-(phenylethynyl)-pyridine (MPEP) potentiates ketamine and heroin reward as assessed by acquisition, extinction, and reinstatement of conditioned place preference in the rat. *Eur J Pharmacol* 606:94-101.

[117] Popik P, Wrobel M (2002) Morphine conditioned reward is inhibited by MPEP, the mGluR5 antagonist. *Neuropharmacology* 43:1210-1217.

[118] Berridge KC, Robinson TE (1998) What is the role of dopamine in reward: hedonic impact, reward learning, or incentive salience? *Brain Res Brain Res Rev* 28:309-369.

[119] O'Connor EC, Crombag HS, Mead AN, Stephens DN (2010) The mGluR5 antagonist MTEP dissociates the acquisition of predictive and incentive motivational properties of reward-paired stimuli in mice. *Neuropsychopharmacology* 35:1807-1817.

[120] Brown RM, Mustafa S, Ayoub MA, Dodd PR, Pfleger KD, Lawrence AJ (2012) mGlu5 receptor functional interactions and addiction. *Front Pharmacol* 3:84.

[121] Koob GF (2013) Negative reinforcement in drug addiction: the darkness within. *Curr Opin Neurobiol* 23:559-563.

[122] Khantzian EJ (1997) The self-medication hypothesis of substance use disorders: a reconsideration and recent applications. *Harv Rev Psychiatry* 4:231-244.

[123] Stewart SH, Karp J, Pihl RO, Peterson RA (1997) Anxiety sensitivity and self-reported reasons for drug use. *J Subst Abuse* 9:223-240.

[124] Koob GF, Le Moal M (2001) Drug addiction, dysregulation of reward, and allostasis. *Neuropsychopharmacology* 24:97-129.

[125] Anthony JC, Petronis KR (1995) Early-onset drug use and risk of later drug problems. *Drug Alcohol Depend* 40:9-15.

[126] Koob GF (2003) Neuroadaptive mechanisms of addiction: studies on the extended amygdala. *Eur Neuropsychopharmacol* 13:442-452.

[127] Swanson CJ, Bures M, Johnson MP, Linden AM, Monn JA, Schoepp DD (2005) Metabotropic glutamate receptors as novel targets for anxiety and stress disorders. *Nat Rev Drug Discov* 4:131-144.

[128] Barr AM, Markou A, Phillips AG (2002) A 'crash' course on psychostimulant withdrawal as a model of depression. *Trends Pharmacol Sci* 23:475-482.

[129] Ettenberg A, Raven MA, Danluck DA, Necessary BD (1999) Evidence for opponent-process actions of intravenous cocaine. *Pharmacol Biochem Behav* 64:507-512.

[130] Barr AM, Phillips AG (1999) Withdrawal following repeated exposure to d-amphetamine decreases responding for a sucrose solution as measured by a progressive ratio schedule of reinforcement. *Psychopharmacology* (Berl) 141:99-106.

[131] McGregor C, Srisurapanont M, Jittiwutikarn J, Laobhripatr S, Wongtan T, White JM (2005) The nature, time course and severity of methamphetamine withdrawal. *Addiction* 100:1320-1329.

[132] Ahmed SH, Kenny PJ, Koob GF, Markou A (2002) Neurobiological evidence for hedonic allostasis associated with escalating cocaine use. *Nat Neurosci* 5:625-626.

[133] Ahmed SH, Koob GF (1999) Long-lasting increase in the set point for cocaine self-administration after escalation in rats. *Psychopharmacology* (Berl) 146:303-312.

[134] Kitamura O, Wee S, Specio SE, Koob GF, Pulvirenti L (2006) Escalation of methamphetamine self-administration in rats: a dose-effect function. *Psychopharmacology* (Berl) 186:48-53.

[135] Paterson NE, Markou A (2003) Increased motivation for self-administered cocaine after escalated cocaine intake. *Neuroreport* 14:2229-2232.

[136] Wee S, Mandyam CD, Lekic DM, Koob GF (2008) Alpha 1-noradrenergic system role in increased motivation for cocaine intake in rats with prolonged access. *Eur Neuropsychopharmacol* 18:303-311.

[137] Hao Y, Martin-Fardon R, Weiss F (2010) Behavioral and functional evidence of metabotropic glutamate receptor 2/3 and metabotropic glutamate receptor 5 dysregulation in cocaine-escalated rats: factor in the transition to dependence. *Biol Psychiatry* 68:240-248.

[138] Jang CG, Whitfield T, Schulteis G, Koob GF, Wee S (2013) A dysphoric-like state during early withdrawal from extended access to methamphetamine self-administration in rats. *Psychopharmacology* (Berl) 225:753-763.

[139] Weiss F (2010) Advances in Animal Models of Relapse for Addiction Research. In: *Advances in the Neuroscience of Addiction*, 2nd Ed (Kuhn CM, Koob GF, eds). Boca Raton (FL).

[140] Aujla H, Martin-Fardon R, Weiss F (2008) Rats with extended access to cocaine exhibit increased stress reactivity and sensitivity to the anxiolytic-like effects of the mGluR2/3 agonist LY379268 during abstinence. *Neuropsychopharmacology* 33:1818-1826.

[141] Spooren W, Gasparini F (2004) mGlu5 receptor antagonists: a novel class of anxiolytics? *Drug News Perspect* 17:251-257.

[142] Kovacevic T, Skelin I, Minuzzi L, Rosa-Neto P, Diksic M (2012) Reduced metabotropic glutamate receptor 5 in the Flinders Sensitive Line of rats, an animal model of depression: an autoradiographic study. *Brain Res Bull* 87:406-412.

[143] Hashimoto K (2009) Emerging role of glutamate in the pathophysiology of major depressive disorder. *Brain Res Rev* 61:105-123.

[144] Deschwanden A, Karolewicz B, Feyissa AM, Treyer V, Ametamey SM, Johayem A, Burger C, Auberson YP, Sovago J, Stockmeier CA, Buck A, Hasler G (2011) Reduced metabotropic glutamate receptor 5 density in major depression determined by [^{11}C]ABP688 PET and postmortem study. *Am J Psychiatry* 168:727-734.

[145] Matrisciano F, Caruso A, Orlando R, Marchiafava M, Bruno V, Battaglia G, Gruber SH, Melchiorri D, Tatarelli R, Girardi P, Mathe AA, Nicoletti F (2008) Defective group-II metaboropic glutamate receptors in the hippocampus of spontaneously depressed rats. *Neuropharmacology* 55:525-531.

[146] Parelkar NK, Wang JQ (2008) Upregulation of metabotropic glutamate receptor 8 mRNA expression in the rat forebrain after repeated amphetamine administration. *Neurosci Lett* 433:250-254.

[147] Parrott AC (1995) Stress modulation over the day in cigarette smokers. *Addiction* 90:233-244.

[148] Kassel JD, Stroud LR, Paronis CA (2003) Smoking, stress, and negative affect: correlation, causation, and context across stages of smoking. *Psychol Bull* 129:270-304.

[149] Epping-Jordan MP, Watkins SS, Koob GF, Markou A (1998) Dramatic decreases in brain reward function during nicotine withdrawal. *Nature* 393:76-79.

[150] Kenny PJ, Gasparini F, Markou A (2003) Group II metabotropic and alpha-amino-3-hydroxy-5-methyl-4-isoxazole propionate (AMPA)/kainate glutamate receptors regulate the deficit in brain reward function associated with nicotine withdrawal in rats. *J Pharmacol Exp Ther* 306:1068-1076.

[151] Sinha R (2008) Chronic stress, drug use, and vulnerability to addiction. *Ann NY Acad Sci* 1141:105-130.

[152] Jonkman S, Risbrough VB, Geyer MA, Markou A (2008) Spontaneous nicotine withdrawal potentiates the effects of stress in rats. *Neuropsychopharmacology* 33:2131-2138.

[153] Watkins SS, Stinus L, Koob GF, Markou A (2000) Reward and somatic changes during precipitated nicotine withdrawal in rats: centrally and peripherally mediated effects. *J Pharmacol Exp Ther* 292:1053-1064.

[154] Hogle JM, Kaye JT, Curtin JJ (2010) Nicotine withdrawal increases threat-induced anxiety but not fear: neuroadaptation in human addiction. *Biol Psychiatry* 68:719-725.

[155] Helton DR, Tizzano JP, Monn JA, Schoepp DD, Kallman MJ (1997) LY354740: a metabotropic glutamate receptor agonist which ameliorates symptoms of nicotine withdrawal in rats. *Neuropharmacology* 36:1511-1516.

[156] Tsai G, Gastfriend DR, Coyle JT (1995) The glutamatergic basis of human alcoholism. *Am J Psychiatry* 152:332-340.

[157] Goodman JS, Gilman A (1958) The aliphatic alcohols. In: *The Pharmacological Basis of Therapeutics*, pp 98-122. New York: Macmillan.

[158] Grant KA, Valverius P, Hudspith M, Tabakoff B (1990) Ethanol withdrawal seizures and the NMDA receptor complex. *Eur J Pharmacol* 176:289-296.

[159] Engberg G, Hajos M (1992) Alcohol withdrawal reaction as a result of adaptive changes of excitatory amino acid receptors. *Naunyn Schmiedebergs Arch Pharmacol* 346:437-441.

[160] Roberto M, Schweitzer P, Madamba SG, Stouffer DG, Parsons LH, Siggins GR (2004) Acute and chronic ethanol alter glutamatergic transmission in rat central amygdala: an in vitro and in vivo analysis. *J Neurosci* 24:1594-1603.

[161] Hoffman PL, Grant KA, Snell LD, Reinlib L, Iorio K, Tabakoff B (1992) NMDA receptors: role in ethanol withdrawal seizures. *Ann NY Acad Sci* 654:52-60.

[162] Luz J, Griggio MA, Plapler H, De-Meo-Bancher M, Carvalho-Kosmiskas JV (1996) Effects of ethanol on energy balance of rats and the inappropriateness of intraperitoneal injection. *Alcohol* 13:575-580.

[163] O'Dell LE, Roberts AJ, Smith RT, Koob GF (2004) Enhanced alcohol self-administration after intermittent versus continuous alcohol vapor exposure. *Alcohol Clin Exp Res* 28:1676-1682.

[164] Braconi S, Sidhpura N, Aujla H, Martin-Fardon R, Weiss F, Ciccocioppo R (2010) Revisiting intragastric ethanol intubation as a dependence induction method for studies of ethanol reward and motivation in rats. *Alcohol Clin Exp Res* 34:538-544.

[165] Becker HC (1994) Positive relationship between the number of prior ethanol withdrawal episodes and the severity of subsequent withdrawal seizures. *Psychopharmacology* (Berl) 116:26-32.

[166] Becker HC, Diaz-Granados JL, Hale RL (1997) Exacerbation of ethanol withdrawal seizures in mice with a history of multiple withdrawal experience. *Pharmacol Biochem Behav* 57:179-183.

[167] Wong RK, Bianchi R, Taylor GW, Merlin LR (1999) Role of metabotropic glutamate receptors in epilepsy. *Adv Neurol* 79:685-698.

[168] Olive MF, Becker HC (2008) Effects of the mGluR2/3 agonist LY379268 and the mGluR5 antagonist MPEP on handling-induced convulsions during ethanol withdrawal in mice. *Alcohol* 42:191-197.

[169] Preuss UW, Koller G, Bahlmann M, Zill P, Soyka M, Bondy B (2002) No association between metabotropic glutamate receptors 7 and 8 (mGlur7 and mGlur8) gene polymorphisms and withdrawal seizures and delirium tremens in alcohol-dependent individuals. *Alcohol Alcohol* 37:174-178.

[170] Overstreet DH, Knapp DJ, Breese GR (2005) Pharmacological modulation of repeated ethanol withdrawal-induced anxiety-like behavior differs in alcohol-preferring P and Sprague-Dawley rats. *Pharmacol Biochem Behav* 81:122-130.

[171] Overstreet DH, Knapp DJ, Breese GR (2002) Accentuated decrease in social interaction in rats subjected to repeated ethanol withdrawals. *Alcohol Clin Exp Res* 26:1259-1268.

[172] Breese GR, Knapp DJ, Overstreet DH (2004) Stress sensitization of ethanol withdrawal-induced reduction in social interaction: inhibition by CRF-1 and benzodiazepine receptor antagonists and a 5-HT1A-receptor agonist. *Neuropsychopharmacology* 29:470-482.

[173] Knapp DJ, Overstreet DH, Breese GR (2007) Baclofen blocks expression and sensitization of anxiety-like behavior in an animal model of repeated stress and ethanol withdrawal. *Alcohol Clin Exp Res* 31:582-595.

[174] Huang MM, Overstreet DH, Knapp DJ, Angel R, Wills TA, Navarro M, Rivier J, Vale W, Breese GR (2010) Corticotropin-releasing factor (CRF) sensitization of ethanol withdrawal-induced anxiety-like behavior is brain site specific and mediated by CRF-1 receptors: relation to stress-induced sensitization. *J Pharmacol Exp Ther* 332:298-307.

[175] Roberts AJ, Heyser CJ, Cole M, Griffin P, Koob GF (2000) Excessive ethanol drinking following a history of dependence: animal model of allostasis. *Neuropsychopharmacology* 22:581-594.

[176] Griffin WC 3d, Haun HL, Hazelbaker CL, Ramachandra VS, Becker HC (2013) Increased extracellular glutamate in the nucleus accumbens promotes excessive ethanol drinking in ethanol dependent mice. *Neuropsychopharmacology*.

[177] Way EL, Loh HH, Shen FH (1969) Simultaneous quantitative assessment of morphine tolerance and physical dependence. *J Pharmacol Exp Ther* 167:1-8.

[178] Schaefer GJ, Michael RP (1986) Changes in response rates and reinforcement thresholds for intracranial self-stimulation during morphine withdrawal. *Pharmacol Biochem Behav* 25:1263-1269.

[179] Schulteis G, Markou A, Gold LH, Stinus L, Koob GF (1994) Relative sensitivity to naloxone of multiple indices of opiate withdrawal: a quantitative dose-response analysis. *J Pharmacol Exp Ther* 271:1391-1398.

[180] Watanabe T, Nakagawa T, Yamamoto R, Maeda A, Minami M, Satoh M (2002) Involvement of glutamate receptors within the central nucleus of the amygdala in naloxone-precipitated morphine withdrawal-induced conditioned place aversion in rats. *Jpn J Pharmacol* 88:399-406.

[181] Fundytus ME, Coderre TJ (1994) Effect of activity at metabotropic, as well as ionotropic (NMDA), glutamate receptors on morphine dependence. *Br J Pharmacol* 113:1215-1220.

[182] Fundytus ME, Ritchie J, Coderre TJ (1997) Attenuation of morphine withdrawal symptoms by subtype-selective metabotropic glutamate receptor antagonists. *Br J Pharmacol* 120:1015-1020.

[183] Popik P, Kozela E, Pilc A (2000) Selective agonist of group II glutamate metabotropic receptors, LY354740, inhibits tolerance to analgesic effects of morphine in mice. *Br J Pharmacol* 130:1425-1431.

[184] Kozela E, Pilc A, Popik P (2003) Inhibitory effects of MPEP, an mGluR5 antagonist, and memantine, an N-methyl-D-aspartate receptor antagonist, on morphine antinociceptive tolerance in mice. *Psychopharmacology* (Berl) 165:245-251.

[185] Rasmussen K, Hsu MA, Vandergriff J (2004) The selective mGlu2/3 receptor antagonist LY341495 exacerbates behavioral signs of morphine withdrawal and morphine-withdrawal-induced activation of locus coeruleus neurons. *Neuropharmacology* 46:620-628.

[186] O'Brien CP, McLellan AT (1996) Myths about the treatment of addiction. *Lancet* 347:237-240.

[187] O'Brien CP, Childress AR, Ehrman R, Robbins SJ (1998) Conditioning factors in drug abuse: can they explain compulsion? *J Psychopharmacol* 12:15-22.

[188] Shaham Y, Shalev U, Lu L, De Wit H, Stewart J (2003) The reinstatement model of drug relapse: history, methodology and major findings. *Psychopharmacology* (Berl) 168:3-20.

[189] Laberg JC (1986) Alcohol and expectancy: subjective, psychophysiological and behavioral responses to alcohol stimuli in severely, moderately and non-dependent drinkers. *Br J Addict* 81:797-808.

[190] Greeley JD, Swift W, Prescott J, Heather N (1993) Reactivity to alcohol-related cues in heavy and light drinkers. *J Stud Alcohol* 54:359-368.

[191] Smolka MN, Buhler M, Klein S, Zimmermann U, Mann K, Heinz A, Braus DF (2006) Severity of nicotine dependence modulates cue-induced brain activity in regions involved in motor preparation and imagery. *Psychopharmacology* (Berl) 184:577-588.

[192] Childress AR, Hole AV, Ehrman RN, Robbins SJ, McLellan AT, O'Brien CP (1993) Cue reactivity and cue reactivity interventions in drug dependence. *NIDA Res Monogr* 137:73-95.

[193] de Wit H, Stewart J (1983) Drug reinstatement of heroin-reinforced responding in the rat. *Psychopharmacology* (Berl) 79:29-31.

[194] Lujan R, Roberts JD, Shigemoto R, Ohishi H, Somogyi P (1997) Differential plasma membrane distribution of metabotropic glutamate receptors mGluR1 alpha, mGluR2 and mGluR5, relative to neurotransmitter release sites. *J Chem Neuroanat* 13:219-241.

[195] Petralia RS, Wang YX, Niedzielski AS, Wenthold RJ (1996) The metabotropic glutamate receptors, mGluR2 and mGluR3, show unique postsynaptic, presynaptic and glial localizations. *Neuroscience* 71:949-976.

[196] Moran MM, McFarland K, Melendez RI, Kalivas PW, Seamans JK (2005) Cystine/glutamate exchange regulates metabotropic glutamate receptor presynaptic inhibition of excitatory transmission and vulnerability to cocaine seeking. *J Neurosci* 25:6389-6393.

[197] Berglind WJ, Whitfield TW, Jr., LaLumiere RT, Kalivas PW, McGinty JF (2009) A single intra-PFC infusion of BDNF prevents cocaine-induced alterations in extracellular glutamate within the nucleus accumbens. *J Neurosci* 29:3715-3719.

[198] Baker DA, Shen H, Kalivas PW (2002) Cystine/glutamate exchange serves as the source for extracellular glutamate: modifications by repeated cocaine administration. *Amino Acids* 23:161-162.

[199] Scalley RD, Conner CS (1978) Acetaminophen poisoning: a case report of the use of acetylcysteine. *Am J Hosp Pharm* 35:964-967.

[200] Kufahl PR, Moore E, Barabas P, Halstengard C, Olive MF (2013) N-Acetylcysteine and its use in the treatment of addictive disorders. In: *Horizons in Neuroscience Research* (Costa A, Villalba E, eds), pp 1-34. New York: Nova Science Publishers.

[201] Miller BR, Dorner JL, Shou M, Sari Y, Barton SJ, Sengelaub DR, Kennedy RT, Rebec GV (2008) Up-regulation of GLT1 expression increases glutamate uptake and attenuates the Huntington's disease phenotype in the R6/2 mouse. *Neuroscience* 153:329-337.

[202] Bannai S (1986) Exchange of cystine and glutamate across plasma membrane of human fibroblasts. *J Biol Chem* 261:2256-2263.

[203] Murphy TH, Miyamoto M, Sastre A, Schnaar RL, Coyle JT (1989) Glutamate toxicity in a neuronal cell line involves inhibition of cystine transport leading to oxidative stress. *Neuron* 2:1547-1558.

[204] Lee B, Platt DM, Rowlett JK, Adewale AS, Spealman RD (2005) Attenuation of behavioral effects of cocaine by the metabotropic glutamate receptor 5 antagonist 2-methyl-6-(phenylethynyl)-pyridine in squirrel monkeys: comparison with dizocilpine. *J Pharmacol Exp Ther* 312:1232-1240.

[205] Backstrom P, Hyytia P (2006) Ionotropic and metabotropic glutamate receptor antagonism attenuates cue-induced cocaine seeking. *Neuropsychopharmacology* 31:778-786.

[206] Iso Y, Grajkowska E, Wroblewski JT, Davis J, Goeders NE, Johnson KM, Sanker S, Roth BL, Tueckmantel W, Kozikowski AP (2006) Synthesis and structure-activity relationships of 3-[(2-methyl-1,3-thiazol-4-yl)ethynyl]pyridine analogues as potent, noncompetitive metabotropic glutamate receptor subtype 5 antagonists; search for cocaine medications. *J Med Chem* 49:1080-1100.

[207] Watterson LR, Kufahl PR, Nemirovsky NE, Sewalia K, Hood LE, Olive MF (2013) Attenuation of reinstatement of methamphetamine-, sucrose-, and food-seeking behavior in rats by fenobam, a metabotropic glutamate receptor 5 negative allosteric modulator. *Psychopharmacology* (Berl) 225:151-159.

[208] Bradbury MJ, Campbell U, Giracello D, Chapman D, King C, Tehrani L, Cosford ND, Anderson J, Varney MA, Strack AM (2005) Metabotropic glutamate receptor mGlu5 is a mediator of appetite and energy balance in rats and mice. *J Pharmacol Exp Ther* 313:395-402.

[209] Keck TM, Zou MF, Bi GH, Zhang HY, Wang XF, Yang HJ, Srivastava R, Gardner EL, Xi ZX, Newman AH (2013) A novel mGluR5 antagonist, MFZ 10-7, inhibits cocaine-taking and cocaine-seeking behavior in rats. *Addict Biol,* in press.

[210] Knackstedt LA, Trantham-Davidson HL, Schwendt M (2013) The role of ventral and dorsal striatum mGluR5 in relapse to cocaine-seeking and extinction learning. *Addict Biol,* in press.

[211] Wang X, Moussawi K, Knackstedt L, Shen H, Kalivas PW (2013) Role of mGluR5 neurotransmission in reinstated cocaine-seeking. *Addict Biol* 18:40-49.

[212] Olive MF (2009) Metabotropic glutamate receptor ligands as potential therapeutics for addiction. *Curr Drug Abuse Rev* 2:83-98.

[213] Peters J, Kalivas PW (2006) The group II metabotropic glutamate receptor agonist, LY379268, inhibits both cocaine- and food-seeking behavior in rats. *Psychopharmacology* (Berl) 186:143-149.

[214] Aultman JM, Moghaddam B (2001) Distinct contributions of glutamate and dopamine receptors to temporal aspects of rodent working memory using a clinically relevant task. *Psychopharmacology* (Berl) 153:353-364.

[215] Reichel CM, Schwendt M, McGinty JF, Olive MF, See RE (2011) Loss of object recognition memory produced by extended access to methamphetamine self-administration is reversed by positive allosteric modulation of metabotropic glutamate receptor 5. *Neuropsychopharmacology* 36:782-792.

[216] Reissner KJ, Kalivas PW (2010) Using glutamate homeostasis as a target for treating addictive disorders. *Behav Pharmacol* 21:514-522.

[217] Moussawi K, Zhou W, Shen H, Reichel CM, See RE, Carr DB, Kalivas PW (2011) Reversing cocaine-induced synaptic potentiation provides enduring protection from relapse. *Proc Natl Acad Sci U S A* 108:385-390.

[218] Reichel CM, Moussawi K, Do PH, Kalivas PW, See RE (2011) Chronic N-acetylcysteine during abstinence or extinction after cocaine self-administration produces enduring reductions in drug seeking. *J Pharmacol Exp Ther* 337:487-493.

[219] Amen SL, Piacentine LB, Ahmad ME, Li SJ, Mantsch JR, Risinger RC, Baker DA (2011) Repeated N-acetylcysteine reduces cocaine seeking in rodents and craving in cocaine-dependent humans. *Neuropsychopharmacology* 36:871-878.

[220] LaRowe SD, Myrick H, Hedden S, Mardikian P, Saladin M, McRae A, Brady K, Kalivas PW, Malcolm R (2007) Is cocaine desire reduced by N-acetylcysteine? *Am J Psychiatry* 164:1115-1117.

[221] Mardikian PN, LaRowe SD, Hedden S, Kalivas PW, Malcolm RJ (2007) An open-label trial of N-acetylcysteine for the treatment of cocaine dependence: a pilot study. *Prog Neuropsychopharmacol Biol Psychiatry* 31:389-394.

[222] LaRowe SD, Kalivas PW, Nicholas JS, Randall PK, Mardikian PN, Malcolm RJ (2013) A double-blind placebo-controlled trial of N-acetylcysteine in the treatment of cocaine dependence. *Am J Addict* 22:443-452.

[223] Grant JE, Odlaug BL, Kim SW (2010) A double-blind, placebo-controlled study of N-acetylcysteine plus naltrexone for methamphetamine dependence. Eur *Neuropsychopharmacol* 20:823-828.

[224] Sari Y, Smith KD, Ali PK, Rebec GV (2009) Upregulation of GLT1 attenuates cue-induced reinstatement of cocaine-seeking behavior in rats. *J Neurosci* 29:9239-9243.

[225] Sondheimer I, Knackstedt LA (2011) Ceftriaxone prevents the induction of cocaine sensitization and produces enduring attenuation of cue- and cocaine-primed reinstatement of cocaine-seeking. *Behav Brain Res* 225:252-258.

[226] Trantham-Davidson H, LaLumiere RT, Reissner KJ, Kalivas PW, Knackstedt LA (2012) Ceftriaxone normalizes nucleus accumbens synaptic transmission, glutamate transport, and export following cocaine self-administration and extinction training. *J Neurosci* 32:12406-12410.

[227] Abulseoud OA, Miller JD, Wu J, Choi DS, Holschneider DP (2012) Ceftriaxone upregulates the glutamate transporter in medial prefrontal cortex and blocks reinstatement of methamphetamine seeking in a condition place preference paradigm. *Brain Res* 1456:14-21.

[228] Koob GF, Le Moal M (2008) Addiction and the brain antireward system. *Annu Rev Psychol* 59:29-53.

[229] Mantsch JR, Yuferov V, Mathieu-Kia AM, Ho A, Kreek MJ (2004) Effects of extended access to high versus low cocaine doses on self-administration, cocaine-induced reinstatement and brain mRNA levels in rats. *Psychopharmacology* (Berl) 175:26-36.

[230] Ahmed SH, Cador M (2006) Dissociation of psychomotor sensitization from compulsive cocaine consumption. *Neuropsychopharmacology* 31:563-571.

[231] Knackstedt LA, Kalivas PW (2007) Extended access to cocaine self-administration enhances drug-primed reinstatement but not behavioral sensitization. *J Pharmacol Exp Ther* 322:1103-1109.

[232] Rogers JL, De Santis S, See RE (2008) Extended methamphetamine self-administration enhances reinstatement of drug seeking and impairs novel object recognition in rats. *Psychopharmacology* (Berl) 199:615-624.

[233] Mantsch JR, Yuferov V, Mathieu-Kia AM, Ho A, Kreek MJ (2003) Neuroendocrine alterations in a high-dose, extended-access rat self-administration model of escalating cocaine use. *Psychoneuroendocrinology* 28:836-862.

[234] Kufahl PR, Watterson LR, Nemirovsky NE, Hood LE, Villa A, Halstengard C, Zautra N, Olive MF (2013) Attenuation of methamphetamine seeking by the mGluR2/3 agonist LY379268 in rats with histories of restricted and escalated self-administration. *Neuropharmacology* 66:290-301.

[235] Martin-Fardon R, Weiss F (2012) (-)-2-oxa-4-aminobicylco[3.1.0]hexane-4,6-dicarboxylic acid (LY379268) and 3-[(2-methyl-1,3-thiazol-4-yl)ethynyl]piperidine (MTEP) similarly attenuate stress-induced reinstatement of cocaine seeking. *Addict Biol* 17:557-564.

[236] Simonyi A, Schachtman TR, Christoffersen GR (2005) The role of metabotropic glutamate receptor 5 in learning and memory processes. *Drug News Perspect* 18:353-361.

[237] Simonyi A, Schachtman TR, Christoffersen GR (2010) Metabotropic glutamate receptor subtype 5 antagonism in learning and memory. *Eur J Pharmacol* 639:17-25.

[238] Ayala JE, Chen Y, Banko JL, Sheffler DJ, Williams R, Telk AN, Watson NL, Xiang Z, Zhang Y, Jones PJ, Lindsley CW, Olive MF, Conn PJ (2009) mGluR5 positive allosteric modulators facilitate both hippocampal LTP and LTD and enhance spatial learning. *Neuropsychopharmacology* 34:2057-2071.

[239] Kinney GG, O'Brien JA, Lemaire W, Burno M, Bickel DJ, Clements MK, Chen TB, Wisnoski DD, Lindsley CW, Tiller PR, Smith S, Jacobson MA, Sur C, Duggan ME,

Pettibone DJ, Conn PJ, Williams DL, Jr. (2005) A novel selective positive allosteric modulator of metabotropic glutamate receptor subtype 5 has in vivo activity and antipsychotic-like effects in rat behavioral models. *J Pharmacol Exp Ther* 313:199-206.

[240] Schlumberger C, Pietraszek M, Gravius A, Klein KU, Greco S, More L, Danysz W (2009) Comparison of the $mGlu_5$ receptor positive allosteric modulator ADX47273 and the $mGlu_{2/3}$ receptor agonist LY354740 in tests for antipsychotic-like activity. *Eur J Pharmacol* 623:73-83.

[241] Gregory KJ, Herman EJ, Ramsey AJ, Hammond AS, Byun NE, Stauffer SR, Manka JT, Jadhav S, Bridges TM, Weaver CD, Niswender CM, Steckler T, Drinkenburg WH, Ahnaou A, Lavreysen H, Macdonald GJ, Bartolomé JM, Mackie C, Hrupka BJ, Caron MG, Daigle TL, Lindsley CW, Conn PJ, Jones CK. (2013) N-Aryl piperazine metabotropic glutamate receptor 5 positive allosteric modulators possess efficacy in preclinical models of NMDA hypofunction and cognitive enhancement. *J Pharmacol Exp Ther* 347:438-457.

[242] Taylor JR, Olausson P, Quinn JJ, Torregrossa MM (2009) Targeting extinction and reconsolidation mechanisms to combat the impact of drug cues on addiction. *Neuropharmacology* 56 Suppl 1:186-195.

[243] Cleva RM, Gass JT, Widholm JJ, Olive MF (2010) Glutamatergic targets for enhancing extinction learning in drug addiction. *Curr Neuropharmacol* 8:394-408.

[244] Cleva RM, Hicks MP, Gass JT, Wischerath KC, Plasters ET, Widholm JJ, Olive MF (2011) mGluR5 positive allosteric modulation enhances extinction learning following cocaine self-administration. *Behav Neurosci* 125:10-19.

[245] Kufahl PR, Nemirovsky NE, Watterson LR, Zautra N, Olive MF (2013) Positive or negative allosteric modulation of metabotropic glutamate receptor 5 (mGluR5) does not alter expression of behavioral sensitization to methamphetamine. *F1000 Research 2.*

[246] Kufahl PR, Hood LE, Nemirovsky NE, Barabas P, Halstengard C, Villa A, Moore E, Watterson LR, Olive MF (2012) Positive allosteric modulation of mGluR5 accelerates extinction learning but not relearning following methamphetamine self-administration. *Front Pharmacol* 3:194.

[247] Widholm JJ, Gass JT, Cleva RM, Olive MF (2011) The mGluR5 positive allosteric modulator CDPPB does not alter extinction or contextual reinstatement of methamphetamine-seeking behavior in rats. *J Addict Res Ther* S1:004.

[248] Kufahl PR, Olive MF (2011) Investigating methamphetamine craving using the extinction-reinstatement model in the rat. *J Addict Res Ther* S1:003.

[249] Bossert JM, Marchant NJ, Calu DJ, Shaham Y (2013) The reinstatement model of drug relapse: recent neurobiological findings, emerging research topics, and translational research. *Psychopharmacology* (Berl) 229:453-476.

[250] Xue YX, Luo YX, Wu P, Shi HS, Xue LF, Chen C, Zhu WL, Ding ZB, Bao YP, Shi J, Epstein DH, Shaham Y, Lu L (2012) A memory retrieval-extinction procedure to prevent drug craving and relapse. *Science* 336:241-245.

[251] Millan EZ, Milligan-Saville J, McNally GP (2013) Memory retrieval, extinction, and reinstatement of alcohol seeking. *Neurobiol Learn Mem* 101:26-32.

[252] McNally GP (2013) Extinction of drug seeking: Neural circuits and approaches to augmentation. *Neuropharmacology*.

[253] Li X, Xi ZX, Markou A (2013) Metabotropic glutamate 7 (mGlu7) receptor: a target for medication development for the treatment of cocaine dependence. *Neuropharmacology* 66:12-23.

[254] Sukoff Rizzo SJ, Leonard SK, Gilbert A, Dollings P, Smith DL, Zhang MY, Di L, Platt BJ, Neal S, Dwyer JM, Bender CN, Zhang J, Lock T, Kowal D, Kramer A, Randall A, Huselton C, Vishwanathan K, Tse SY, Butera J, Ring RH, Rosenzweig-Lipson S, Hughes ZA, Dunlop J (2011) The metabotropic glutamate receptor 7 allosteric modulator AMN082: a monoaminergic agent in disguise? *J Pharmacol Exp Ther* 338:345-352.

[255] Kufahl PR, Barabas P, Halstengard C, Olive MF (2012) Potential use of antidepressants as therapies for drug use disorders. In: *Frontiers in Clinical Drug Research - CNS and Neurological Disorders* (Atta-ur-Rahman, ed), pp 3-37. United Arab Emirates: Bentham Science.

[256] Dravolina OA, Zakharova ES, Shekunova EV, Zvartau EE, Danysz W, Bespalov AY (2007) mGlu1 receptor blockade attenuates cue- and nicotine-induced reinstatement of extinguished nicotine self-administration behavior in rats. *Neuropharmacology* 52:263-269.

[257] Chiamulera C, Borgo C, Falchetto S, Valerio E, Tessari M (1996) Nicotine reinstatement of nicotine self-administration after long-term extinction. *Psychopharmacology* (Berl) 127:102-107.

[258] Shaham Y, Adamson LK, Grocki S, Corrigall WA (1997) Reinstatement and spontaneous recovery of nicotine seeking in rats. *Psychopharmacology* (Berl) 130:396-403.

[259] Buczek Y, Le AD, Wang A, Stewart J, Shaham Y (1999) Stress reinstates nicotine seeking but not sucrose solution seeking in rats. *Psychopharmacology* (Berl) 144:183-188.

[260] Plaza-Zabala A, Martin-Garcia E, de Lecea L, Maldonado R, Berrendero F (2010) Hypocretins regulate the anxiogenic-like effects of nicotine and induce reinstatement of nicotine-seeking behavior. *J Neurosci* 30:2300-2310.

[261] Ramirez-Nino AM, D'Souza MS, Markou A (2013) N-acetylcysteine decreased nicotine self-administration and cue-induced reinstatement of nicotine seeking in rats: comparison with the effects of N-acetylcysteine on food responding and food seeking. *Psychopharmacology* (Berl) 225:473-482.

[262] Knackstedt LA, LaRowe S, Mardikian P, Malcolm R, Upadhyaya H, Hedden S, Markou A, Kalivas PW (2009) The role of cystine-glutamate exchange in nicotine dependence in rats and humans. *Biol Psychiatry* 65:841-845.

[263] Schmaal L, Berk L, Hulstijn KP, Cousijn J, Wiers RW, van den Brink W (2011) Efficacy of N-acetylcysteine in the treatment of nicotine dependence: a double-blind placebo-controlled pilot study. *Eur Addict Res* 17:211-216.

[264] Alajaji M, Bowers MS, Knackstedt L, Damaj MI (2013) Effects of the beta-lactam antibiotic ceftriaxone on nicotine withdrawal and nicotine-induced reinstatement of preference in mice. *Psychopharmacology* (Berl) 228:419-426.

[265] Paterson NE, Markou A (2004) Prolonged nicotine dependence associated with extended access to nicotine self-administration in rats. *Psychopharmacology* (Berl) 173:64-72.

[266] Fowler CD, Lu Q, Johnson PM, Marks MJ, Kenny PJ (2011) Habenular α5 nicotinic receptor subunit signalling controls nicotine intake. *Nature* 471:597-601.

[267] Cohen A, Koob GF, George O (2012) Robust escalation of nicotine intake with extended access to nicotine self-administration and intermittent periods of abstinence. *Neuropsychopharmacology* 37:2153-2160.

[268] Gilpin NW, Whitaker AM, Baynes B, Abdel AY, Weil MT, George O (2013) Nicotine vapor inhalation escalates nicotine self-administration. *Addict Biol*, in press.

[269] Koob GF, Buck CL, Cohen A, Edwards S, Park PE, Schlosburg JE, Schmeichel B, Vendruscolo LF, Wade CL, Whitfield TW Jr, George O (2013) Addiction as a stress surfeit disorder. *Neuropharmacology*, in press.

[270] Schumann G, Johann M, Frank J, Preuss U, Dahmen N, Laucht M, Rietschel M, Rujescu D, Lourdusamy A, Clarke TK, Krause K, Dyer A, Depner M, Wellek S, Treutlein J, Szegedi A, Giegling I, Cichon S, Blomeyer D, Heinz A, Heath S, Lathrop M, Wodarz N, Soyka M, Spanagel R, Mann K (2008) Systematic analysis of glutamatergic neurotransmission genes in alcohol dependence and adolescent risky drinking behavior. *Arch Gen Psychiatry* 65:826-838.

[271] Gass JT, Sinclair CM, Cleva RM, Widholm JJ, Olive MF (2011) Alcohol-seeking behavior is associated with increased glutamate transmission in basolateral amygdala and nucleus accumbens as measured by glutamate-oxidase-coated biosensors. *Addict Biol* 16:215-228.

[272] Sinclair CM, Cleva RM, Hood LE, Olive MF, Gass JT (2012) mGluR5 receptors in the basolateral amygdala and nucleus accumbens regulate cue-induced reinstatement of ethanol-seeking behavior. *Pharmacol Biochem Behav* 101:329-335.

[273] Zhao Y, Dayas CV, Aujla H, Baptista MA, Martin-Fardon R, Weiss F (2006) Activation of group II metabotropic glutamate receptors attenuates both stress and cue-induced ethanol-seeking and modulates c-fos expression in the hippocampus and amygdala. *J Neurosci* 26:9967-9974.

[274] Bossert JM, Poles GC, Sheffler-Collins SI, Ghitza UE (2006) The mGluR2/3 agonist LY379268 attenuates context- and discrete cue-induced reinstatement of sucrose seeking but not sucrose self-administration in rats. *Behav Brain Res* 173:148-152.

[275] Backstrom P, Hyytia P (2005) Suppression of alcohol self-administration and cue-induced reinstatement of alcohol seeking by the mGlu2/3 receptor agonist LY379268 and the mGlu8 receptor agonist (S)-3,4-DCPG. *Eur J Pharmacol* 528:110-118.

[276] Rodd ZA, McKinzie DL, Bell RL, McQueen VK, Murphy JM, Schoepp DD, McBride WJ (2006) The metabotropic glutamate 2/3 receptor agonist LY404039 reduces alcohol-seeking but not alcohol self-administration in alcohol-preferring (P) rats. *Behav Brain Res* 171:207-215.

[277] Adams CL, Cowen MS, Short JL, Lawrence AJ (2008) Combined antagonism of glutamate mGlu5 and adenosine A2A receptors interact to regulate alcohol-seeking in rats. *Int J Neuropsychopharmacol* 11:229-241.

[278] Adams CL, Short JL, Lawrence AJ (2010) Cue-conditioned alcohol seeking in rats following abstinence: involvement of metabotropic glutamate 5 receptors. *Br J Pharmacol* 159:534-542.

[279] Kufahl PR, Martin-Fardon R, Weiss F (2011) Enhanced sensitivity to attenuation of conditioned reinstatement by the mGluR2/3 agonist LY379268 and increased functional

activity of mGluR2/3 in rats with a history of ethanol dependence. *Neuropsychopharmacology* 36:2762-2773.

[280] Meinhardt MW, Hansson AC, Perreau-Lenz S, Bauder-Wenz C, Stahlin O, Heilig M, Harper C, Drescher KU, Spanagel R, Sommer WH (2013) Rescue of infralimbic mGluR2 deficit restores control over drug-seeking behavior in alcohol dependence. *J Neurosci* 33:2794-2806.

[281] Kennedy WK, Leloux M, Kutscher EC, Price PL, Morstad AE, Carnahan RM (2010) Acamprosate. *Expert Opin Drug Metab Toxicol* 6:363-380.

[282] Lhuintre JP, Moore N, Tran G, Steru L, Langrenon S, Daoust M, Parot P, Ladure P, Libert C, Boismare F, et al. (1990) Acamprosate appears to decrease alcohol intake in weaned alcoholics. *Alcohol Alcohol* 25:613-622.

[283] Mason BJ, Heyser CJ (2010) Acamprosate: a prototypic neuromodulator in the treatment of alcohol dependence. *CNS Neurol Disord Drug Targets* 9:23-32.

[284] Mason BJ, Heyser CJ (2010) The neurobiology, clinical efficacy and safety of acamprosate in the treatment of alcohol dependence. *Expert Opin Drug Saf* 9:177-188.

[285] Anton RF, O'Malley SS, Ciraulo DA, Cisler RA, Couper D, Donovan DM, Gastfriend DR, Hosking JD, Johnson BA, LoCastro JS, Longabaugh R, Mason BJ, Mattson ME, Miller WR, Pettinati HM, Randall CL, Swift R, Weiss RD, Williams LD, Zweben A (2006) Combined pharmacotherapies and behavioral interventions for alcohol dependence: the COMBINE study: a randomized controlled trial. *JAMA* 295:2003-2017.

[286] Kampman KM, Pettinati HM, Lynch KG, Xie H, Dackis C, Oslin DW, Sparkman T, Sharkoski T, O'Brien CP (2009) Initiating acamprosate within-detoxification versus post-detoxification in the treatment of alcohol dependence. *Addict Behav* 34:581-586.

[287] Ferreira Seiva FR, Amauchi JF, Ribeiro Rocha KK, Souza GA, Ebaid GX, Burneiko RM, Novelli EL (2009) Effects of N-acetylcysteine on alcohol abstinence and alcohol-induced adverse effects in rats. *Alcohol* 43:127-135.

[288] Seiva FR, Amauchi JF, Rocha KK, Ebaid GX, Souza G, Fernandes AA, Cataneo AC, Novelli EL (2009) Alcoholism and alcohol abstinence: N-acetylcysteine to improve energy expenditure, myocardial oxidative stress, and energy metabolism in alcoholic heart disease. *Alcohol* 43:649-656.

[289] Sari Y, Sakai M, Weedman JM, Rebec GV, Bell RL (2011) Ceftriaxone, a beta-lactam antibiotic, reduces ethanol consumption in alcohol-preferring rats. *Alcohol Alcohol* 46:239-246.

[290] Qrunfleh AM, Alazizi A, Sari Y (2013) Ceftriaxone, a beta-lactam antibiotic, attenuates relapse-like ethanol-drinking behavior in alcohol-preferring rats. *J Psychopharmacol* 27:541-549.

[291] LaLumiere RT, Kalivas PW (2008) Glutamate release in the nucleus accumbens core is necessary for heroin seeking. *J Neurosci* 28:3170-3177.

[292] Bossert JM, Busch RF, Gray SM (2005) The novel mGluR2/3 agonist LY379268 attenuates cue-induced reinstatement of heroin seeking. *Neuroreport* 16:1013-1016.

[293] Bossert JM, Gray SM, Lu L, Shaham Y (2006) Activation of group II metabotropic glutamate receptors in the nucleus accumbens shell attenuates context-induced relapse to heroin seeking. *Neuropsychopharmacology* 31:2197-2209.

[294] Zhou W, Liu H, Zhang F, Tang S, Zhu H, Lai M, Kalivas PW (2007) Role of acetylcholine transmission in nucleus accumbens and ventral tegmental area in heroin-seeking induced by conditioned cues. *Neuroscience* 144:1209-1218.

[295] Zhou W, Kalivas PW (2008) N-acetylcysteine reduces extinction responding and induces enduring reductions in cue- and heroin-induced drug-seeking. *Biol Psychiatry* 63:338-340.

[296] Levenga J, Hayashi S, de Vrij FM, Koekkoek SK, van der Linde HC, Nieuwenhuizen I, Song C, Buijsen RA, Pop AS, Gomezmancilla B, Nelson DL, Willemsen R, Gasparini F, Oostra BA (2011) AFQ056, a new mGluR5 antagonist for treatment of fragile X syndrome. *Neurobiol Dis* 42:311-317.

[297] Novartis (2013) Study to evaluate the effect of AFQ056 in obsessive compulsive disorder (OCD) patients resistant to selective serotonin reuptake inhibitor (SSRI) therapy. *ClinicalTrials.gov, 3/14/2013 Edition*. Bethesda, MD: National Library of Medicine.

[298] Addex (2013) ADX10059 as an add-on therapy to proton pump inhibitors (PPIs) in patients with gastroesophageal reflux (GERD). *ClinicalTrials.gov, 7/13/2012 Edition*. Bethesda, MD: National Library of Medicine.

[299] Bergink V, Westenberg HG (2005) Metabotropic glutamate II receptor agonists in panic disorder: a double blind clinical trial with LY354740. *Int Clin Psychopharmacol* 20:291-293.

[300] Kellner M, Muhtz C, Stark K, Yassouridis A, Arlt J, Wiedemann K (2005) Effects of a metabotropic glutamate(2/3) receptor agonist (LY544344/LY354740) on panic anxiety induced by cholecystokinin tetrapeptide in healthy humans: preliminary results. *Psychopharmacology* (Berl) 179:310-315.

[301] Rorick-Kehn LM, Perkins EJ, Knitowski KM, Hart JC, Johnson BG, Schoepp DD, McKinzie DL (2006) Improved bioavailability of the mGlu2/3 receptor agonist LY354740 using a prodrug strategy: in vivo pharmacology of LY544344. *J Pharmacol Exp Ther* 316:905-913.

[302] Eriksson AH, Varma MV, Perkins EJ, Zimmerman CL (2010) The intestinal absorption of a prodrug of the mGlu2/3 receptor agonist LY354740 is mediated by PEPT1: in situ rat intestinal perfusion studies. *J Pharm Sci* 99:1574-1581.

[303] Hashimoto K, Malchow B, Falkai P, Schmitt A (2013) Glutamate modulators as potential therapeutic drugs in schizophrenia and affective disorders. *Eur Arch Psychiatry Clin Neurosci* 263:367-377.

[304] Berry-Kravis E, Hessl D, Coffey S, Hervey C, Schneider A, Yuhas J, Hutchison J, Snape M, Tranfaglia M, Nguyen DV, Hagerman R (2009) A pilot open label, single dose trial of fenobam in adults with fragile X syndrome. *J Med Genet* 46:266-271.

[305] Rylander D, Iderberg H, Li Q, Dekundy A, Zhang J, Li H, Baishen R, Danysz W, Bezard E, Cenci MA (2010) A mGluR5 antagonist under clinical development improves L-DOPA-induced dyskinesia in parkinsonian rats and monkeys. *Neurobiol Dis* 39:352-361.

[306] Cavallone L (2013) Pharmacokinetics and side effects of the mGlu5 antagonist fenobam in adult healthy volunteers. *ClinicalTrials.gov, 3/2/2013 Edition*. Bethesda, MD: National Library of Medicine.

[307] Breese GR, Sinha R, Heilig M (2011) Chronic alcohol neuroadaptation and stress contribute to susceptibility for alcohol craving and relapse. *Pharmacol Ther* 129:149-171.

[308] Ulrich-Lai YM, Herman JP (2009) Neural regulation of endocrine and autonomic stress responses. *Nat Rev Neurosci* 10:397-409.

[309] Herman JP, Figueiredo H, Mueller NK, Ulrich-Lai Y, Ostrander MM, Choi DC, Cullinan WE (2003) Central mechanisms of stress integration: hierarchical circuitry controlling hypothalamo-pituitary-adrenocortical responsiveness. Front *Neuroendocrinol* 24:151-180.

[310] Taylor SB, Conrad CD, Olive MF. Stress and maze performance in rodents. In: *The Maze Book: Your Guide Book to Theories, Practices and Protocols for Testing Rodent Cognition*. (Bimonte-Nelson HA, ed): Humana Press, in press.

[311] de Kloet ER, Vreugdenhil E, Oitzl MS, Joels M (1998) Brain corticosteroid receptor balance in health and disease. *Endocrine Rev* 19:269-301.

[312] Piazza PV, Deroche V, Deminière JM, Maccari S, Le Moal M, Simon H (1993) Corticosterone in the range of stress-induced levels possesses reinforcing properties: implications for sensation-seeking behaviors. *Proc Natl Acad Sci USA* 90:11738-11742.

[313] Fahlke C, Hurd E, Hansen S (1996) Facilitation of ethanol consumption by intracerebroventricular infusions of corticosterone. *Psychopharmacology* (Berl) 127:133-139.

[314] Goeders NE (1997) A neuroendocrine role in cocaine reinforcement. *Psychoneuroendocrinology* 22:237-259.

[315] Piazza PV, Le Moal M (1997) Glucocorticoids as a biological substrate of reward: physiological and pathophysiological implications. *Brain Res Rev* 25:359-372.

[316] Koob GF (2010) The role of CRF and CRF-related peptides in the dark side of addiction. *Brain Res* 1314:3-14.

[317] Cullinan WE, Herman JP, Watson SJ (1993) Ventral subicular interaction with the hypothalamic paraventricular nucleus: Evidence for a relay in the bed nucleus of the stria terminalis. *J Comp Neurol* 332:1-20.

[318] Jankord R, Herman JP (2008) Limbic regulation of hypothalamo-pituitary-adrenocortical function during acute and chronic stress. *Ann NY Acad Sci* 1148:64-73.

[319] Diorio D, Viau V, Meaney MJ (1993) The role of the medial prefrontal cortex (cingulate gyrus) in the regulation of hypothalamic-pituitary-adrenal responses to stress. *J Neurosci* 13:3839-3847.

[320] Figueiredo HF, Bruestle A, Bodie B, Dolgas CM, Herman JP (2003) The medial prefrontal cortex differentially regulates stress-induced c-fos expression in the forebrain depending on type of stressor. *Eur J Neurosci* 18:2357-2364.

[321] Crane JW, Ebner K, Day TA (2003) Medial prefrontal cortex suppression of the hypothalamic–pituitary–adrenal axis response to a physical stressor, systemic delivery of interleukin-1β. Eur *J Neurosci* 17:1473-1481.

[322] Radley JJ, Arias CM, Sawchenko PE (2006) Regional differentiation of the medial prefrontal cortex in regulating adaptive responses to acute emotional stress. *J Neurosci* 26:12967-12976.

[323] Dong HW, Petrovich GD, Swanson LW (2001) Topography of projections from amygdala to bed nuclei of the stria terminalis. *Brain Res Brain Res Rev* 38:192-246.

[324] Feldman S, Conforti N, Itzik A, Weidenfeld J (1994) Differential effect of amygdaloid lesions on CRF-41, ACTH and corticosterone responses following neural stimuli. *Brain Res* 658:21-26.

[325] Sawchenko PE, Li HY, Ericsson A (2000) Circuits and mechanisms governing hypothalamic responses to stress: a tale of two paradigms. *Prog Brain Res* 122:61-78.

[326] Bhatnagar S, Vining C, Denski KAI (2004) Regulation of chronic stress-induced changes in hypothalamic-pituitary-adrenal activity by the basolateral amygdala. *Ann NY Acad Sci* 1032:315-319.

[327] Radley JJ, Gosselink KL, Sawchenko PE (2009) A discrete GABAergic relay mediates medial prefrontal cortical inhibition of the neuroendocrine stress response. *J Neurosci* 29:7330-7340.

[328] Makino S, Gold PW, Schulkin J (1994) Corticosterone effects on corticotropin-releasing hormone mRNA in the central nucleus of the amygdala and the parvocellular region of the paraventricular nucleus of the hypothalamus. *Brain Res* 640:105-112.

[329] Olive MF, Koenig HN, Nannini MA, Hodge CW (2002) Elevated extracellular CRF levels in the bed nucleus of the stria terminalis during ethanol withdrawal and reduction by subsequent ethanol intake. *Pharmacol Biochem Behav* 72:213-220.

[330] Overstreet DH, Knapp DJ, Breese GR (2004) Modulation of multiple ethanol withdrawal-induced anxiety-like behavior by CRF and CRF1 receptors. *Pharmacol Biochem Behav* 77:405-413.

[331] Funk CK, O'Dell LE, Crawford EF, Koob GF (2006) Corticotropin-releasing factor within the central nucleus of the amygdala mediates enhanced ethanol self-administration in withdrawn, ethanol-dependent rats. *J Neurosci* 26:11324-11332.

[332] Ulrich-Lai YM, Xie W, Meij JT, Dolgas CM, Yu L, Herman JP (2006) Limbic and HPA axis function in an animal model of chronic neuropathic pain. *Physiol Behav* 88:67-76.

[333] Ostrander MM, Ulrich-Lai YM, Choi DC, Richtand NM, Herman JP (2006) Hypoactivity of the hypothalamo-pituitary-adrenocortical axis during recovery from chronic variable stress. *Endocrinology* 147:2008-2017.

[334] Rivier C, Bruhn T, Vale W (1984) Effect of ethanol on the hypothalamic-pituitary-adrenal axis in the rat: role of corticotropin-releasing factor (CRF). *J Pharmacol Exp Ther* 229:127-131.

[335] Merlo Pich E, Lorang M, Yeganeh M, Rodriguez de Fonseca F, Raber J, Koob GF, Weiss F (1995) Increase of extracellular corticotropin-releasing factor-like immunoreactivity levels in the amygdala of awake rats during restraint stress and ethanol withdrawal as measured by microdialysis. *J Neurosci* 15:5439-5447.

[336] Rasmussen DD, Boldt BM, Bryant CA, Mitton DR, Larsen SA, Wilkinson CW (2000) Chronic daily ethanol and withdrawal: 1. Long-term changes in the hypothalamo-pituitary-adrenal axis. *Alcohol Clin Exp Res* 24:1836-1849.

[337] Koob G, Kreek MJ (2007) Stress, dysregulation of drug reward pathways, and the transition to drug dependence. *Am J Psychiatry* 164:1149-1159.

[338] Roberto M, Cruz MT, Gilpin NW, Sabino V, Schweitzer P, Bajo M, Cottone P, Madamba SG, Stouffer DG, Zorrilla EP, Koob GF, Siggins GR, Parsons LH (2010) Corticotropin releasing factor-induced amygdala gamma-aminobutyric acid release plays a key role in alcohol dependence. *Biol Psychiatry* 67:831-839.

[339] Darlington DN, Miyamoto M, Keil LC, Dallman MF (1989) Paraventricular stimulation with glutamate elicits bradycardia and pituitary responses. *Am J Physiol* 256:R112-119.

[340] van den Pol AN, Wuarin JP, Dudek FE (1990) Glutamate, the dominant excitatory transmitter in neuroendocrine regulation. *Science* 250:1276-1278.

[341] Herman JP, Mueller NK, Figueiredo H (2004) Role of GABA and glutamate circuitry in hypothalamo-pituitary-adrenocortical stress integration. *Ann NY Acad Sci* 1018:35-45.

[342] Makara GB, Stark E (1975) Effect of intraventricular glutamate on ACTH release. *Neuroendocrinology* 18:213-216.

[343] Joanny P, Steinberg J, Oliver C, Grino M (1997) Glutamate and N-methyl-D-aspartate stimulate rat hypothalamic corticotropin-releasing factor secretion in vitro. *Journal of Neuroendocrinology* 9:93-97.

[344] Ziegler DR, Herman JP (2000) Local integration of glutamate signaling in the hypothalamic paraventricular region: regulation of glucocorticoid stress responses. *Endocrinology* 141:4801-4804.

[345] Johnson MP, Kelly G, Chamberlain M (2001) Changes in rat serum corticosterone after treatment with metabotropic glutamate receptor agonists or antagonists. *J Neuroendocrinol* 13:670-677.

[346] Scaccianoce S, Matrisciano F, Del Bianco P, Caricasole A, Di Giorgi Gerevini V, Cappuccio I, Melchiorri D, Battaglia G, Nicoletti F (2003) Endogenous activation of group-II metabotropic glutamate receptors inhibits the hypothalamic-pituitary-adrenocortical axis. *Neuropharmacology* 44:555-561.

[347] Ohishi H, Shigemoto R, Nakanishi S, Mizuno N (1993) Distribution of the mRNA for a metabotropic glutamate receptor (mGluR3) in the rat brain: an in situ hybridization study. *J Comp Neurol* 335:252-266.

[348] Ohishi H, Shigemoto R, Nakanishi S, Mizuno N (1993) Distribution of the messenger RNA for a metabotropic glutamate receptor, mGluR2, in the central nervous system of the rat. *Neuroscience* 53:1009-1018.

[349] Meeker RB, Greenwood RS, Hayward JN (1994) Glutamate receptors in the rat hypothalamus and pituitary. *Endocrinology* 134:621-629.

[350] Ohishi H, Akazawa C, Shigemoto R, Nakanishi S, Mizuno N (1995) Distributions of the mRNAs for L-2-amino-4-phosphonobutyrate-sensitive metabotropic glutamate receptors, mGluR4 and mGluR7, in the rat brain. *J Comp Neurol* 360:555-570.

[351] Wright RA, Arnold MB, Wheeler WJ, Ornstein PL, Schoepp DD (2001) [^{3}H]LY341495 binding to group II metabotropic glutamate receptors in rat brain. *J Pharmacol Exp Ther* 298:453-460.

[352] Gosnell HB, Silberman Y, Grueter BA, Duvoisin RM, Raber J, Winder DG (2011) mGluR8 modulates excitatory transmission in the bed nucleus of the stria terminalis in a stress-dependent manner. *Neuropsychopharmacology* 36:1599-1607.

[353] Bradbury MJ, Giracello DR, Chapman DF, Holtz G, Schaffhauser H, Rao SP, Varney MA, Anderson JJ (2003) Metabotropic glutamate receptor 5 antagonist-induced stimulation of hypothalamic-pituitary-adrenal axis activity: interaction with serotonergic systems. *Neuropharmacology* 44:562-572.

[354] Herman JP, Eyigor O, Ziegler DR, Jennes L (2000) Expression of ionotropic glutamate receptor subunit mRNAs in the hypothalamic paraventricular nucleus of the rat. *J Comp Neurol* 422:352-362.

[355] Eyigor O, Minbay Z, Cavusoglu I, Jennes L (2005) Localization of kainate receptor subunit GluR5-immunoreactive cells in the rat hypothalamus. *Brain Res Mol Brain Res* 136:38-44.

[356] Mitsukawa K, Mombereau C, Lotscher E, Uzunov DP, van der Putten H, Flor PJ, Cryan JF (2006) Metabotropic glutamate receptor subtype 7 ablation causes dysregulation of the HPA axis and increases hippocampal BDNF protein levels: implications for stress-related psychiatric disorders. *Neuropsychopharmacology* 31:1112-1122.

[357] Wieronska JM, Branski P, Szewczyk B, Palucha A, Papp M, Gruca P, Moryl E, Pilc A (2001) Changes in the expression of metabotropic glutamate receptor 5 (mGluR5) in the rat hippocampus in an animal model of depression. *Pol J Pharmacol* 53:659-662.

[358] Iyo AH, Feyissa AM, Chandran A, Austin MC, Regunathan S, Karolewicz B (2010) Chronic corticosterone administration down-regulates metabotropic glutamate receptor 5 protein expression in the rat hippocampus. *Neuroscience* 169:1567-1574.

[359] Neugebauer V, Zinebi F, Russell R, Gallagher JP, Shinnick-Gallagher P (2000) Cocaine and kindling alter the sensitivity of group II and III metabotropic glutamate receptors in the central amygdala. *J Neurophysiol* 84:759-770.

[360] Schwendt M, Reichel CM, See RE (2012) Extinction-dependent alterations in corticostriatal mGluR2/3 and mGluR7 receptors following chronic methamphetamine self-administration in rats. *PLoS One* 7:e34299.

[361] Rassnick S, Heinrichs SC, Britton KT, Koob GF (1993) Microinjection of a corticotropin-releasing factor antagonist into the central nucleus of the amygdala reverses anxiogenic-like effects of ethanol withdrawal. *Brain Res* 605:25-32.

[362] Basso AM, Spina M, Rivier J, Vale W, Koob GF (1999) Corticotropin-releasing factor antagonist attenuates the "anxiogenic-like" effect in the defensive burying paradigm but not in the elevated plus-maze following chronic cocaine in rats. *Psychopharmacology* (Berl) 145:21-30.

[363] Spooren WP (2000) Anxiolytic-like effects of the prototypical metabotropic glutamate receptor 5 antagonist 2-methyl-6-(phenylethynyl)pyridine in rodents. *J Pharmacol Exp Ther* 295:1267-1275.

[364] Schulz B (2001) The metabotropic glutamate receptor antagonist 2-methyl-6-(phenylethynyl)-pyridine (MPEP) blocks fear conditioning in rats. *Neuropharmacology* 41:1-7.

[365] Tatarczynska E (2001) Potential anxiolytic- and antidepressant-like effects of MPEP, a potent, selective and systemically active mGlu5 receptor antagonist. *Br J Pharmacol* 132:1423-1430.

[366] Spooren WP, Schoeffter P, Gasparini F, Kuhn R, Gentsch C (2002) Pharmacological and endocrinological characterisation of stress-induced hyperthermia in singly housed mice using classical and candidate anxiolytics (LY314582, MPEP and NKP608). *Eur J Pharmacol* 435:161-170.

[367] Rolls ET (1996) A theory of hippocampal function in memory. *Hippocampus* 6:601-620.

[368] Johnson MP, Kelly G, Chamberlain M (2001) Changes in rat serum corticosterone after treatment with metabotropic glutamate receptor agonists or antagonists. *J Neuroendocrinol* 13:670-677.

[369] Monn JA (1997) Design, synthesis, and pharmacological characterization of (+)-2-aminobicyclo[3.1.0]hexane-2,6-dicarboxylic acid (LY354740): a potent, selective, and

orally active group 2 metabotropic glutamate receptor agonist possessing anticonvulsant and anxiolytic properties. *J Med Chem* 40:528-537.

[370] Benvenga MJ, Overshiner CD, Monn JA, Leander JD (1999) Disinhibitory effects of LY354740, a new mGluR2 agonist, on behaviors suppressed by electric shock, in rats and pigeons. *Drug Dev Res* 47:37-44.

[371] Klodzinska A (1999) Potential anti-anxiety, anti-addictive effects of LY354740, a selective group II glutamate metabotropic receptors agonist in animal models. *Neuropharmacology* 38:1831-1839.

[372] Moore NA, Rees G, Monn JA (1999) Effects of the group II metabotropic glutamate receptor agonist, LY354740 on schedule-controlled behaviour in rats. *Behav Pharmacol* 10:319-325.

[373] Shekhar A, Keim SR (2000) LY354740, a potent group II metabotropic glutamate receptor agonist prevents lactate-induced panic-like response in panic-prone rats. *Neuropharmacology* 39:1139-1146.

[374] Ferris P, Seward E, Dawson GR (2001) Interactions between LY354740, a group II metabotropic agonist and the $GABA_A$-benzodiazepine receptor complex in the rat elevated plus-maze. *Journal of Psychopharmacology* 15:76-82.

[375] Tizzano JP, Griffey KI, Schoepp DD (2002) The anxiolytic action of mGlu2/3 receptor agonist, LY354740, in the fear-potentiated startle model in rats is mechanistically distinct from diazepam. *Pharmacol Biochem Behav* 73:367-374.

[376] Grillon C, Cordova J, Levine LR, Morgan CA (2003) Anxiolytic effects of a novel group II metabotropic glutamate receptor agonist (LY354740) in the fear-potentiated startle paradigm in humans. *Psychopharmacology* (Berl) 168:446-454.

[377] Swanson CJ, Perry KW, Schoepp DD (2004) The mGlu2/3 receptor agonist, LY354740, blocks immobilization-induced increases in noradrenaline and dopamine release in the rat medial prefrontal cortex. *J Neurochem* 88:194-202.

[378] Schoepp DD, Wright RA, Levine LR, Gaydos B, Potter WZ (2003) LY354740, an mGlu2/3 receptor agonist as a novel approach to treat anxiety/stress. *Stress* 6:189-197.

[379] Schoepp DD (2004) Case study: utility of metabotropic glutamate agonists in psychiatric illness. *5th World Congress on Stress.*

[380] Chojnacka-Wojcik E, Tatarczynska E, Pilc A (1996) Anxiolytic-like effects of metabotropic glutamate antagonist (RS)-[alpha]-methylserine-O-phosphate in rats. *Pol J Pharmacol* 48:507-509.

[381] Chojnacka-Wojcik E, Tatarczynska E, Pilc A (1997) The anxiolytic-like effect of metabotropic glutamate receptor antagonists after intrahippocampal injection in rats. *Eur J Pharmacol* 319:153-156.

[382] Tatarczynska E (2002) Anxiolytic- and antidepressant-like effects of group III metabotropic glutamate agonist (1S,3R,4S)-1-aminocyclopentane-1,3,4-tricarboxylic acid (ACPT-I) in rats. *Pol J Pharmacol* 54:707-710.

[383] Masugi M (1999) Metabotropic glutamate receptor subtype 7 ablation causes deficit in fear response and conditioned taste aversion. *J Neurosci* 19:955-963.

[384] Linden AM (2002) Increased anxiety-related behavior in mice deficient for metabotropic glutamate 8 (mGlu8) receptor. *Neuropharmacology* 43:251-259.

[385] Cryan JF (2003) Antidepressant and anxiolytic-like effects in mice lacking the group III metabotropic glutamate receptor mGluR7. *Eur J Neurosci* 17:2409-2417.

[386] Adinoff B, Risher-Flowers D, De Jong J, Ravitz B, Bone GH, Nutt DJ, Roehrich L, Martin PR, Linnoila M (1991) Disturbances of hypothalamic-pituitary-adrenal axis functioning during ethanol withdrawal in six men. *Am J Psychiatry* 148:1023-1025.

[387] Vescovi PP, Coiro V, Volpi R, Passeri M (1992) Diurnal variations in plasma ACTH, cortisol and beta-endorphin levels in cocaine addicts. *Hormone Res* 37:221-224.

[388] Ehrenreich H, Schuck J, Stender N, Pilz J, Gefeller O, Schilling L, Poser W, Kaw S (1997) Endocrine and hemodynamic effects of stress versus systemic CRF in alcoholics during early and medium term abstinence. *Alcohol Clin Exp Res* 21:1285-1293.

[389] Adinoff B, Junghanns K, Kiefer F, Krishnan-Sarin S (2005) Suppression of the HPA axis stress-response: implications for relapse. *Alcohol Clin Exp Res* 29:1351-5.

[390] Sinha R, Fox HC, Hong KA, Bergquist K, Bhagwagar Z, Siedlarz KM (2008) Enhanced negative emotion and alcohol craving, and altered physiological responses following stress and cue exposure in alcohol dependent individuals. *Neuropsychopharmacology* 34:1198-1208.

[391] Hughes JR (1992) Tobacco withdrawal in self-quitters. *J Consult Clin Psychol* 60:689-697.

[392] Kouri EM, Pope HG Jr, Lukas SE (1999) Changes in aggressive behavior during withdrawal from long-term marijuana use. *Psychopharmacology* (Berl) 143:302-308.

[393] Mulvaney FD, Alterman AI, Boardman CR, Kampman K (1999) Cocaine abstinence symptomatology and treatment attrition. *J Subst Abuse Treat* 16:129-135.

[394] Budney AJ, Hughes JR (2006) The cannabis withdrawal syndrome. *Curr Opin Psychiatry* 19:233-238.

[395] Baker TB, Brandon TH, Chassin L (2004) Motivational influences on cigarette smoking. *Annu Rev Psychol* 55:463-491.

[396] Dodge R, Sindelar J, Sinha R (2005) The role of depression symptoms in predicting drug abstinence in outpatient substance abuse treatment. *J Subst Abuse Treat* 28:189-196.

[397] Sinha R (2007) The role of stress in addiction relapse. *Curr Psychiatry Rep* 9:388-395.

[398] Paliwal P, Hyman SM, Sinha R (2008) Craving predicts time to cocaine relapse: further validation of the Now and Brief versions of the cocaine craving questionnaire. *Drug Alcohol Depend* 93:252-259.

[399] Harris GC, Altomare K, Aston-Jones G (2001) Preference for a cocaine-associated environment is attenuated by augmented accumbal serotonin in cocaine withdrawn rats. *Psychopharmacology* (Berl) 156:14-22.

[400] El Hage C, Rappeneau V, Etievant A, Morel AL, Scarna H, Zimmer L, Berod A (2012) Enhanced anxiety observed in cocaine withdrawn rats is associated with altered reactivity of the dorsomedial prefrontal cortex. *PLoS One* 7:16.

[401] Paine TA, Jackman SL, Olmstead MC (2002) Cocaine-induced anxiety: alleviation by diazepam, but not buspirone, dimenhydrinate or diphenhydramine. *Behav Pharmacol* 13:511-523.

[402] Rudoy CA, Van Bockstaele EJ (2007) Betaxolol, a selective beta$_1$-adrenergic receptor antagonist, diminishes anxiety-like behavior during early withdrawal from chronic cocaine administration in rats. *Prog Neuropsychopharmacol Biol Psychiatry* 31:1119-1129.

Index

A

B

C

D

E

F

G

J

K

L

M

N

O

P

Q

R

S

T

U

V

W

Y